Preface

This book, as its title declares, is about the epistemology of quantum physics. However, we use "epistemology" in a rather loose and generic sense to mean things and aspects related mainly to the interpretation and justification of the scientific theory in its technical form. Accordingly, the contents of the book are essentially of non-formal nature[1] although the formalism of quantum mechanics will also be investigated (rather briefly and superficially) inline with the needs and requirements of the epistemological investigation and considerations. Therefore, the reader of this book should ideally be familiar with (at least) the basic concepts and formulations of quantum mechanics.

Our original plan was to write a complete book about the quantum theory both in its formalism and in its interpretation (or epistemology to be more general). However, this plan was abandoned due to the existence of many excellent texts dedicated to (or focused on) the formalism of this theory and hence we restricted our attention in the present book on the interpretative (or epistemological) aspects which in our view require more investigation and discussion to clarify and rectify many controversial (and even some non-controversial) issues. Moreover, a thorough treatment that includes both the formalism and epistemology of the quantum theory requires more than one book (which is not within our current plan).

The reader should note that a general scientific and mathematical background (at the undergraduate level) is required to understand the book properly and appreciate its contents. The reader should also note that the present book, unlike our previous books, does not contain questions or exercises or solved problems (thanks to its non-formal nature which we indicated earlier). However, it is like our previous books in style and favorable characteristics (such as clarity, graduality and intensive cross referencing).

The book, in our view, is particularly useful to those who have special interest in the interpretative aspects of quantum theory and the philosophy of science although it should be useful even to those who are interested in the purely-scientific and technical aspects of the quantum theory since the contents of the book should broaden the understanding of these aspects and provide them with qualitative and interpretative dimensions (as well as the added benefit of the brief and marginal investigation of the formalism of quantum mechanics).

Taha Sochi
London, June 2022

---

[1] We may also describe it as: non-technical or contemplative or philosophical or interpretative or qualitative or descriptive nature (with epistemology being the focus of interest).

Contents

Preface 1

Table of Contents 2

Nomenclature 6

1 Preliminaries **8**
   1.1 General Remarks 8
   1.2 Introducing Quantum Physics 10
   1.3 General Factors to Consider in Assessing Science 12
      1.3.1 Humanism 12
      1.3.2 Scale of Observer and Observed 13
      1.3.3 Inertness of Observation 15
   1.4 Comparison between Wave and Particle 16
   1.5 Comparison between Massive and Massless Objects 17
   1.6 Double-Slit Experiment 17
      1.6.1 Dual Nature 19
      1.6.2 Effect of Observation and Knowledge 19
      1.6.3 Self-Interference 20
      1.6.4 Delayed Cause 22
      1.6.5 Memory 22
      1.6.6 Semi-Randomness 22
      1.6.7 Other Features 22
   1.7 The Principle of Causality 23
   1.8 The Principle of Locality 27
   1.9 Relationship between Causality and Locality 31
   1.10 Realism and Idealism 31
   1.11 Consciousness and its Role in Science 32

2 Scientific Theory and its Interpretation **34**
   2.1 Objectives of Science 34
   2.2 Logic in Science 34
   2.3 Epistemology 35
   2.4 The Epistemological Principles of Science 35
      2.4.1 The Principles of Reality and Truth 35
      2.4.2 The Principle of Causality 36
      2.4.3 The Principle of Non-Uniqueness of Science 36
      2.4.4 The Principle of Economy 37
      2.4.5 The Principle of Intuitivity 38
   2.5 Eligibility Criteria of Scientific Theory 38
      2.5.1 Logicality 39
      2.5.2 Compliance with the Principles of Reality and Truth 42
      2.5.3 Compliance with the Principle of Causality 44
      2.5.4 Physicality 44
      2.5.5 Testability 45
   2.6 Validity Criteria of Scientific Theory 45
      2.6.1 Scientific Evidence 46
      2.6.2 Compliance with the Established Theories 47
   2.7 Optimality Criteria of Scientific Theory 48

	2.7.1	Simplicity	48
	2.7.2	Empiricism	48
	2.7.3	Realisticity	49
	2.7.4	Harmony	49
	2.7.5	Objectivity	49
	2.7.6	Quantitativity	50
	2.7.7	Predictivity	50
	2.7.8	Interpretativity	50
2.8	Relationship between Formalism and Interpretation		51
2.9	Eligibility Criteria of Scientific Interpretation		53
	2.9.1	Logicality	53
	2.9.2	Compliance with the Principles of Reality and Truth	53
	2.9.3	Compliance with the Principle of Causality	53
	2.9.4	Physicality	53
	2.9.5	Testability	54
	2.9.6	Comprehensibility	56
	2.9.7	Thoroughness	56
	2.9.8	Rationalization	57
2.10	Validity Criteria of Scientific Interpretation		57
2.11	Optimality Criteria of Scientific Interpretation		58
2.12	Role and Importance of Interpretation		58
2.13	Existence and Uniqueness of Scientific Interpretation		59
2.14	Evolution and Achievements of Science		60
2.15	Freewill and Science		62
	2.15.1	Freewill and Physicality	62
	2.15.2	Freewill and Indeterminism	63
	2.15.3	Freewill and Causality	64

3 Quantum versus Classical — 65
- 3.1 Comparison between Classical and Quantum Phenomena — 65
- 3.2 Comparison between Classical and Quantum Waves — 66
- 3.3 Comparison between Classical and Quantum Particles — 68
- 3.4 Comparison between Classical and Quantum Measurement — 69
- 3.5 Comparison between Classical and Quantum Theories — 70
- 3.6 Relationship between Our Views to Classical and Quantum Worlds — 71
- 3.7 Relationship between Classical and Quantum Physics — 72
- 3.8 Border between Classical and Quantum Domains — 75

4 Formalism of Quantum Theory — 77
- 4.1 Wavefunction — 77
- 4.2 The Axiomatic Framework of Quantum Mechanics — 80
- 4.3 Schrodinger's Equation — 83
- 4.4 Operators in Quantum Theory — 85
- 4.5 Quantum Observables — 87
- 4.6 Angular Momentum and Spin — 88
  - 4.6.1 SG Type Experiments — 90
- 4.7 The Principles of Quantum Theory — 91
  - 4.7.1 The Uncertainty Principle — 92
  - 4.7.2 The Correspondence Principle — 96
  - 4.7.3 The Complementarity Principle — 99
  - 4.7.4 Relationships between the Principles of Quantum Theory — 101
- 4.8 The Dynamics of Quantum Mechanics — 102

	4.9 The Basic Elements of Formalism	103

5 Characteristic Features of Quantum Theory — 104
- 5.1 Quantization — 104
- 5.2 Indeterminism — 105
  - 5.2.1 Superposition — 106
  - 5.2.2 Probabilisticity — 107
  - 5.2.3 Uncertainty — 108
- 5.3 Duality — 109
  - 5.3.1 Particle-Wave Duality — 109
  - 5.3.2 Formalism Duality — 113
  - 5.3.3 Classic-Quantum Duality — 113
- 5.4 Quantum Interference — 113
- 5.5 Quantum Tunneling and Scattering — 115
- 5.6 Quantum Entanglement — 117
- 5.7 Quantum Measurement and Collapse of Wavefunction — 120
- 5.8 Linearity — 124
- 5.9 Complementarity — 125
- 5.10 Other Features — 125

6 Technical Assessment of Quantum Theory — 127
- 6.1 Wavefunction — 128
- 6.2 The Axiomatic Framework of Quantum Mechanics — 130
- 6.3 Schrodinger's Equation — 130
- 6.4 Indeterminism — 132
- 6.5 Quantum Interference — 134
- 6.6 The Uncertainty Principle — 134
  - 6.6.1 Justification of the Uncertainty Principle — 136
  - 6.6.2 The Uncertainty Principle and Creation — 137
  - 6.6.3 Uncertainties about the Uncertainty Principle — 138
  - 6.6.4 Relationship between the Uncertainty Principle and Other Principles — 138
- 6.7 The Correspondence Principle — 139
- 6.8 The Complementarity Principle — 141
  - 6.8.1 Slit-Grid Experiment — 143
- 6.9 Quantum Tunneling and Scattering — 144
- 6.10 Quantum Entanglement — 146
  - 6.10.1 Threatening Features of Entanglement — 147
  - 6.10.2 Front Lines of Entanglement — 149
- 6.11 Quantum Measurement and Collapse of Wavefunction — 150
- 6.12 Conservation Principles in Quantum Mechanics — 154
- 6.13 Characteristics of Double-Slit Experiment — 155
  - 6.13.1 Dual Nature — 156
  - 6.13.2 Effect of Observation and Knowledge — 157

7 General Assessment of Quantum Theory — 160
- 7.1 Historical Evolution — 160
- 7.2 Classic-Quantum Duality — 161
- 7.3 Validity Domain — 162
- 7.4 Mathematical Apparatus — 165
- 7.5 Logicality — 165
- 7.6 Correctness — 166
- 7.7 Optimality — 167

	7.8	Experiments in Quantum Physics	167
	7.9	Tone	169
	7.10	Formalism-Interpretation Confusion	170
	7.11	Inconsistencies and Question Marks	170
	7.12	Problem with Causality	173
	7.13	Problem with Locality	174
	7.14	Problem with Realism	174
	7.15	Difficulty of Interpretation	176

8 Schools of Interpretation — 180
- 8.1 Copenhagen Interpretation — 184
- 8.2 Hidden-Variable Interpretation — 188
- 8.3 Many-Worlds Interpretation — 193
- 8.4 Transactional Interpretation — 199
- 8.5 Consistent Histories Interpretation — 200
- 8.6 Our Interpretation — 201

9 Challenges to Quantum Theory — 205
- 9.1 Paradoxes — 206
  - 9.1.1 Bubble Paradox — 207
  - 9.1.2 Schrodinger's Cat Paradox — 207
  - 9.1.3 Wigner's Friend Paradox — 211
  - 9.1.4 Suicide Paradox — 211
  - 9.1.5 Other Paradoxes — 212
- 9.2 Incompatibility — 212
  - 9.2.1 Clash with Other Theories — 213
  - 9.2.2 Inconsistency with Other Theories — 213
- 9.3 Incompleteness — 213
  - 9.3.1 The EPR Argument — 214
  - 9.3.2 Bell's Theorem — 216
  - 9.3.3 Bell's Experiments — 221

10 Appendix — 224
- 10.1 Slit-Grid Experiment — 224
- 10.2 The Original EPR Argument — 224
  - 10.2.1 The Essence of the Original EPR Argument — 225
  - 10.2.2 Realism — 226
  - 10.2.3 Completeness and Correctness — 227
  - 10.2.4 Other Issues about the Original EPR Argument — 228
- 10.3 Proofs and Arguments in Support of Bell's Theorem — 229
  - 10.3.1 A Proof of Bell's Theorem — 229
  - 10.3.2 Another Proof of Bell's Theorem — 237
  - 10.3.3 General Argument in Support of Bell's Theorem — 239
  - 10.3.4 Geometric Argument in Support of Bell's Theorem — 240
- 10.4 Aspect's Experiment — 241

Epilogue — 243

References — 246

Index — 247

Author Notes — 256

Nomenclature

In the following list, we define the common symbols, notations and abbreviations which are used in the book as a quick reference for the reader.

$\langle \; \rangle$	average (or mean or expectation) value
$[\hat{A}, \hat{B}]$	commutator of operators $\hat{A}$ and $\hat{B}$
$\sim$	comparable in size
* (asterisk)	complex conjugate
$\cdot$ (dot)	dot product
∇	nabla differential operator
ˆ (hat)	operator
3D	3-dimensional
a, **b**, **c**	unit vectors (in 3D space)
c	the characteristic speed of light in vacuum ($= 299792458$ m/s)
det	determinant
dV	infinitesimal volume element
E	energy
Eq., Eqs.	Equation, Equations
h	Planck's constant ($\simeq 6.62607 \times 10^{-34}$ J s)
$\hbar$	reduced Planck's constant ($= h/2\pi$)
H	Hamiltonian
$\hat{H}$	Hamiltonian operator
i	imaginary unit
iff	if and only if
LHV	local hidden-variable
m	mass
O	observable
$\hat{O}$	operator representing the observable O
p	magnitude of momentum
p	momentum
$\hat{\mathbf{p}}$	quantum operator for momentum ($-i\hbar\nabla$)
p_x, p_y, p_z	components of 3D momentum vector in x, y, z directions
P	probability
r	position vector
$\hat{\mathbf{r}}$	quantum operator for position
r, θ, ϕ	spherical coordinates
$\hat{S}_x, \hat{S}_y, \hat{S}_z$	operators representing components of spin in x, y, z directions
$\hat{S}_\theta$	operator representing component of spin in θ direction
SG	Stern-Gerlach
t	time
T	kinetic energy
$\hat{T}$	kinetic energy operator
v	speed
V	volume
V	potential energy
$\hat{V}$	potential energy operator
x, y, z	coordinates of 3D space (usually rectangular Cartesian)
α, β	spin-up state, spin-down state
α_x, β_z	spin-up state along x axis, spin-down state along z axis
Δ	uncertainty

θ	angle between two directions or vectors (possibly polar angle)
λ	wavelength
λ	hidden variable(s)
μ	expectation value (for the product of spins)
μ_{LHV}	expectation value (for the product of spins) according to local hidden-variable theories
μ_{qm}	expectation value (for the product of spins) according to quantum mechanics
ν	frequency
ρ	probability density
ψ	wavefunction
$|\psi|^2$	probability density (in quantum mechanics)

Chapter 1
Preliminaries

In this chapter we provide general clarifying remarks about issues related to the basic terminology and conventions used in the writing of the book (to avoid potential confusion in the future). We also provide a set of brief notes about quantum physics to introduce the subject and familiarize the readers with its basic features and elements. Similar preparatory investigations related to quantum physics (such as the double-slit experiment and comparison between wave and particle) are also included. We also include some general discussions about certain aspects required in our future investigations (e.g. the principles of causality and locality).

1.1 General Remarks

In this section we present a number of general remarks (related to the conventions, terminology and commonly occurring issues in this book) outlined in the following points:

• As declared in the title of the book and indicated in the Preface, this book is about the epistemology of quantum physics (rather than the formal and technical aspects of this theory) and hence some of the investigated issues require prior familiarity with some technical aspects and details that may not be found in this book (although they usually should be outlined somewhere in the book). Therefore, the unfamiliar reader should refer (if necessary) to the literature. We also refer the reader to the literature for any required knowledge that is not directly related to quantum mechanics (e.g. general physics or calculus) although we may casually explain some of these requirements.[2]

• The terms "formalism" and "interpretation" are used extensively in this book (as well as in the literature of quantum theory). Broadly speaking, the formalism of a scientific theory is the theory in its technical and rigorous scientific form (which is usually expressed by well-defined terminology and symbols and cast into strict qualitative or/and quantitative relations), while the interpretation of a scientific theory is a rational explanation and justification to the formalism that can be envisaged as a model that we can understand and appreciate within our classic and macroscopic experiences.

• We use "classical" (in terms like classical theory or classical physics or classical mechanics) to oppose "quantum" although the term can be used to oppose other non-classical theories and branches of physics such as Lorentz mechanics. Yes, if we use this term in a different meaning then we will notify the reader about this exceptional use.

• Our investigation in this book is restricted to "classical" quantum mechanics[3] (with no combination with other branches like Lorentz mechanics or electromagnetism) and hence we have no interest, for instance, in quantum field theory or quantum electrodynamics (apart from some casual references).

• We use expressions like "quantum scale" to refer to the scale of phenomena that should be investigated and described by quantum physics (rather than by classical physics). Likewise, expressions like "quantum particle" and "quantum system" should have similar interpretations. So, "quantum *something*" in such expressions should mean something that belongs to the quantum world and hence its investigation requires the use of quantum theory rather than classical theory. We also use "macroscopic" as an alternative to "classical" and "microscopic" as an alternative to "quantum".[4]

---

[2] In fact, we tried to make the book as complete and self-contained as possible (considering the level and scope of the book) although this may not be the cases with regard to the formalism of quantum mechanics.

[3] We may also call it "elementary quantum mechanics" or "elementary quantum theory".

[4] Although "microscopic" may be used conventionally (or conveniently) to a scale between "macroscopic" and "quantum" (or sub-microscopic), we depart from this convention. So, our criterion for being "microscopic" is the applicability of quantum mechanics primarily (which is typically represented by the atomic and sub-atomic scale).

1.1 General Remarks

- The terms "quantum theory", "quantum physics" and "quantum mechanics" are generally used in similar meanings although the context occasionally dictates what is more appropriate (usually based on considerations related to specificity and generality).[5]
- We may use vector terms (like momentum) to refer to the magnitude of vector quantities. For example, we may say "the momentum p" where p is the magnitude (which is a scalar) of the momentum (which is a vector).
- We use "thinker" to mean any living being (which in reality is a human being) that is capable of abstract thinking.
- We use "rational" in expressions like "rational intellectual process" or "rational knowledge" to refer to types of intellectual activities and products whose objective is to understand and describe reality and hence they are based on rational thinking and logical reasoning. This is to exclude other types of intellectual activities and products such as art and literature.
- We use the words "metaphysics" and "metaphysical" to refer to types of knowledge and objects that do not belong to the physical (or natural) world. For example, "atom" is physical while "angel" is metaphysical.
- We generally rely on the context to lift the ambiguity of some commonly occurring expressions. For example, Schrodinger's equation has a time-dependent form and a time-independent form and hence we may use "Schrodinger's equation" without the qualifiers "time-dependent" or "time-independent" where the context is sufficient to identify the meaning. This similarly applies to the wavefunction where we use the same symbol (i.e. ψ) for the time-dependent and time-independent where the context [or the argument associated with ψ such as $\psi(x,t)$] should lift any ambiguity (noting that ψ is used generically or generally and hence it can include either or both).
- We use "quantum system" very often generically to mean any physical object or configuration that is subject to the rules of quantum physics, and hence it has very wide (and even multiple) meaning that includes things like "wavepacket" and "single particle" (as well as composite quantum objects).
- Considering that "observable" means something that can be observed and measured, in many contexts we use "observation" and "measurement" (and their derivatives) interchangeably or use one to refer to both (noting that observation is a kind of measurement and vice versa).
- In some cases and contexts "observer" should mean the entire observation system (as opposite to the observed phenomenon) and hence it includes the measuring equipment as well as the (human) observer.
- Despite the obvious difference between "observation" and "experiment" in their primary meanings (as passive and active interventions), they may be used interchangeably as well as using one to mean both.
- We use "superluminal speed" to mean speed that exceeds the characteristic speed of light in vacuum (i.e. c) regardless of being finite or infinite and regardless of being measurable/detectable or not (although for more clarity we may indicate in some contexts "infinity" as a distinct possibility).
- We use "action at a distance" in two meanings: interaction between spatially-separated objects with no intervening physical signal that communicates this interaction, and interaction between spatially-separated objects with an intervening physical signal whose speed is superluminal (including infinite). We refer to the former with "spatial action at a distance" and refer to the latter with "temporal action at a distance".
- Regarding our view about special relativity and "Lorentz mechanics" (which we refer to frequently) and their relation, we refer the reader to our book "The Mechanics of Lorentz Transformations" where we differentiate between "Lorentz mechanics" which is the (accepted and empirically-endorsed) formalism and "special relativity" which is the (rejected and logically-inconsistent) interpretation of "Lorentz mechanics".[6] It is noteworthy that we generally use "Lorentzian" in place of "relativistic" unless special

---

[5] In fact, "quantum mechanics" strongly suggests the formal and technical aspect of the quantum theory (which is rather loose and general and can include interpretation for instance) and hence "quantum mechanics" is commonly used to refer to the formalism of the theory. Regarding "quantum physics", it is between "quantum theory" and "quantum mechanics" in generality and specificity although it strongly suggests the scientific aspect of the theory. However, it should be noted that "physics" in "quantum physics" may refer to the actual laws that govern the outside world and in this sense it is different from the theory.

[6] As pointed out in our book "The Mechanics of Lorentz Transformations", special relativity takes the credit for the success of Lorentz mechanics (while Lorentz mechanics takes the blame for the failures and inconsistencies of special relativity). The culprit of this is the confusion between formalism and interpretation and the lack of understanding that the successes belong to the formalism (i.e. Lorentz mechanics) and the failures belong to the interpretation (i.e. special relativity).

relativity is intended specifically (rather than Lorentz mechanics). The reason is our rejection of special relativity as an illogical interpretation of the formalism based on the Lorentz transformations.

- Considering that the book is written primarily to individuals of general scientific education (at the level of undergraduate in science) rather than to specialists in this filed, clarity is a primary factor in the style of writing. Accordingly, we generally use very simple language and presentation methods with many cross referencing as well as some occasional repetition. We hope this will make the book more useful and of broader audience (although it may become boring to some occasionally).[7]
- For the convenience of the readers of the electronic versions of this book, all references and cross references are hyperlinked (although they are not highlighted, with colors for example, to avoid potential distortion).
- The prefix "eigen" (which precedes words like function or vector or value or state) is a German word meaning "own" or "proper" or "characteristic" and is commonly used in mathematics (e.g. linear algebra) and quantum mechanics. Although we assume that knowledge of the technical meaning and significance of this prefix is a preliminary background requirement, in the following we explain briefly the mathematical concept of "eigenvalue equation" (also known as "eigenequation") which is the pivot of the "eigen" terminology. In summary, an eigenvalue equation is a mathematical relation of the form:

$$\hat{O}\psi = C\psi \tag{1}$$

where the operator $\hat{O}$ acts on a function ψ to reproduce ψ but scaled by a constant C. Accordingly, ψ is an eigenfunction of this equation and C is its eigenvalue (corresponding to the operator $\hat{O}$).

- Words like "dominant" or "mainstream" or "common" or "rare" (or ... etc.) may be used in this book to describe an opinion or view or interpretation or theory (or ... etc.). However, it should be noted that descriptions (or rather judgments) like these represent our opinion and feeling based on our inspection of the literature (as well as on similar opinions, feelings and judgments that we find in the literature); otherwise we are not pollsters or statisticians. We should also note that "dominance" and its alike vary with time and hence what was dominant 4 or 5 decades ago may not be dominant now. So, the reader should treat such judgments (whether from us or from other authors) with caution and skepticism.

1.2 Introducing Quantum Physics

In the following points we summarize some of the basic properties of quantum physics and outline its nature (noting that most of these properties will be investigated more thoroughly later on):[8]

1. In physics, the word "quantum" means a discrete physical entity or property that can take only certain (countable) values. The word is usually associated with tiny objects (at atomic and sub-atomic scale) and hence it suggests "tininess" as well as "discreteness".[9]
2. The (primary and typical) domain of quantum physics is the atomic and sub-atomic physical systems (which we call quantum systems).[10] However, the validity of quantum physics is extended (rather theoretically and hypothetically) by the correspondence principle (see for instance § 4.7.2) to classical systems although in reality this does not have a practical value in most cases (such as in replacing or

---

[7] I should confess that quantum theory (in its formalism and interpretation) is riddled with subtleties and ambiguities (as well as many controversies) and hence it is (unlike some other theories and branches of science) not easy to understand and appreciate without persistent effort and direct experience and work on it (and sometimes even this may not be sufficient). Nevertheless, I hope I succeeded overall in my effort and achieved (at least) some of my objectives.

[8] Considering the introductory nature of this section, the presented materials are not sufficiently detailed or sufficiently rigorous. Also, some points may depend on (or represent) certain interpretations.

[9] It is noteworthy in this context that in the early days of quantum mechanics the subject was referred to or associated with "atomism" (which also suggests both "discreteness" and "tininess").

[10] As indicated earlier, "system" here (and in similar contexts) should be understood in its general sense which should include individual objects (e.g. a single particle). Also, whether "sub-atomic" should include all scales smaller than atomic scale or it may not apply to scales that are smaller than a certain sub-atomic scale is not clear. In fact, the applicability of quantum physics to scales much smaller than atomic size is questionable due to the lack of observational evidence at these scales (noting that such scales are beyond our observational capabilities and hence even their existence, let alone their physics, is not clear). We should also note that "atomic" should be understood as a typical scale and hence it should include molecular scale (possibly with some restrictions on some "giant" molecules).

1.2 Introducing Quantum Physics

reproducing Newton's laws) since in classical systems classical physics is employed (with no need for quantum physics and even with no realistic possibility of using it or benefiting from it).[11]

3. There are two main types of formulation of quantum physics: wave mechanics and matrix mechanics (see § 5.3.2). They are both valid and usable.
4. From a mathematical viewpoint, quantum mechanics may be described as a combination of Hilbert spaces (of finite or infinite dimensionality) and linear algebra (with some use of other mathematical branches and techniques like calculus and complex numbers).[12]
5. A quantum system is physically represented by its wavefunction (or state function; see § 4.1 and § 6.1) which may be described as a complex-valued probability amplitude (meaning that the probabilities for the possible outcomes of measurements conducted on the system are derived from it as will be clarified next). Accordingly, the physical state of a quantum system is represented mathematically by its wavefunction (in Hilbert space).[13]
6. The tempo-spatial evolution of the wavefunction is governed deterministically by the Schrodinger equation (see § 4.3).
7. The wavefunction ψ has no physical meaning (i.e. in itself and on its own).[14] Yes, the square of its modulus (i.e. $|\psi|^2 = \psi\psi^*$ where the asterisk means complex conjugate) is physically meaningful (i.e. it has physical implications and consequences in a probabilistic sense).
8. According to the probabilistic interpretation of quantum mechanics (which we may describe as the standard and commonly-accepted interpretation of quantum mechanics), the intensity of the wavefunction of a quantum object (where this intensity is given by the square of its modulus $|\psi(\mathbf{r},t)|^2$) represents the probability density for the presence of the object in a tiny volume of space around a given point $\mathbf{r}$ in space at a given instant t in time.[15]
9. In a given physical setting, only certain physical quantities are observable where the outcome of observation is generally probabilistic in nature (as outlined already) and may have an intrinsic uncertainty.
10. If two observables can (in principle) be known precisely at the same time then they are *compatible*, and if they cannot then they are *incompatible* and hence they are subject to the so-called uncertainty principle (see for instance § 4.7.1 and § 6.6).[16]
11. The wavefunction is supposed to collapse following (or rather by) observation/measurement (where the meaning of "collapse" is rather fuzzy and controversial as will be investigated later on; see for instance § 5.7 and § 6.11).[17]

---

[11] In fact, the extension by the correspondence principle is practically restricted to quantum systems of large quantum numbers whose results can be obtained from classical physics (see for instance § 4.7.2 and § 6.7). The theory can also be combined with other branches of physics (e.g. electromagnetism) and hence it has applications in macroscopic systems. The theory may also provide a conceptual or statistical basis for theories of macroscopic systems. This issue (and related issues) will be investigated in detail later on.

[12] Hilbert space may be defined concisely as an inner product (and complete) vector space. Hilbert space may be seen as a generalization of the concept of Euclidean space (of finite dimensionality) to include spaces of infinite dimensionality. Also see footnote [165] on page 79. We should also note that "linear algebra" here should be understood in its more general formulation than what is provided in elementary linear algebra texts.

[13] Being complex-valued means that the use of complex numbers and variables is a necessity in quantum mechanics. This is unlike the use of complex numbers and variables in other branches of physics which is no more than a convenient choice or option. We also note that it may be more appropriate to use "amplitude" for the modulus of the wavefunction rather than for the wavefunction itself. It should also be noted that we have time-dependent wavefunction $\psi(x,y,z,t)$ which has temporal dependency as well as spatial dependency and time-independent wavefunction $\psi(x,y,z)$ which has only spatial dependency (where we use the same symbol ψ for both but possibly with different arguments).

[14] We may also say: it is not physically observable.

[15] We assume ψ to be normalized to unity (as required by probability). This applies to similar occurrences in this book. We should also note that other physical quantities (such as the expectation values) can also be obtained from the wavefunction (see § 4.5).

[16] Incompatible observables may also be called complementary or conjugate observables or canonically conjugates (i.e. to each other). We note that compatible observables possess a complete set of simultaneous eigenfunctions while incompatible observables do not.

[17] "Collapse" broadly means a transition (or reduction) from a superposition state to a definite state. It should be noted that "collapse" is essentially an interpretative aspect of quantum theory although it is generally included (axiomatically) in the commonly-accepted formulation of quantum mechanics. Accordingly, collapse is denied by certain schools of formulation and interpretation of quantum physics. We should also note that other terms (such as "reduction") are used

12. The collapse of wavefunction is generally indeterministic and is governed by the postulate of measurement (rather than by the Schrodinger equation which is deterministic).

Finally, it is useful to be aware of the fact (which will be concluded and repeated again and again) that quantum theory is very successful in formalizing the (quantum) physical phenomena but it is not so in capturing the "spirit" behind these phenomena and reflecting the essence of the physical laws that govern theses phenomena. So, from the perspective of "conquering" the world (which is one of the main objectives of science; see § 2.1) the theory is very successful (at least so far) but from the perspective of understanding the world (which is another main objective of science) it is much less successful (at least so far). It seems as if the quantum theory describes the physical phenomena by their symptoms rather than by their physical essence.[18] The awareness of this important fact (or rather assessment) at this early stage in the book should make the reader more able to appreciate many aspects of the quantum theory and its virtues and vices. So, the reader should keep in mind that quantum theory is brilliant empirically and dull epistemologically.

1.3 General Factors to Consider in Assessing Science

These are general factors that should be taken into account when assessing any scientific theory and its interpretation (or in fact any type of rational knowledge). The main and most common of these factors are briefly investigated in the following subsections.

1.3.1 Humanism

Science is a product of mankind and hence it is characterized by the features of our cognitive system and how we think and interact with our environment (i.e. Nature). In fact, "humanism" does not only represent the "species" factor that enters in the determination and identification of the scientific process and knowledge, but it also includes many other factors such as cultural, social and personal factors. This, for instance, could partly explain the fact that we can have more than one correct scientific theory for describing and formulating a single physical phenomenon. As we will see (refer for instance to § 2.4.3), any physical phenomenon can be correctly described and formulated by more than one scientific theory without violating the rules of logic or the principles of reality and truth (see § 2.4.1 and § 2.5.1). This is because any type of human knowledge is not really an ideal image-reflection (or image extraction) of reality but it is rather an interaction[19] between the thinker and his environment (i.e. Nature) and hence this process (as well as its outcome which is the "extracted image") is affected by many humanist factors. In other words, the acquired knowledge (or the "extracted image") is actually an artistic impression of the thinker by Nature (or the organism by environment). Thus, it is more like an artistic impressionist painting than a high-definition photograph (or mirror-reflection).

So in brief, *correct* science does not represent a pure image and mirror-reflection of reality (or absolute truth) but it is a mix of reflection (or discovery) and invention (or fabrication) where this process and its outcome are affected and determined by many humanist factors (in the broadest sense of "humanist" as indicated above). As a result, the progress (or rather the development or evolution whether positively or negatively) of science is a continuous process that fundamentally follows our cultural and social evolution (as well as other types of evolution including our evolution as a species although this type of evolution is of very different nature and on very different time scale). Accordingly, we should not search for (or expect to reach) a final version of science or an ultimate scientific theory. In fact, science will reach its final *static*

---

in place of "collapse". So, at this stage in the book we can describe "collapse" generically and briefly as: the change in the wavefunction caused by measurement.

[18] In fact, most of the modern "scientific" advances (particularly in the quantum field) are basically of technological, rather than scientific, nature. In other words, they represent empirical and practical advances rather than conceptual and epistemological ones. The success of quantum mechanics empirically and practically and its failure conceptually and epistemologically should indicate that conquering the world can be easier than understanding it. This conclusion may also be drawn from our experiences with other branches of science and types of knowledge.

[19] It should be noted that this sort of "interaction" is not supposed (as far as humanism is concerned) to interfere with the observed phenomenon and affect its nature (unlike the interaction that we will meet in § 1.3.3).

state when humanity reaches its ultimate *static* state (by extinction or by conversion to another species which may halt this continuously-evolving process).

1.3.2 Scale of Observer and Observed

This factor may be seen as originating from the previous factor (although we consider it separately due to its specificity and particular importance to quantum theory). The essence of this factor is that because we are a species of a given size, our sensory and conceptual experiences are acquired from the natural world (as perceived by us) on the scale that is comparable to our size (and in fact it is processed by our "humanist" nature which is the product of an evolutionary process on this scale and size). This means that the validity of the patterns and models (including concepts and paradigms) that we acquire from our past physical experiences and we use as elements and prototypes in our scientific theories is generally restricted to certain scale and size and hence these patterns and models are not necessarily valid as physical models to the world on other scales and sizes.

This is based on the fact that the patterns and models (that are acquired and developed as abstractions from the physical world) are generally scale-dependent and hence some patterns and models may be compatible with certain scales but not with other scales. In other words, not all patterns and models are valid for any scale and size (or not all patterns and models are valid for describing the physical world at any scale and size). For example, the physical pattern or model of "wave" is acquired from our physical experiences at classical (or macroscopic) scale and hence the "wave" model may not be sensible to describe physical objects and phenomena at pico or femto scales because these scales were beyond our reach during the acquisition and development of the "wave" model and hence this model may not reflect the nature of the world at these scales. This similarly applies to many other concepts, models and patterns like "color" or "brittleness".[20]

In the following points we provide more clarifications about the "scale of observer and observed" as a factor that should be considered in the assessment of scientific theories:

1. "Scale" is more general than "size" and hence it includes for instance speed (i.e. how fast or slow) and intensity (i.e. how strong or weak a certain physical property like electric charge). So in brief, "scale" is about the magnitude of a physical attribute (whether size or something else). Yes, for quantum physics "size" is the primary attribute of the "scale" factor (also see point 3).
2. As the title of the present subsection indicates, the "scale" factor is actually about the *relative* magnitude as determined by the scale of the observer in comparison to the scale of the observed phenomenon. This is because the scale of the observer should primarily determine the scale of the observed phenomenon to which the patterns and models of the observer are appropriate or not.
3. The "scale" factor is what makes quantum physics and Lorentz mechanics special among scientific branches and theories. This is because the extreme smallness of quantum objects and the extreme fastness of Lorentzian objects make them unusual and hence our physical models and patterns (which we developed and acquired during our long evolutionary history dealing with macroscopic and slow objects) are not suitable to describe and quantify quantum and Lorentzian phenomena. Accordingly, classical physics (as opposite to both quantum and Lorentzian physics) which is based on our inherited (macroscopic and slow) models and patterns becomes an invalid scientific theory for describing and quantifying quantum and Lorentzian phenomena. In brief, the "scale" factor (in size for quantum and in speed for Lorentzian) is the main factor that invalidates classical physics in these fields.

In fact, we believe that this should similarly apply to other scientific theories and branches such as astronomy, astrophysics and cosmology where the scale in space and time (as well as other properties and attributes) is unusual and hence it should not be considered classical or macroscopic (see point 7 of § 7.3). Accordingly, we may need to revise our principles, laws, assumptions, concepts, models, etc. that we are currently using in the investigation of subjects like astronomy and cosmology because these (classical/macroscopic) principles ... etc. may not be appropriate to use due to the scale factor.

---

[20] In fact, if the invalidity of "wave" microscopically is suspicious the invalidity of "color" or "brittleness" (and their alike) should be obvious because they are fundamentally macroscopic. For instance, having red and green electrons or brittle and ductile atoms is obviously nonsensical.

In fact, the failures[21] in these fields (even by some supposedly "non-classical" theories like general relativity) could be a sign for the failure of our (classical/macroscopic) principles ... etc. by the scale factor.

4. As we will see, the failure of the interpretations of quantum mechanics and the inconsistencies they have (which sometimes reach even the formalism) originate partially from using the macroscopic concepts and patterns to describe quantum phenomena. In our view, the scale factor is one of the primary reasons for the epistemological failure of quantum theory (as reflected in the difficulty of finding an acceptable interpretation to its formalism despite the empirical success of this formalism within the limitations of indeterminism). Also see for instance § 7.6 and § 8.6.

5. The subjective nature (to a certain extent) of quantum theory (since the observer seems to be involved in the observed physical phenomenon through measurement) may be partly explicable by the "scale" factor (as well as humanism). Also see § 1.3.3, § 2.7.5 and § 7.9.

6. The scale factor could have an effect even on some of our most fundamental and seemingly-intuitive concepts like space and time or our philosophical and epistemological patterns and paradigms such as our concept about physical reality and its nature. For example, the paradigms of "space" and "time" may not be applicable or appropriate (at least in their exact sense) at very large or very small scales (such as the scales of Universe and nucleon). Similarly, "macroscopic realism" may not be exactly applicable to "microscopic realism" (see for instance § 1.10 and § 8.6). As we will see, the epistemology itself could be scale dependent.

7. Can the scale factor affect logic (and hence we may have modified or different versions of logic for worlds and realities of different scales)? This issue will be investigated later on (see for instance point 10 of § 2.5.1 as well as § 8.6).

8. The aforementioned fact that our patterns and models are scale-dependent should lead to an obvious intrinsic limitation in our ability to describe and interpret (in a realistic way) physical phenomena incommensurate to our scale. In fact, there are many examples in physics about this limitation and how we use (justifiably or unjustifiably) scale-limited patterns and models approximately or inappropriately to describe and interpret phenomena that are beyond our scale because we have no other (more realistic) choice. This is seen vividly in quantum physics where the concepts of "particle" and "wave" (which are essentially macroscopic) for instance are used to describe quantum objects and systems despite the scale-limited nature of these concepts (as can be inferred from the use of things like "duality" and "complementarity" to remove obvious inconsistencies resulted from the use of these concepts noting that there is no such duality or complementarity in real particles and waves and from a classical perspective). Similarly, models like electron spin or electron circulation around the nucleus (which suggest macroscopic-type mechanical models of these quantum phenomena) are used despite their limitation and misleading nature (by the confession of quantum physicists themselves).[22] In fact, this highlights an essential limitation of science in general. After all, we are *humans* and of certain physical *scale* and hence we have no direct access to entities that are beyond our familiar experiences which are acquired and shaped according to our type and scale. Therefore, we should be content to use these scale-limited models and patterns (but cautiously) as approximate prototypes to investigate these entities.[23] The danger, however, emerges when these models and patterns are treated (recklessly) as realistic and exact prototypes (and this sort of recklessness seems to be common in science as in all other aspects of human life).

We should finally note that quantum physics provides vivid examples for this type of limitation (i.e.

---

[21] These failures are demonstrated, for instance, by the need for dark energy or dark matter or creation or nonsensical consequences like travel in time ... etc. which modern physics (in these subjects in particular) is full of.

[22] A more elaborate example of the limitations of classical-type models in describing the quantum world can be found in the many failures of the classically-based (and even semi-classically-based) models of the electron and the nonsensical and contradictory consequences that these models lead to. However, the quantum mechanical description of the behavior of electron (despite the failure of the proposed *visual* models about its nature) is very successful despite its phenomenological nature and lack of *classical sensibility*.

[23] If we feel confused when we move from one city to another city or from one society to another society (of different culture), then how can we expect to have full understanding when we move from one world to another world (e.g. from macroscopic to microscopic).

limitation by scale) because we are generally aware of our limitation in the quantum field. However, in our view (which we outlined earlier) this limitation is more common and widespread than what the physicists may think. For example, this limitation could or should apply to astronomy and cosmology as much as it applies to quantum physics because in astronomy and cosmology we also deal with unfamiliar scale-dependent phenomena (although this time the scale limitation arises from other scale factors such as being very large and extensive in space and time).

1.3.3 Inertness of Observation

This means that observation is presumably a passive process whose role is to capture an already-existing phenomenon without affecting (or altering or disturbing) the observed phenomenon by the observation process. In fact, the root of this is the presumed independence of the observer from the observed which any conventional (or classical) observation process is based on. This factor (if accepted as a factor for assessing scientific theories) could lead to the rejection of the outcome of any observation that alters the observed phenomenon since such an observation is not really an observation but it is a mix of interaction and observation (or rather an observation of an interaction of the observer with the observed) and hence it cannot be an honest (or exact or precise) representation and description of the supposedly-observed phenomenon. For example, if we want to measure the momentum of an electron by bouncing a photon off it (where the impact of the photon affects the momentum of the electron without the possibility of distinguishing or isolating the effect of impact) then what is actually measured is not the momentum of the electron (immediately prior to the impact which is the objective of the measurement) but the altered or modified momentum.

More clarifications about the inertness (or passivity) of observation and its role as a factor that should be considered in assessing scientific theories are given in the following points:

1. The effect of this factor (i.e. the inertness of observation) will appear in assessing certain aspects of quantum theory (e.g. in the theory of quantum measurement and the collapse of wavefunction or the uncertainty principle and its background) where the distinction between the classical view and the quantum view will become apparent. See for instance 4.7.1, § 5.7, § 6.6 and § 6.11.
2. The "independence of the observer from the observed" (which inertness is supposed to be based on) may be considered as another principle of reality and truth (or a derivative of one of these principles; see § 2.4.1) where the reality is supposed to exist in a completely determined and independent form regardless of the observer and the process of observation, and hence the role of observation is to capture and reveal this preexisting and completely-determined reality, as required by inertness, rather than capturing a different reality (i.e. the reality after alteration) or even worse by creating the reality (by the observation process) that to be captured.
3. Inertness of observation should be seen as a necessary condition for any acceptable *objective* observation whose purpose and aim is to capture an image of the reality (which is supposedly independent of the observer and observation according to the *objectivity* constraint). This is because without inertness what is actually captured by observation is not the independent and original reality but the influenced and modified reality (which is a different reality from the independent and original reality which the observer is supposed to capture objectively).
4. In its strict sense and idealistic form, inertness of observation may be seen as an ideal situation (at least for observation of quantum systems) and hence it may not be achievable practically although the effect of interfering (i.e. of the observer with the observed) can, in general, be minimized. However, there are some examples of certain types of observation (e.g. by delayed choice or by entanglement) that supposedly eliminate any disturbance to the observed by observation.[24] Anyway, in an approximate sense and from a practical perspective inertness is certainly achievable when the disturbance is minimal

---

[24] In delayed choice observation the observation occurs (indirectly) after the occurrence of the phenomenon (see § 1.6.2 and § 6.13.1), while in entanglement the observation (or measurement) of an entangled object will give information (i.e. observation) of the other entangled object and hence the other entangled object is "observed" inertly (see § 5.6 and § 6.10). However, it should be noted that the inertness of observation in delayed choice observations may be questionable (at least) in some cases such as when the knowledge supposedly has a back-in-time effect (see for instance § 1.6.2 and § 6.13.2).

and beyond the capability of the experimental resolution. For example, if we measure the momentum of an energetic heavy atom by the impact of a microwave photon then the disturbance is minimal and hence the observation is practically (although not idealistically) inert. So, inertness is achievable (at least approximately) in quantum systems as well as in classical systems.

5. Referring to our discussion in § 1.3.1, inertness in a strict and idealistic sense is not achievable but from another perspective or sense that is: the assumption of complete pre-determination of reality and its total independence of the observer is not realistic because observation is an observer-dependent process and hence its outcome depends on the nature and properties of the observer and his observation equipment (e.g. his cognitive system, his cultural and social background, his time and location, his background knowledge, type and resolution of his apparatus, etc.). As indicated earlier, human (rational) knowledge is not an ideal image-reflection of reality but rather an artistic impression of reality since knowledge contains an element of creation and invention and it is not a pure revelation and discovery. In brief, even if reality is totally independent and determinate ontologically, it is not so epistemologically. The error in the idealistic view of realism is that it confuses ontological realism (which is absolutely legitimate and acceptable as a philosophical choice even in its extreme and idealistic form) with epistemological realism (which is absolutely legitimate and acceptable as an epistemological choice and even as a necessity but not in its extreme and idealistic form which is untenable due to the fact that observation requires an observer and hence the outcome of observation depends in its nature and details on the nature and properties of the observer and his observation equipment). Also see § 1.10, § 7.14 and § 8.6.

1.4 Comparison between Wave and Particle

"Wave" and "particle" are classical concepts used in the quantum theory. Therefore, in the following comparison we generally follow a classical approach and hence some of the following properties may not fit exactly with the quantum theory. The purpose of this is to induce our intuition (or common sense) about these paradigms which were conceptualized primarily in classical physics and were imported later to quantum physics. We outline this comparison in the following points:

1. **Localization**: particles are localized objects and hence they are well-defined spatially while waves are not localized and hence they are diffuse (or dispersed or delocalized) in space. This is demonstrated, for instance, by the following properties:
    - Particles have well-defined trajectories while waves do not.
    - Waves can pass each other (since they can simultaneously occupy the same space thanks to their diffuse nature) while particles cannot (since they are localized and hence no more than one particle can occupy the same position at the same time).
    - Waves experience diffraction (essentially as a consequence of delocalization and lack of well-defined trajectory) while particles do not.[25]
2. **Mass and Energy**: primarily, particles have mass while waves have energy (noting the link between mass and energy through certain relations like Poincare's mass-energy equivalence relation).
3. **Physical properties**: although the distinction between particles and waves by mass and energy may be the most prominent of its kind, there are similar distinctions based on physical properties other than mass and energy. For instance, particles have charge,[26] volume and geometric shape but waves do not have. On the other hand, waves have amplitude, frequency and wavelength but particles do not have.
4. **Superposition**: waves superimpose on each other while particles do not. This is demonstrated, for instance, by properties like interference which is a distinctive wave property.[27]

---

[25] We note that diffraction may be seen as a consequence of another property of waves. In fact, we can even put diffraction as an independent point (i.e. not under localization or any other point). Anyway, this is not an important issue as long as we agree that diffraction is a wave property.

[26] Even neutral particles have charge, i.e. zero. We may also characterize particles by "chargeability" (i.e. being chargeable) and hence avoid the zero-charge "embarrassment".

[27] In fact, superposition may be seen as a consequence of delocalization since waves are spatially diffuse and hence they can share the same space where they interact with each other (constructively and destructively) to produce phenomena

5. **Spatial intensity**: as a result of superposition and interference (in waves) the intensity of waves and particles are fundamentally different where the intensity of waves (i.e. where their strength is high/low) is subject to interference while the intensity (or concentration) of particles is not (and hence it is subject to simple accumulation).[28]

We should finally note that in quantum mechanics the paradigm of "wavepacket"[29] is used to model quantum objects since quantum objects generally demonstrate dual particle-wave behavior and hence they (like wavepacket) are neither particles strictly and plainly (since they are diffuse and delocalized) nor waves strictly and plainly (since they vanish far away from where they are supposed to be and hence they are comparatively localized). As we know (from Fourier analysis), the wavepacket is made from a superposition of waves of varying wavelengths and amplitudes.

1.5 Comparison between Massive and Massless Objects

In this section we make a comparison between massive and massless objects. This comparison is outlined in the following points (noting that these points represent general guidelines rather than strict rules and conditions and they are primarily about quantum objects):

1. Massive objects demonstrate particle properties primarily and wave properties secondarily while massless objects are the opposite (see § 1.4).
2. The wave nature of massive objects is demonstrated as probability (or matter) waves while the wave nature of massless objects is demonstrated as real (or classical or physical) waves (see § 3.2).[30] Also see point 6 of § 5.3.1 and point 4 of § 7.11.
3. Massless objects are intrinsically Lorentzian (i.e. "relativistic") while massive objects may or may not be so (depending on their speed).
4. The difference in nature between massive and massless objects may also be inferred from other physical properties and formulations, e.g. from the expression of momentum p for these types of objects where $p = mv$ for massive objects[31] (noting that massive object has mass m) while $p = E/c$ for massless objects (noting that massless object has energy but not mass); see § 1.4.

1.6 Double-Slit Experiment

It is common in the literature of quantum physics to start the investigation of the subject by introducing this experiment and evaluating its significance. In fact, this experiment represents an ideal prototype for the behavior of quantum objects and their perplexing aspects and attributes which cannot be explained or envisaged by using classical physics or common sense or familiar concepts and patterns. Hence, understanding this experiment and appreciating its significance will save a lot of time and energy in going through many details when we discuss other quantum observations and examine other quantum characteristics and systems (which are usually more complicated in their nature, rationale, setting, procedure, behavior, etc.). So, by just referring to the double-slit experiment in these contexts we can invoke all

---

like interference.

[28] We made "spatial intensity" an independent point in this comparison despite being a consequence of superposition and interference (in waves) because of its special importance in quantum theory since it determines the probability (or statistical) distribution of quantum objects.

[29] Broadly speaking, "wavepacket" is a relatively-localized wave disturbance made of a superposition of waves of different properties (e.g. frequency).

[30] As we see, this comparison should lead to the conclusion that the particle/wave nature in these types of objects is not exactly the same and hence the supposed particle-wave duality (see for instance § 1.6.1 and § 5.3.1) in the two types is not the same. For instance, when we consider the (massive) electron we see that its wave nature is actually a *non-physical* "probability wave" (which effectively determines its statistical distribution in space) while when we consider the (massless) photon we see that its wave nature is actually a physical "electromagnetic wave" (which is made of electric and magnetic fields that are physically observable and experimentally measurable).

[31] For simplicity, we ignored the Lorentz factor (which may be justified either by considering the classical case or by embedding the factor in m). This could also be more general from the perspective of considering classical and Lorentzian mechanics.

the important distinctive features that characterize quantum phenomena and systems and retrieve essential conceptual experiences needed for analyzing, interpreting and appreciating quantum phenomena and their physics. Hence, for this reason (as well as for other similar reasons) we listed this experiment in our preliminaries despite the fact that it could be seen (rather rightly) as a minor subject and an insignificant detail.

Double-slit experiment is well known in classical physics and it dates back (at least) to the days of Thomas Young at the beginning of the nineteenth century. As we know, the objective of the classical double-slit experiment is to demonstrate the wave property of visible light. However, double-slit experiment in its quantum version is identical (in its setting and procedure) to the classical version apart from the use (or rather the consideration) of quantum objects that are usually seen as particles (e.g. electrons).[32] In fact, even light (in its quantum conceptualization as photons) can be used in the quantum double-slit experiment (as will be seen below especially in single-particle version of this experiment where the quantum objects are released and detected individually one at a time). The main objective of the quantum version of this experiment is to reveal the particle-wave duality of quantum objects (including light). To be more general and inclusive we may say: the main objective (among other objectives) of the quantum version of this experiment is to reveal the nature of quantum objects and if they are waves or particles or both or something else. However, as we will see this experiment will also raise questions and highlight problematic issues about the nature of quantum phenomena and objects such as the role of the observer knowledge on the behavior of quantum objects and their self-interference.

In the following subsections we outline the important features of the quantum double-slit experiment and try to assess its significance. However, before we start this investigation we should note that there are some delicate details about this experiment that cannot be determined definitely from the literature (i.e. by their nature they can be available only to the experimenters). As we are not experimenters, we generally trust (or pretend to trust) the experimenters in what they report and rely on their preliminary interpretation and justification (since they have direct first hand experience) although we may need occasionally to do some (rather reasonable) guesswork.

We may also need to express some reservations on certain aspects of the reported details. For example, we may put a question mark on the accuracy (or efficiency) of the detector and if it can be guaranteed to be 100% reliable (i.e. it can detect every single particle). In fact, 100% reliability (or at least high reliability) is required in such experiment if we have to accept some of the claimed results and conclusions (e.g. self-interference). We may also put a question mark on the practical possibility of emitting single particles (e.g. electrons or photons) at least in a highly reliable and systematic way (and at least in some of the reported experiments). Similarly, the mechanism used for localizing the particles on the screen is generally unknown to us (noting that the nature of this mechanism could be significance on the analysis and interpretation of the results).[33]

We should also note that there are many different types, variants and elaborations to the double-slit experiment where specific modifications are introduced on the setting and configuration of this experiment in its original form to test certain quantum mechanical aspects (such as the effect of knowledge on the interference pattern). In fact, the double-slit brand of experiments is one of the richest categories of experiments in quantum physics and in physics in general (as a quick inspection of the research literature should reveal). Apart from its experimental value (which is directly related to the formalism of quantum theory and its rules and principles) the double-slit experiment (in its various versions and modifications) is commonly used as a test bed for examining and assessing the interpretative aspects of quantum mechanics

---

[32] Being identical to the classical version should give us more reasons for not providing a description of its setting and procedure in this book since these details can be found in any physical text about this subject. However, we should note that the double-slit experiment for electrons (and similar quantum particles) may not be achievable in its classic (or photon of visible light) version due to the very short wavelength of the "electron wave" although similar experiments based on the same principles and similar results are achievable (in fact a double-slit experiment with electrons was performed in the 1960s but the setting is more complicated where electrostatic "lenses" were used to magnify the fringes of the interference pattern). Hence, in some parts of the discussion about the double-slit experiment (e.g. for electrons) the "double-slit" setting is considered a prototype rather than an exact configuration.

[33] For example, the concept of "particle" may require a certain level of resolution (as well as a certain detection mechanism) that is not available for the employed detectors.

and this should add an extra value to this experiment. Also see § 2.10 and § 7.8.

Interestingly, despite its exceptional importance, the analysis in the literature of this experiment is generally of qualitative and descriptive (rather than quantitative and mathematical) nature. For example, there is no sufficient formal analysis (e.g. in terms of wavefunction and how it collapses to produce the interference pattern) that can be used to justify rigorously and formally the various perplexing aspects of this experiment (e.g. the semi-random nature of the buildup of the interference pattern which will be discussed later). As we will see, the analysis of this experiment is largely based on qualitative (or semi-qualitative) models and prototypes (such as pilot waves and wavepacket) which are essentially approximate and imitate (rather than exact and rigorous) models.

1.6.1 Dual Nature

The main feature of this experiment is that the quantum objects behave as waves in their propagation (and hence they demonstrate interference pattern) and behave as particles in their detection (and hence they hit the detection screen at specific points). The interference pattern requires the opening of both slits and hence this pattern is destroyed when only one slit is open. This can be easily explained by the wave properties of the quantum objects as in the classical version of this experiment which involves waves. So, if the quantum objects are electrons then we can simply explain this feature by saying: electrons have a wave nature (like light) and hence the interference pattern requires the opening of both slits. Similarly, hitting the detection screen at specific points can be easily explained by particle properties since particles are localized entities. However, what is perplexing (or seemingly so) is this dual behavior as if the objects have dual particle-wave nature which is classically incomprehensible. As we will see, there are proposals (such as pilot waves, i.e. particles guided by waves) to rationalize this dual behavior (and dual nature) and make sense of it although these proposals are not completely satisfactory.

In brief, the dual nature feature of quantum objects (with all its aspects and demonstrations) may be explained rather easily and generically by claiming that these objects have both wave properties and particle properties. The unfamiliarity of this unusual particle-wave duality may be justified broadly by appealing to the scale factor (see § 1.3.2) where the strangeness of this duality is because our patterns and models have evolved in a macroscopic world where physical objects behave either as waves or as particles. In other words, there is no contradiction or inconsistency in having physical objects that behave as waves and as particles (regardless of the actual physical realization and mechanism of this duality and how it occurs and is envisaged) although we are not familiar with this duality in our daily life and in classical physics (which is generally based on our experiences in daily life).

As indicated above, there are specific suggestions in the literature about the nature of this duality and the potential physical realization and mechanism behind it and how it can be envisaged. These suggestions may be used (if needed) to rationalize this dual behavior and give it a sensible and tangible form. For example, the models of "pilot wave" and "wavepacket" may offer to some people rather intuitive and simple explanations to some of the features of this duality although these models are not totally convincing (since they cannot explain and rationalize all the features of this duality) and hence they should be treated as approximates and imitates rather than entirely-realistic and exact models. Anyway, the issue of particle-wave duality will be investigated in more details in the future (see for instance § 5.3.1 and § 6.13.1).

1.6.2 Effect of Observation and Knowledge

Another feature is that the interference pattern is destroyed when both slits are open but the slit from which each object passes was identified (i.e. we know the trajectories of the objects). This feature is not easy to explain because it seems as if the quantum objects know what we know about them and hence they adjust their behavior accordingly, i.e. they behave like waves (that interfere) when we do not know their trajectories and behave like particles (with no interference) when we know their trajectories. However, before we discuss this issue further we remark that (at least) some of the experimental procedures that are used to (allegedly) demonstrate the effect of observation and knowledge are questionable (although we will pretend to accept these claims).

There are two main situations to consider in this regard: in one situation (call it SA) our observation of the trajectories interfere with the objects and in the other situation (call it SB) our observation does not interfere at all. An example of SA is the use of an intrusive method of observation that has a direct physical impact on the objects such as by hitting the objects with a photon or by intercepting them with an electric field, while an example of SB (and in fact the best example) is the so-called delayed choice methods (see point 4 of § 1.3.3) where the trajectories are identified after the objects pass the double-slit and hit the display screen (where the pattern appears) and hence the objects (and hence the "trajectories") are not affected at all by observation (or at least that what should be if we believe in causality and the impossibility of having a cause after its effect, i.e. delayed cause; see § 1.6.4).[34]

In fact, SA may be easily explained and justified by claiming: it is reasonable that the method of observation affects the objects and changes their nature (i.e. from waves to particles) regardless of the actual physical mechanism of this change and how it occurs (and regardless of the magnitude of the effect and if it is negligible or not). However, SB is very hard to explain because it seems as if the objects know (possibly in advance) what we know (and hence they adjust their behavior accordingly). This means that there are actually several strange aspects that require justification (considering delayed choice case as an instance of SB):

- The behavior of the objects depends on our knowledge (as if reality is determined or affected by our consciousness or awareness which my be seen as an extreme form of subjectivity and idealism; see § 1.11).
- The objects know what we know (as if they are intelligent creatures and they have the ability to read our mind).
- The objects can predict what we will know (as if they can foresee the future). Alternatively, the cause of the particular behavior (which is our knowledge) follows the caused (which is the behavior of these objects as waves or particles); see § 1.6.4 as well as § 1.7.

All these aspects are difficult to explain and rationalize classically and intuitively. The effect of observation and knowledge will be investigated in § 6.13.2 from a more general perspective (i.e. not restricted to the double-slit experiment).

1.6.3 Self-Interference

Another important quantum feature of the double-slit experiment is that the interference pattern builds up when we fire a single object (e.g. single electron) at a time as if the object interferes with itself.[35] In fact, "self-interference" is not an accurate term to represent this phenomenon because what we actually have in this sort of experiment (i.e. when the objects are fired one at a time) is a distribution of particles according to an interference pattern that belongs to the entire experiment over its entire history. So, it seems as if there is an already-existing interference pattern according to which the particles are statistically distributed (rather than a pattern generated by the individual objects as "self-interference" suggests). Hence we may ask: is it possible for instance that the interference pattern belongs to (and possibly is generated by) the setting and configuration of the experiment and not to (or by) the fired quantum objects?

This possibility seems entirely odd and counter-intuitive and hence it cannot offer a sensible explanation. This possibility can also be refuted by the fact that when both slits are open the setting is the same whether or not we observe the path of the object and hence the effect of observation and knowledge (see § 1.6.2) cannot be explained unless the consciousness of the observer is regarded as part of the setting which seems nonsensical (see § 1.11). Moreover, if the interference pattern belongs to (or is generated by) the setting then why it should be affected by the attributes of the fired objects (e.g. their energy).

Anyway, the statistical distribution of particles according to this interference pattern cannot be explained because why these particles should follow this statistical distribution pattern noting that neither total determinism nor total randomness can explain this (because it is random from the aspect of the individual

---

[34] It should be noted that (at least) some of the delayed choice methods are questionable.
[35] "Self-interference" may also be called "single-particle interference" or other similar terms. We should also repeat our previous reservation about the achievability of this type (i.e. single-shot) of experiment at least persistently (considering the technical difficulties of controlling the emission of objects to be one at a time).

1.6.3 Self-Interference

particles and deterministic from the aspect of the pattern that includes all the particles contributing to this pattern). In fact, the failure to explain this semi-random (or semi-deterministic) behavior extends to other possibilities (e.g. pilot waves which will be investigated next) and is not restricted to this possibility. Moreover, this failure should extend even to the double-slit experiment in itself regardless of being conducted in a single-shot manner or not (see § 1.6.6).

Another possible explanation is the aforementioned pilot wave model. However, even this model cannot rationalize this behavior. In fact, the pilot wave model cannot rationalize this behavior of the double-slit experiment in itself regardless of being conducted in a single-shot manner or not. For example, the pilot wave model cannot explain the semi-random nature of this behavior (as indicated already; also see § 1.6.6). Moreover, if we note that the interference of waves requires interaction of two objects (i.e. waves) at the same time and in a common space, then the idea of self-interference (whether by pilot waves or something else) is bizarre because when the path length (i.e. across the slits) of the supposed wave is different then there is no possibility of the wave to interact with itself unless we assume that the speed of the wave is path-dependent (i.e. it is faster along the longer path) which is nonsensical noting that the path difference in this case should vary randomly (even if we accept that the speed of light wave or matter wave is not constant). In other words, how the two parts of the split wave communicate with each other to tune their speeds such that the "particle" will interfere with itself at an exact point on the screen (as required by the particle nature and in accordance with the interference pattern).

In fact, we may need to assume a wave pattern that exists instantly (i.e. with infinite speed) so that it can guide the "particle" to the exact hit point according to the self-interfering pattern. However, if so then we do not only need to assume infinite speed (which most, if not all, physicists reject) but we should also need to assume the existence of this pattern independent of the ejected particles and hence this interference pattern (and thus the supposed wave pattern) should belong to the setting of the experimental device (as seen in the first possibility) and not to the emitted particles, i.e. we do not actually have a dual particle-wave nature of the emitted quantum objects (where particles are guided by their own pilot waves) but we have particles guided by an interference pattern (of a wave) that belongs to (and possibly is generated by) the experimental device and setting (which seems very bizarre and nonsensical as we found earlier). In fact, there are many other problematic aspects and issues in the pilot wave model that we cannot go through because they require excessive (and unnecessary) explanations and details. In brief, if this model is supposed to rationalize this experiment (and its self-interference feature in particular) then it seems to irrationalize more than rationalize.

A third possibility is the wavepacket model where the "wave" nature of the wavepacket generates the interference pattern while the "packet" nature of the wavepacket generates the localization effect of the detected particles. In fact, this possibility is as bad as the previous ones for the same or/and similar reasons. For example, the wavepacket model cannot explain the semi-randomness of this behavior (i.e. why the packets are statistically distributed on the detection screen according to this pattern?). In fact, the "wave" nature of the wavepacket model cannot even explain the interference pattern itself (due to the comparative localization of the "packets") let alone the entire behavior. We may also claim that even the "packet" nature of the wavepacket model cannot explain the localization (due to the uncertainty about the spatial extension of the packet unless we consider its center for instance or tolerate uncertainty in the position of detection due to limited precision of the detector). So again, the wavepacket model seems to irrationalize more than rationalize.

Anyway, there are other serious challenges to all the above possibilities and models (as well as to the other possibilities and models that are proposed in the literature).[36] So, to the best of our knowledge there

---

[36] In fact, we can imagine other possibilities and scenarios (whether in the literature or not) but none of which (in our view) offer a convincing and rational explanation. Also, the above three possibilities can be refuted and contested by other challenges and question marks (but for brevity we do not go through the other details noting that what we mentioned is sufficient for our purpose). It is noteworthy that none of the possibilities seem to be associated with (or supported by) sufficient formal analysis to justify the claimed explanations and interpretations. For example, we do not see how a formal and rigorous treatment (based on the Schrodinger equation) can lead to (or imply or show the consistency of ... etc.) the "pilot waves" or "wavepacket" models and how they supposedly explain the semi-random nature that we indicated above. By the way, apart from not being associated with (or supported by) sufficient formal analysis, the sensibility of the employed models themselves (regardless of their use in the double-slit experiment or something else)

is no tenable explanation and rationalization to the self-interference feature of the double-slit experiment (and indeed to the other features of this experiment) and hence the perplexity of this quantum behavior remains. Therefore, in our view the only possibility (which can offer only a generic and broad justification) is the lack of direct experience with the quantum world (due mainly to the scale factor; see § 1.3.2) which makes any macroscopic (or classically-developed) models like "particle" and "wave" (as well as their "duality") approximates and imitates that may be useful for developing the formalism of quantum theory or providing an educational tool but they are not good enough for providing a deep and intuitive insight into the quantum world and the nature of its objects and phenomena.

1.6.4 Delayed Cause

We may also call this feature "late cause" (or "back-in-time effect" or "back-in-time influence") where a cause creates an effect in its past. It is obvious that according to some possibilities and scenarios in the double-slit experiment and its variants it seems as if a late knowledge or event has an effect on the outcome of the experiment which took place earlier, and this is an instance of delayed cause.[37] As we will see in § 1.7, delayed cause violates the principle of causality (i.e. no effect can occur before its cause) and hence it should be rejected. Accordingly, any experiment or possibility or scenario that requires (for its rationalization and interpretation) the assumption of delayed cause should be rejected or interpreted differently to avoid delayed cause. Also see § 1.7 (and point 12 of that section in particular).

1.6.5 Memory

The build-up of interference pattern in the single-shot version[38] of double-slit experiment (as investigated earlier) may require having a memory that keeps building the pattern over time according to a predetermined and long-lasting model (regardless of if and how we can envisage self-interference) and this should be another feature of oddity and another source of question marks, e.g. to whom (or to which object or setting) this memory belongs?

1.6.6 Semi-Randomness

This is another interesting quantum feature that can be seen in the double-slit experiment (in its various shapes and forms) as well as in other types of quantum experiments. As indicated in § 1.6.3, there is an element of randomness in the double-slit experiment which makes it semi-random (or semi-deterministic). This is because the individual particles are distributed indeterministically while the interference pattern (according to which these particles are distributed) is deterministic. So, how the individual particles manage to find their positions according to this pattern (especially in the single-shot version of this experiment)?

1.6.7 Other Features

There are other features of the double-slit experiment that are puzzling or potentially problematic or at least they require special attention such as the possibility of action at a distance (which will be investigated later; see for instance § 1.8 and § 7.13). However, we think the above investigation is sufficient (noting that the other features of the double-slit experiment are generally quantum mechanical features and hence they will be necessarily investigated in this book within other contexts or experiments or quantum properties).

---

lead sometimes to nonsensical or counter-intuitive consequences (e.g. the phase and group velocities of the wavepacket may move in opposite directions) which casts a shadow on the sensibility and validity of these models in themselves.

[37] In fact, the supposedly delayed cause can be something other than knowledge (as indicated already) and can occur in quantum experiments other than double-slit experiment (in its different variants).

[38] In fact, "single-shot" is not a necessity; what is required is gradual build-up.

1.7 The Principle of Causality

The essence of the principle of causality is the claim of an *intrinsic* association between a given event (or events) called cause and another given event (or events) called effect (where we use "claim" to indicate that causality cannot in general be proved, i.e. it is essentially a postulate or a hypothesis). This association makes the occurrence of the effect inevitable when the cause occurs (also see the upcoming point 1). In fact, there are many aspects in the principle of causality that deserve inspection and attention. However, due to the limited space and scope we will discuss (briefly) only some of these aspects, i.e. those aspects that are most relevant to our subsequent discussion and investigation of quantum physics.

One aspect is the origin of this principle and its justification. In our view, the origin of the causality principle is the (intuitive) denial of creation and annihilation either because they are impossible or because they are unimaginable (or at least they are not familiar phenomena). In fact, this should justify the principle in itself although individual and specific causality relations still require further (and specific) origins and justifications. Yes, the desire of obtaining persistent and consistent patterns (i.e. the perpetual nature of the observed association) should also play a role in the determination of specific causal relations.[39] So, the principle itself is based (in our view) on the "intuitive" denial of creation and annihilation, while the instances of this principle are based (in their specificity) on special reasons (usually obtained by observation) in association with this principle and the desire to have persistent patterns.

A second aspect is whether the principle of causality is based on induction or deduction.[40] This seems to be a controversial issue and a matter of philosophical debate and choice. However, it should be noted that the generalization of the results obtained (supposedly) by induction still requires a sort of deduction (even if the origin is presumed to be induction). It should also be noted that the individual and specific causality relations in science are largely and primarily based on induction since science is a form of experimental or observational type of knowledge and hence induction should play the major role in extracting and formulating scientific relations and principles although it should still need (especially in its philosophical and epistemological aspects) deduction and inference.

A third aspect is the chronology of causal relations, i.e. the aforementioned "intrinsic association" is hierarchical in nature where the cause is supposed to produce the effect in a certain chronological order between the two (i.e. they occur simultaneously or the effect follows the cause in time). Accordingly, this hierarchy does not allow the occurrence of the effect before the occurrence of the cause (i.e. the effect cannot temporally precedes its cause). In fact, this is important for the appreciation of the causality principle and analyzing its origin and roots. In our view, this (temporally-limited) hierarchy should also be linked to the denial of creation because in the time period between the occurrence of effect and the occurrence of cause the effect has no cause, i.e. it is a creation that should be denied. To put it differently we can say: once the "effect" emerges (i.e. appears before the "cause") there is no need for a cause anymore and hence this principle is not needed (and hence not applicable) for correlating this particular "effect" to this particular "cause" and justifying the emergence of this "effect" (although the effect should still need in general a non-delayed cause to explain its emergence). As indicated earlier (see § 1.6.4) and will be seen next (see point 12), we call such alleged causal relations (i.e. where the effect supposedly precedes its cause) "delayed cause" relations (although the "delayed cause" is not really a cause and the relations are not really causal).

A fourth aspect is the justification of the *intrinsic* nature of this association. In our view, the internal sense of "constructional" relation between the cause and the effect is what justifies this intrinsic nature. In other words, the repetitive association in itself (even if it is perpetual) does not imply causality unless we have a "sense" or "feeling" or "intuition" of a constructional and inherent relation between the cause and the effect (where this feeling normally originates from our past experiences).[41] For example, if two

---

[39] The indicated desire should originate from our mental habit for consistency and rationality which is part of our identity as rational creatures and intellectual thinkers.

[40] In gross terms, induction is where we go from special to general, while deduction is where we go from general to special. Alternatively, induction is based on gathering and generalizing observations and hence it represents (in some sense) empiricism, while deduction is based on reasoning and hence it represents (in some sense) rationalism. It should be noted that "deduction" does not necessarily imply "proof" (as it might be suggested linguistically).

[41] This is inline with what we indicated earlier that causality is essentially postulated rather than proved. So, the ultimate

1.7 The Principle of Causality

unrelated persons living in remotely-separated accommodations happened to wake up everyday at 7:00 o'clock in the morning (or one at 7:00 o'clock and one at 7:30), no rational person would believe that there is any causal relation between these events. On the other hand, we may conclude a causal relation between two events from a single observation based on our sense or feeling of this inherent relation (as well as relying on our past experience and knowledge in association with this principle and the desire of having persistent and consistent patterns that rationalize our observations). This should indicate that even if we believe that the origin of the principle of causality is induction, the specific causal relations require more than induction (or observation) since they are based on our internal sense and rational thinking that is based on our overall past experience and knowledge (i.e. these relations require deduction and intuition as much as they require induction and observation).[42]

In the following points we will try to investigate some issues related to the principle of causality:

1. "Perpetual association" in the above context should mean that the effect must occur when the cause occurs and the effect cannot occur without the occurrence of cause (where we consider in this formulation "necessary and sufficient" causal relations although causality in itself could be more general such as when the effect has more than one cause). In fact, in the latter example the cause can be considered as "one of the possible causes" and hence the uniqueness of cause can be maintained (e.g. the cause of heat is "one of fire and friction"). Alternatively, each specific effect is correlated specifically to its cause (e.g. the cause of "heat of fire" is fire while the cause of "heat of friction" is friction) and hence uniqueness is maintained.

2. As indicated above, the causality relation at the empirical level is no more than an association (continuously-repeated in the past and supposedly continuously-repeated in the future) between two events (or sets of events). The distinction between the cause and the effect in this relation is decided either by the chronology of the events (i.e. the first-occurring is the cause and the second-occurring is the effect) or by the dependency of the events (i.e. the independently-occurring is the cause and the dependently-occurring is the effect).[43] However, we should note that chronology and dependency may not be available in some causal relations (e.g. simultaneous events not under our control) and hence we should rely on other means (e.g. guess and intuition) to distinguish between cause and effect.

3. Whether the principle of causality is restricted to the macroscopic world or it is valid even in the quantum world seems to be a matter that can be debated and disputed (as quantum theory may suggest noting for instance the probabilistic nature of quantum phenomena which may be seen as a violation of causality at least in its strict sense). However, (in our view) as a requirement of rationality and consistency (especially if the principle originates from deduction) the principle should be general and hence it should apply to the quantum world as to the classical world. Yes, the specific causal relations may be less obvious with regard to the quantum world since we have no familiar experience or direct observation of the quantum world and its phenomena and their relations. Nevertheless, most of the specific quantum causal relations are observed and deduced indirectly through macroscopic phenomena (represented typically by the reactions of the measuring devices which are macroscopic) and hence the macroscopic relations should be rationally projected onto the microscopic relations. Yes, an issue may be raised (and will be discussed later; see point 13) about the extent of the validity of

---

root of causality is a "psychological" factor motivated and inspired by the desire of consistency and rationality. Otherwise, no one can prove that B is the effect of the cause A (especially when A and B occur simultaneously) since "causality" is an abstract philosophical/epistemological concept (rather than a scientific concept or anything that can be proved experimentally or theoretically).

[42] The role of past experience and knowledge should justify, in part, the "nonsensical" nature of many old "superstitious" beliefs (e.g. about the spiritual or supernatural nature of astronomical events and their relations to deities and angels as seen for instance in astrology) despite the fact that the people who created and embraced these beliefs (motivated by certain associations) are not less intelligent than us (noting that they are just a few hundred or a few thousand years back in time and hence on the scale of natural evolution they are genetically identical, or at least very similar, to us).

[43] "Independently-occurring" means it occurs by choice and "dependently-occurring" means it occurs involuntary (i.e. when the cause occurs it occurs with no control over its occurrence). For example, we are free to set fire or not but we are not free to have heat from this fire because heat is intrinsic to fire. We should also note that the causality itself is justified by the above-indicated "sense of constructional relation" while the distinction (i.e. which is which) is made by the chronology or dependency (although the chronology and dependency should also contribute to the formation of this sense).

1.7 The Principle of Causality

the principle of causality and possible restrictions on it at the quantum level to justify for instance the probabilistic nature of quantum behavior.

4. Regardless of its origin, its validity and its domain of validity (and regardless of anything else), the principle of causality should be a required ingredient for any consistent intellectual theory (whether scientific or philosophical or epistemological) because the aforementioned pattern of "perpetual association" (when it occurs) cannot be explained rationally and systematically without this principle. Yes, the missing deductive part of the theory (i.e. the cause "generates" the effect) requires justification which rationality should be able to explain in general. So in brief, we may have a free choice about the origin, validity, domain of validity, etc. of this principle but we have no choice about the need to this principle (at least broadly) for creating rational propositions and theories. In other words, causality is an essential and indispensable element in our rationality kit.

5. Regarding the relation between the principle of causality and locality, there is no necessity for the cause and effect to be in the same place.[44] So, causal relations between two things which are spatially-separated (and hence they are not in a direct contact) is acceptable and reasonable (at least in principle and in general and according to the common consensus). However, in such cases of spatial-separation is there a necessity for a sort of communication between the cause and the effect (which may be envisaged by the existence of a sort of medium or field that facilitates the communication, e.g. by electromagnetic signals) to justify and explain the influence exerted by the cause on the effect? The potential necessity of a sort of communication could be justified by the impossibility (or at least non-imaginability) of exerting such influence without the mediation of a sort of an agent in-between that transfers this influence. In fact, the essence of this is the possibility (or impossibility) of "action at a distance" in its spatial (rather than temporal) sense regardless of any temporal aspect.[45]

Anyway, in our view **spatial action at a distance** and its possibility or impossibility is a philosophical and epistemological (rather than scientific) issue which is subject to free choice (i.e. there is no logical inconsistency in the notion of "spatial action at a distance"). However, if spatial action at a distance was allowed or used in any theory (whether scientific or not), it should be (like anything else) compliant with all the criteria and requirements of eligibility and validity (i.e. spatial action at a distance should not lead to any inconsistency or irrationality).[46] Nevertheless, spatial action at a distance may affect physicality (if it is used in a scientific theory) by claiming that it is non-physical (or it is metaphysical), although this is a matter of opinion and preference (since in this situation we will be on the very edge of science). Regarding **temporal action at a distance**, it should be logically and physically acceptable (despite its denial by special relativists) with no potential violation of the causality principle (assuming communication between the cause and caused). More investigation about these issues will follow (see for instance § 1.8, § 1.9, § 7.12 and § 7.13).

6. Regarding the relation between the principle of causality and simultaneity, we can repeat (or rather imitate) the discussion and argumentation of the previous point (noting that locality is about space and simultaneity is about time and hence they represent the two faces of the same space-time concept). In fact, locality and simultaneity are not really separate or independent issues (due to the intimate link between space and time especially in science and in modern physics in particular, as indicated already). Anyway, the important point to note in this context is that instantaneous action at a distance (whether spatial or temporal where in the latter case it requires infinite speed of communication) may be seen as non-physical. Nevertheless (and regardless of its physicality or not), it may not be needed necessarily (as an assumption) in its strict sense if we assume immeasurably-high finite speed (which is

---

[44] "Locality" here means "being in the same place" and hence it is rather different from "locality" used in the "principle of locality" (see § 1.8) as well as subsequently. In fact, "locality" in this book (as well as in the literature) may mean the principle of locality (or the content of this principle) and may mean being local (where the context usually determines unambiguously which is which). Also, being in the same place (in the strict and literal sense) should be impossible (at least in the familiar and common cases) even if the two are co-local.

[45] Spatial action at a distance (or spatial non-locality) means causal relations without physical communication between the cause and effect, while temporal action at a distance (or temporal non-locality) means causal relations (usually) with physical communication between the cause and effect but with violation of certain speed limits (e.g. by being superluminal). See § 1.1.

[46] The criteria of eligibility and validity of scientific theories and their interpretations will be investigated in chapter 2.

practically infinite although strictly it is not). Yes, it may be needed in some cases for specific reasons such as avoiding the violation of some conservation principles (see for instance point 14 of § 6.11).

7. We should also draw the attention to the relation between the issue of action at a distance (especially temporally) and the issue of delayed cause (i.e. when the effect precedes its cause in time) which will be investigated later (see point 12; also see § 1.6.4). For instance, if temporal action at a distance is not allowed (as special relativists claim) then any remote influence that supposedly violates the imposed speed limit should imply violation of causality because it means the occurrence of the effect before the signal can reach it, and hence if we note that the actual cause (or at least a condition for the action of the cause) in this case is the arrival of signal then the effect occurred before its cause (which is an instance of delayed cause).[47]

8. As suggested earlier, the justification for embracing the principle of causality is its ability to provide a basis for predictability, rationality, consistency (as well as other desired factors). These factors should also explain the need for the principle of causality as an epistemological necessity. In more detail, even if we assume that the causality principle is an ontological or philosophical choice rather than a necessity (see for instance § 2.4.2), it should be obvious that this principle is an epistemological necessity (at least in principle regardless of its potential limitations) because no science (or rational knowledge to be more general) can be built without this principle (in some shape and form) although the details of the extension and limitation of this principle and its exact nature may be subject to debate and dispute. The reason for this necessity is based (as indicated above) on the requirement of predictability, rationality, consistency, ... etc. that any rational theory needs.

9. Does indeterminism (in general) affect causality? In fact, this is a rather generic and ambiguous question. Hence, there is no definite answer to this question (as it stands) because some types of indeterminism affect causality while other types do not. For example, the indeterminism of "having B after or before A" should violate the supposed causality of A to B, while "having B or C after A" may not violate the supposed causality of A to B due for instance to the possibility of a "hidden variable" that restricts the causality.[48] So, in principle indeterminism may or may not affect causality. Yes, indeterminism affects (epistemologically) our knowledge of causality since it represents a form of ignorance about causes and effects and hence it should compromise causality (epistemologically) in its strict and completely-deterministic sense (although in general it does not necessarily negate causality altogether).[49]

10. Referring to the previous point, what makes indeterminism (potentially) violate causality? Again, this is a rather generic and ambiguous question. However, we can say (in a rather generic and broad tenor): determinism implies predictability with certainty while indeterminism implies unpredictability (or rather probabilistic predictability with uncertainty as it is the case with quantum mechanical phenomena). The association or bond between determinism and causality is based on the repetitivity of occurrence that distinguishes causal relations (i.e. the effect always follows the cause) and this pattern of repetitivity is seemingly violated when the outcome is probabilistic or uncertain (i.e. indeterministic). In brief, according to causality: "identical preliminaries lead to identical consequences" while according to indeterminism: "identical preliminaries do not lead to identical consequences".

11. Does freewill contradict causality (at least in its strict sense)? This issue will be investigated in detail in § 2.15. However, this issue is related to the issue of indeterminism (which was investigated in the previous points) because freewill implies indeterminism and uncertainty since with freewill the outcome of events (that have seemingly similar or identical conditions and circumstances) is indeterminate and

---

[47] In fact, we are assuming the signal to be subject to the speed limit; otherwise the violation occurs by the signal itself and its superluminal nature.

[48] In fact, this example is not very tight but it should give an idea about what we mean.

[49] In this regard it may be argued that causality in its strict and familiar form implies the definiteness of the cause and effect and hence probabilistic causal relations or "probabilistic causality" (as in quantum mechanics; see § 5.2.2) should violate causality (in its strict sense) because even though we have a relation between a cause and an effect the indeterminacy (in the outcome) should violate the causality since the correlation between a specific cause and a specific effect has no cause and cannot be explained causally. So, in such cases what we actually have is "partial causality" where part of the association is explained causally and the other part is not and hence causality is violated "partially" (and thus it is violated). Also see § 6.4.

uncertain.

12. As indicated earlier, delayed cause (i.e. when the cause supposedly occurs after its effect) should obviously violate the causality principle, i.e. the delayed cause cannot be really a cause for its alleged effect. This is because in the time between the occurrence of effect and the occurrence of cause the effect has no cause (which is a violation of causality since the occurrence of effect in this time has no cause). We may also say: if the "effect" occurs then there should be no need for the delayed "cause" (or indeed any delayed "cause") since the "effect" is already in existence. Yes, a delayed cause may be needed for the continuation of the existence of effect but this is a different story because the cause then is for the continuation of the existence of effect (rather than for the existence itself) and hence it is not delayed at all. Also see § 1.6.4.

13. As indicated earlier, we should consider partial causality, i.e. a cause may or may not produce a certain effect or a cause may produce probabilistically an effect, within a list of possible effects (like the effects in quantum phenomena). Accordingly, we may ask: can causal relations be subject to an uncertainty (similar to the uncertainty of physical quantities like position and momentum which is embedded in the uncertainty principle)? This uncertainty (if accepted) should explain the limited form of causality seen in quantum phenomena where a cause may lead (statistically and alternately) to a number of possible effects. In fact, this could address the issue of indeterminism as a challenge to causality because if partial causality is allowed then the indeterminism can be accommodated within it. Anyway, despite the fact that partial causality is not acceptable classically (since causality in its classical sense is completely deterministic) it may be accepted as a last resort (in quantum mechanics for instance) if strict and deterministic causality cannot be maintained noting that partial causality meets (although not totally and ideally) the justification criteria set in point 8. Nevertheless, strict and deterministic causality should be maintained as long as possible and as much as possible (as it is the case at classical level).

To conclude, causality in essence is a philosophical (or ontological) and epistemological paradigm rather than a scientific paradigm. The reason for not being scientific in essence is that science is about observation, and what is actually observed in (presumably) causal relations is the association (or correlation or correspondence) between phenomena rather than the causality itself (i.e. causality is a philosophically and epistemologically added value to this association).[50] Nevertheless, it is an epistemological necessity (and possibly ontological necessity as well although this is not of concern to science) for building and rationalizing science (see § 2.4.2). Accordingly, any epistemologically consistent and rational scientific theory should comply with the principle of causality and hence this principle should not be violated in science (as well as in other forms of rational knowledge). Yes, if it cannot be maintained thoroughly, strictly and deterministically (as in quantum mechanics for example where it is likely that strict and deterministic causal relations do not hold), it should still be maintained partially and indeterministically (as much as possible and where and when it is tenable) due to its epistemological necessity to maintain rationality, predictability, etc.

1.8 The Principle of Locality

The essence of this principle (which stems from special relativity) is that: a physical event at a given location cannot have an effect on another event at another location instantaneously or faster than a certain speed limit (i.e. the speed of light c). In other words, when the cause and effect are spatially separated then the effect should occur after the cause following a certain time delay (i.e. between the occurrence of cause and the occurrence of effect). The requirement of a time delay is because the (supposedly-required) signal that carries (or communicates) the influence of the cause to the effect needs a finite time for its propagation from the location of the cause to the location of the effect. However, it should be noted that the "principle of locality" could also mean the ban of superluminal speed in general (regardless of any

---

[50] We note that even if in some causal relations we may distinguish the cause from the caused by chronological order or by independence and dependence we still need this added value for establishing the philosophical and epistemological content to these causal relations.

causal considerations), and this meaning may be implied occasionally where the context should determine the ultimate meaning (see for instance the upcoming point 3).

As indicated, the principle of locality is based on the special relativistic claim (which originates from the second postulate of special relativity) that the characteristic speed of light in vacuum (i.e. c) is the ultimate physical speed and hence no physical signal or information or influence or interaction can be communicated by a faster than light (i.e. superluminal) speed.[51] In the following points we assess this principle (which will also be discussed in the future within other subjects and contexts such as quantum entanglement; see for instance § 5.6 and § 6.10):

1. The principle of locality is based (implicitly and partially) on the assumption that causality is communicated through physical signals (i.e. the denial of spatial action at a distance) and this may be disputed (see § 1.7). Accordingly, if spatial action at a distance is allowed then instantaneous remote interaction should be possible (i.e. infinite speed of interaction is allowed since there is no signal and hence no restriction on its speed) and therefore the principle of locality has no subject. However, if spatial action at a distance is not allowed then instantaneous remote interaction may not be allowed (since the physicality of signal of infinite speed may be questioned), but this should not affect the possibility of violating (as well as respecting) locality. In brief, the principle of locality has a subject (i.e. it could be valid or not) only according to the temporal action at a distance scenario (where remote influences can be exerted only through signals).

2. The second postulate of special relativity (as well as special relativity itself) is questionable. In fact, even if we accept the second postulate it can be claimed that the validity of this postulate is restricted to the speed of light (i.e. electromagnetic signals) and cannot be generalized to all types of signal.[52] This claim should apply even to Lorentz mechanics (i.e. the formalism) which is a well-established theory because even if Lorentz mechanics implies (arguably) the restriction of the ultimate physical speed to c we can still claim that this restriction is related to light and within the theory of Lorentz mechanics and its limitations.

3. The violations (whether definite or tentative) to the principle of locality have been defended by special relativists (or the adherents of locality) by claiming or allegedly-showing that no energy or information can be transferred or exchanged by superluminal speed in these violations.[53] Regardless of the merit of these claims and arguments (and if they are right or wrong and if they can be established in general and in all cases or not)[54] the violation of the generally-accepted speed restriction of special relativity occurs by the mere existence of superluminal speed regardless of transferring or exchanging energy or information (or anything else) or not.

---

[51] In fact, the second postulate of special relativity is essentially about the constancy of the speed of light. However, it was extended and generalized by special relativists (using causality arguments for example) to include all types of signal and to make this speed the ultimate physical speed. More about these issues can be found in our book "The Mechanics of Lorentz Transformations".

[52] To be more general we may say: "restricted to the speed of certain types of signals, i.e. physical signals such as waves of physical fields". This is to address a potential challenge that gravitational waves (assuming their existence is proved, which is questionable, and assuming their speed is the same as the speed of light according to general relativity, which is also questionable) also propagate with the speed of light. We should also note that matter waves (which are different from physical waves) potentially have infinite or immeasurable superluminal speed (see § 3.2) and hence if there are certain speed restrictions on the communication of signals and information then this should be imposed on physical waves and not on matter waves (noting that even for physical waves the speed restriction by c may be imposed only on certain types of them).

[53] A common view (which was indicated above) among special relativists in the recent times about the relation between locality and causality is to restrict locality to causal relationships (i.e. non-locality is allowed if causality is not violated). Accordingly, non-locality that cannot transfer energy or information (or influence to be more general) is allowed while non-locality that can transfer energy or information is banned (and hence the causal implications of special relativity are not violated which means restriction of special relativistic locality constraints to the type of non-locality that affects causality).

[54] In fact, at least some of these claims and arguments can be challenged and possibly refuted directly and straightforwardly. For example, if the collapse of wavefunction is a real and instantaneous physical event (as it is supposed to be at least according to some interpretations) then a causal influence can be exerted superluminally by a measurement on an entangled object A (in a pair of entangled objects) where the effect of collapse on the other entangled object B (in the pair) can have causal influence at the location of B. This means that by observing A we can trigger (superluminally) an event at the location of B.

1.8 The Principle of Locality

In fact, some of these claims and arguments are based on defending the causality (according to the implications of special relativity and within certain cases and circumstances) which supposedly is not violated by this sort of locality violations (i.e. when no energy or information is transferred). Nevertheless, they cannot save the postulate of special relativity (even if they are correct) because saving the "special relativistic" causality requires first and beforehand the validity of special relativity which is based on the validity of its postulate (which is threatened, at least tentatively, by any superluminal speed). So in brief, any legitimacy of locality from special relativistic causality arguments is based ultimately on the validity of special relativity (and its postulates as well as relativistically-based causal arguments) in the first place and this validity is threatened by any type of superluminal speed (at least tentatively which should be sufficient).

We should also note that the impossibility of sending information[55] through superluminal means cannot be established logically (i.e. superluminality is not illogical). What can be established (from the proposed instances and examples) is that no information can be sent by the proposed instances and examples (due allegedly to lack of proper physical mechanisms and means). However, it is always possible that someone in the future comes with a trick or technique that overcomes this difficulty and hence he can send useful information through superluminal means. In fact, we have many examples in the history of science about such alleged impossibilities.[56] One of these examples is the "impossibility" of getting information about the chemical composition of stars (as attributed to the French philosopher Auguste Comte) which was refuted later by the development of spectroscopy. Another example of such "impossibility" (which is from the recent history of quantum theory itself) is about the distinction between the predictions of quantum mechanics and the predictions of hidden-variable theories where it was once believed that the distinction is entirely theoretical and cannot be verified experimentally. However, this "impossibility" was refuted later by the emergence of the Bell theorem and inequality (see § 9.3.2) and the subsequent experimental verification by the work of Aspect and others (see § 9.3.3 and § 10.4). So, as long as there is no logical necessity for the impossibility of sending information superluminally it remains a possibility (and hence a challenge to special relativity).[57] Also see footnotes [54] and [56].

It is noteworthy that the possibility of superluminal speeds is not restricted to quantum entanglement experiments and its alike where sending information in this type of experiments may seem impossible. In fact, there are experiments on devices through which the response to signals across these devices seems instant and hence sending information superluminally through such devices was considered as a practical possibility that can have useful applications (although of limited usefulness at the current time due to size restrictions). Also see point 6.

4. To legitimize the principle of locality (or rather a modified form of it) in a more general way and get rid of its dependence on special relativity (so that the principle becomes more secure), special relativity may be excluded as a basis for this principle. Instead, this principle may be established on the (alleged and seemingly-intuitive) basis that all communications should have finite speeds and hence any non-local communication with infinite speed (as implied seemingly by certain quantum mechanical effects and phenomena like wavefunction collapse and quantum entanglement) should be excluded.

However, the maximum that can be drawn from this form of the principle of locality (assuming it is valid) is the necessity of the existence of a time delay between the cause and effect and this can be realized by signals of finite but immeasurably-high speed (where the possibility of the existence of such signals was indicated in the previous point). In other words, if the cause and effect are communicating through a signal whose speed is immeasurably-high (say 10^{20} m/s) then the principle

---

[55] We use here only "information" to avoid clogging the text; otherwise this should apply to other things like energy or influence.

[56] In fact, we can propose many ideas about transmitting and exchanging information superluminally. For example, if a shadow that moves superluminally passed through point A and triggered a physical action there and next passed through point B then the information of the triggered action at point A at a specific time can be concluded from the speed of the shadow and hence this technique can be used to transfer and exchange information between A and B. In fact, this is a very basic proposal; many more elaborate and tight proposals can be suggested in this regard.

[57] In fact, according to some novel quantum mechanical applications (e.g. in quantum communication and encryption) the possibility of transmission of information superluminally seems real (or at least very likely).

1.8 The Principle of Locality

of locality (or rather its modified form) is not violated although the influence seems non-local due to the immeasurability of the speed of the communication signal. Accordingly, what seems to be non-local quantum effects with infinite speed (e.g. wavefunction collapse) could become local (although it appears non-local) and hence we avoid the violation of the principle of locality which may be a challenge to quantum mechanics (assuming the acceptance of this principle) although other challenges (to both quantum mechanics and locality) could still be posed. Also see point 7 of the present section as well as point 14 of § 6.11.

5. In our view (based on what have been said so far), the principle of locality (in its commonly-accepted relativistic interpretation) is not a scientific necessity (i.e. there is no scientific evidence for its validity or/and generality). Moreover, it is not even a philosophical or epistemological necessity. Accordingly, we can simply reject any argument based on this alleged principle without further ado.

6. Although the main experimental challenge to the principle of locality comes from the experiments of quantum entanglement, there are other types of experiments in which locality seems to have been violated (e.g. devices through which signals are transmitted faster than c). So, even if quantum entanglement experiments were refuted or re-interpreted to avoid non-locality, locality can still be challenged by other experiments and demonstrations.[58] Whether superluminality in these experiments occurs in vacuum or not is irrelevant as far as the principle of locality is concerned although it may have some relevance to special relativity. Also see point 3.

7. It should be noted that (at least) in some situations the violation of locality (assuming it is allowed) must be through infinite speed (i.e. not by finite superluminal speed) if some conservation principles should hold (i.e. strictly at all times). The reader is referred to point 14 of § 6.11 for more clarification (as well as an example). Also see point 4 of the present section.

8. As indicated earlier, we distinguish between spatial non-locality (or spatial action at a distance) and temporal non-locality.[59] Although some people may reject or loathe this distinction we think it is sensible and useful.

9. Some forms of non-locality may require (global) simultaneity and hence the admission of global time and global frame. Accordingly, non-locality could become a threat to special relativity not only through superluminality (i.e. by violating the speed limit c) but also through the admission of global frame which is denied by special relativity. In fact, "global" may not be necessary in this context considering that "privileged" could be sufficient for this threat (since special relativity denies privileged frame). This may apply even to temporal relations more relaxed than simultaneity where (well-defined) locality in some cases may require global or privileged frame (regardless of simultaneity).

To conclude, locality is not a philosophically or epistemologically established principle. Moreover, its theoretical and experimental scientific foundations (which mainly come from special relativity) are not well established due to the challenges to special relativity and the doubts and inconsistencies in its axiomatic and logical structure (as well as lack of other evidence for the validity of this principle). In fact, the principle of locality is challenged (and seemingly violated) in some experiments related, for instance, to quantum entanglement and to superluminal transmission across certain devices. Yes, Lorentz mechanics (which is well established empirically) may be a challenge to superluminality (and hence to non-locality in this sense), but as we outlined elsewhere Lorentz mechanics may require or imply a conditioned and limited (in type of signal as well as in domain) form of locality which cannot be extended and generalized to become a generally-valid principle.[60] Moreover, spatial action at a distance (if accepted) cannot be challenged by Lorentz mechanics or by special relativity (due to the absence of any type of signal and hence no speed restriction can be imposed physically). Hence, we cannot raise locality to the state of (generally-valid) principle (although for convenience and to be inline with the mainstream literature we

---

[58] For example, if quantum entanglement experiments were challenged by the alleged impossibility of exerting causal influences superluminally (e.g. through sending information) this should not apply to this sort of experiments because in these experiments (or at least some of them) information, energy, influence, ... etc. can be transferred superluminally (and this should violate these relativistic causality restrictions).

[59] We note that locality is primarily about space but due to the link between space and time in Lorenz mechanics (and special relativity) and their merge into spacetime the two are linked and hence locality can be limited temporally (in a sensible way) by certain speed restrictions.

[60] In fact, this should in principle apply even to special relativity.

will continue to use "principle" as a label).

Anyway, there is nothing in the formalism of quantum theory (i.e. quantum mechanics) that requires locality in any sense and hence the quantum theory is self-consistent (or logically consistent) with and without this alleged principle.[61] So, any violation of locality (whether definite or tentative) can be accommodated within the quantum theory with no problem (as far as quantum theory is concerned). Any alleged problem raised by a violation of locality then belongs to some other theory (i.e. it is not our problem as quantum physicists) and hence the other theory (especially special relativity) should address and take care of this problem. In other words, a presumed violation of locality (regardless of being a principle or not) by quantum theory (which is overwhelmingly supported by experiment) could be a challenge to the other theory but not to the quantum theory and hence we (as quantum physicists) do not need to worry about such a violation. Also see § 7.13 and § 9.2.1.

1.9 Relationship between Causality and Locality

As seen in the last two sections, the relationship between the principles of causality and locality is intimate. In fact, this relation was investigated sufficiently within the last two sections and hence we have nothing to add apart from some useful remarks which are outlined in the following points:

1. As explained already, violation of locality (i.e. by allowing superluminality) should imply (according to some physicists) violation of causality (at least in some cases). However, this may be the case if we accept special relativity (and its second postulate in particular); otherwise non-locality in itself does not imply non-causality. It should also be obvious that violation of causality does not mean violation of locality, i.e. non-causality in itself does not imply non-locality. So in general, these principles work and apply independently and hence a violation of one is not necessarily a violation of the other although such association may be established in particular cases and for particular reasons (and may be based on particular views).

2. Referring to the previous point, if we use our conceptualization and terminology in distinguishing between spatial and temporal action at a distance (see § 1.1 and § 1.7) then we can say: if causality does not require communication (i.e. by physical signals) then non-locality cannot violate causality (unless we assume communication is required in particular cases). This means that spatial action at a distance cannot violate causality. Therefore, any potential violation of causality by non-locality should be restricted to temporal action at a distance.

1.10 Realism and Idealism

Realism and idealism are two conflicting philosophies that have roots in the oldest forms of philosophical thinking and contemplation. These philosophies have obviously significant impact on science (and on its philosophy and epistemology in particular) especially those branches of physics that are on the edge of our classical world or even beyond. This is represented typically by quantum mechanics and Lorentz mechanics which deal with the non-classical worlds of extremely tiny objects and extremely fast objects which we have no familiarity with or direct experience and hence philosophical contemplation is needed in their interpretation and rationalization. In fact, the subject of realism and idealism is too extensive to be investigated in this section or in this book. However, we feel obliged to outline our opinion in this regard since this is essential for future reference and assessment.

In our view, realism (as represented primarily by the principles of reality and truth; see § 2.4.1) is an ontological choice (and hence we are *free* to embrace realism or not from a philosophical perspective), but it is an epistemological necessity (and hence we *must* embrace realism from this perspective). The reason is that ontological philosophy is essentially a collection of contemplations and impressions (like art) and

---

[61] In fact, the formalism in itself does not require or imply locality or non-locality. So, locality and non-locality essentially belong to the interpretation of the formalism of quantum theory and its experimental evidence. Yes, if the measurement postulate (see for instance § 4.2 and § 5.7) is part of the formalism and it includes the paradigm of "collapse of wavefunction" (or a similar paradigm) then non-locality may be *suggested* by the formalism (since "collapse of wavefunction" *suggests* non-locality).

hence we have almost unlimited freedom to *believe* ontologically in what we believe (or rather in what we want to believe) since this essentially is a subjective and self-centered experience (which may be compared even to religious beliefs and experiences). However, epistemology is a rationality-based and disciplined subject whose objective is to "know the world" by tuning our inner world with the (supposedly real, unique and independent) outside world and hence no rational or sensible epistemology can be established without realism. So, the "art of knowing the world" requires more stringent rules and principles to be "rational" and "sensible" (and hence consistent with "knowledge") and these rules and principles are essentially embedded in realism.

However, although realism is an epistemological necessity, epistemological realism (as opposite to ontological realism) is not acceptable in its extreme and idealistic form which considers (correct and rational) knowledge as an exact and unique image of reality (rather than an artistic impression or painting of reality). As indicated earlier (see for instance point 5 of § 1.3.3), even if we assume that reality is totally independent and determinate ontologically (according to strict ontological realism), it is not so epistemologically. This is due to the fact that observation is an observer-dependent process and hence the outcome of observation depends in its details on the details of the observer and his equipment.[62]

So to summarize, ontological realism (as well as ontological idealism) is a choice, but epistemological realism is a necessity in its moderate form and should be rejected in its extreme form (like epistemological idealism).[63] Thus, epistemologically we could have (to some extent) a determinate and independent reality but our knowledge of this reality is not an exact and unique reflection of it and hence it is possible (and even necessary from some perspectives) to have certain dependencies on the observer/equipment and certain elements of indeterminacy and uncertainty. So, we still have a valid criterion for truth (by the existence of a determinate and independent reality which our knowledge is supposed to reflect to some extent and hence it represents the truth) but we also have a degree of freedom and diversity in the truth (represented for example by the non-uniqueness of science or knowledge to be more general; see § 2.4.3) thanks to the dependency of our knowledge in some details on the observer and his equipment.

1.11 Consciousness and its Role in Science

"Consciousness" is a very big subject that extends to and influences many aspects of life and branches of knowledge (e.g. religion, law, philosophy, biology, psychology, humanities, etc.). In fact, "consciousness" is surrounded by many mysteries, ambiguities, question marks, controversies, etc. (e.g. about its definition, nature, origin, role, limitation, etc.). However, these matters are of little interest to us in this book and hence our objective in this section is to highlight its role in science (and quantum mechanics in particular). In this regard, we adopt a rather generic and intuitive definition of "consciousness" which is essentially based on the daily use of this word (noting that it could have a strict and elaborate technical definition in certain scientific contexts and theories).

Until the emergence of quantum theory no scientist (or at least no respected and renowned scientist) suspected (let alone believed in) the possibility of a role of the consciousness of observer on the observed physical phenomena. In fact, classical physics (and classical science to be more general) is based on total objectivity where the observer (as such) is supposed to be a receptor that is influenced by (but does not influence) the outside world and hence the consciousness of the observer cannot have any role other than "observing" or recording the observed phenomena. The root and justification of this supposition is our internal feeling (or intuition) which makes this supposition not only plausible but even intuitive and

---

[62] Regarding the relationship between ontological realism and epistemological realism, it may be claimed that they are independent of each other and hence we may embrace/reject each independent of our embracement/rejection of the other (and therefore we have four valid possibilities). However, from a sensibility point of view they may not be totally independent of each other and hence we may need to reject some of the four possibilities. In fact, if we consider our view about the necessity of epistemological realism then we should have only two possibilities (both of which seems valid), but again from a sensibility point of view it may be claimed that only one possibility (i.e. ontological realism with epistemological realism but in moderate form at least for the epistemological realism) is valid. Also see footnote [71] on page 36.

[63] Regarding quantum mechanics (which is the subject of interest in this book), we will see later on that if classical-type realism cannot be held (i.e. no tenable form of "classical-like" quantum reality can be maintained) then we can keep a form of partial realism, i.e. realism at classical level and partial realism at quantum level. Also see point 4 of § 8.6.

1.11 Consciousness and its Role in Science

self-evident. In fact, this supposition is partly linked to the independence of reality (see for instance § 1.3.3 and § 2.4.1).

However, with the emergence of quantum theory (and thanks to the rather odd role given to measurement in this theory) the role of consciousness in the physical phenomena emerged (at least) as a sensible possibility for the interpretation of quantum mechanics. As we will see, according to some schools of interpretation the consciousness of observer is credited for the collapse of wavefunction and hence it has an essential and intrinsic role in the physics of quantum phenomena itself (and not only in the "observation" in its classical sense). So in brief, classically consciousness acts like a passive receptor of information from outside world,[64] while quantum mechanically it (possibly) has an active role in the physics of the quantum world through its function in quantum measurement and the collapse of wavefunction (see for instance § 5.7 and § 6.11 as well as § 9.1.3). This alleged role of consciousness in quantum physics does not only violate objectivity (in its classical sense) but even realism (to some extent) where reality is supposed to be independent of the observer (see for instance § 1.10 and § 2.4.1).

We should finally note that our focus in the above discussion is the conventional branches of science and physics in particular (whether classical or not). In fact, there are modern scientific, or allegedly scientific, branches (such as parapsychology) whose very subject of investigation is the possible influence or effect of consciousness (or observation or knowledge) on the physical phenomena in the outside world and possible non-conventional types of interaction between the inner world of observer and the outer world of observed (or physical reality). These non-conventional branches generally do not follow the conventional rules and methods of physics (and even science in general) and hence they should be treated and assessed differently and within their own "scientific" space and domain. Also see footnote [254] on page 114 as well as § 1.6.2 and § 6.13.2.

---

[64] As usual, "classical" here is opposite to "quantum mechanical" and hence it applies in general even to non-quantum branches of modern physics.

Chapter 2
Scientific Theory and its Interpretation

This chapter is dedicated to the investigation and discussion of general aspects related to scientific theories and their interpretations (mostly from epistemological perspectives). This includes, for instance, the general objectives of science, the epistemological principles of science and the eligibility and validity criteria for scientific theories and interpretations. The objective of this is to provide general rules and principles, as well as background knowledge and awareness, required or/and useful in the assessment of scientific theories and their interpretations (particularly in the forthcoming assessment of quantum mechanics and its interpretation).

2.1 Objectives of Science

There are two main objectives to science:[65]
• **Practical** (or pragmatic or empirical or ... etc.) objective which is conquering the world and benefiting from its resources (or **making use** of it).
• **Theoretical** (or conceptual or intellectual or ... etc.) objective which is understanding the world (or **making sense** of it).
In more simple and compact terms, these objectives are about having the ability to predict the behavior of the world (and hence maximizing the benefit from it and minimizing the dangers and risks), and having the ability to rationalize the behavior of the world (and hence having a better understanding of how it works).

It should be obvious (especially when considering quantum theory which is the subject of this book) that the formalism of scientific theory is largely about the first objective while its interpretation is largely about the second objective. It should be similarly obvious that these objectives are not totally independent and hence more/better understanding leads to more/better conquest (and more/better conquest leads to more/better understanding).

Finally, we would like to insist that "understanding" is an objective in itself because we are intellectual species and hence understanding is important to us like conquest (although it may not be regarded by some to be as important as conquest since it does not seem to represent a direct biological need).[66] This should highlight and explain the importance of interpretation (which some scientists regard as redundant and irrelevant). Also see § 2.12.

2.2 Logic in Science

In simple terms, logic is a collection of rules and principles that regulate our (rational) thinking to ensure consistency (or rather self-consistency) and rationality. This means that the rules of logic should be respected in any rational form of thinking and knowledge and this should obviously include science both in its formalism and in its epistemology and interpretation. So in brief, the compliance with logic in science is an absolute necessity and hence any scientific theory that violates logic (explicitly or implicitly, directly or indirectly, formally or epistemologically, etc.) should be rejected without further ado because this violation means that it is inconsistent and irrational (or at least it leads to inconsistent and irrational consequences) and hence it cannot be accepted as a form of rational knowledge.

---

[65] We mean here "science in general" as opposite to individual branches of science which have more specific objectives. We should also note that these main objectives are not restricted to science but they belong to all types of rational knowledge (whether scientific or not).
[66] In fact, we can claim "understanding is as important as conquest" from the perspective of being "humans" (of intellectual needs and demands) rather than just "animals" (with only biological needs and demands).

2.3 Epistemology

We can define (according to our view) an epistemological theory (or framework or system, whether general or specific and particular) as a system of interrelated definitions and conventions (representing and reflecting a presumed reality) that are consistent internally (or intrinsically or logically) and externally (or extrinsically or observationally).[67] This means that the theory accommodates and produces concepts and propositions that are totally consistent and do not lead to contradictory implications and consequences (neither internally nor externally). The ultimate purpose of any epistemological theory is to provide an optimal adaptation of the individuals and groups with their environment (i.e. the presumed reality) and satisfy their needs.[68]

2.4 The Epistemological Principles of Science

In the following subsections we outline the principles that underlie the epistemology of science and hence they provide the required ground for the development of scientific theories and interpretations. However, it should be noted that these principles represent our personal view and proposition. Moreover, we present only those principles that are of primary interest and use to us in this book rather than a comprehensive list of all principles of this kind (also see footnote [69]).

2.4.1 The Principles of Reality and Truth

The epistemological foundation with regard to reality and truth are summarized in the following three basic principles which are pivotal not only to science but to all types of rational knowledge:
- The **existence of reality** which means the existence of a real world beyond and outside the observer where the reality of this world is independent of the observer.
- The **uniqueness of reality** which means that this reality (as identified in the previous point) is unique and hence we have only one reality.
- The **uniqueness of truth** which means that we have only one truth which represents the honest reflection of the (existing and unique) reality (as identified in the previous points).

In the following points we discuss briefly some important issues related to these principles:
1. As indicated earlier (see § 1.3.3), the "independence of the observer from the observed" may be considered as another principle of reality and truth or as a derivative of one of these principles (noting that we indicated this independence in the last part of the first principle).
2. We may claim that these principles (or at least some of them) can be either extracted or justified (ultimately) by the rules of logic.[69] For example, the principle of uniqueness of truth is based in a sense on the logical rule of consistency (or rejection of contradiction) since non-uniqueness of truth leads to inconsistency and contradiction (e.g. "I exist and I do not exist" which is a contradictory statement). Anyway, the rules of logic and the rules of reality and truth work hand in hand to ensure the sensibility and consistency of our propositions regardless of being independent of each other or not. Also see § 2.5.2.
3. It is important to note that we (as scientists) should consider these principles from a purely epistemological perspective rather than from an ontological perspective.[70] Hence, the existence of reality, for instance, should not mean the existence of this alleged reality ontologically and in itself as an outside entity but the existence of this alleged reality epistemologically and for ourselves as an entity that we

---

[67] In fact, this should be seen as a definition for an acceptable (or "correct" or "true") epistemological theory (noting that "epistemology" should be about "true" knowledge rather than alleged knowledge).
[68] We note that "needs" should include non-biological as well as biological (or physical) needs although the needs (of organism) are generally biological in their roots and origin.
[69] It is important to remark (while referring to logic) that we could have suggested (as the first epistemological principle of science) the "principle of logicality" (i.e. the necessity of any scientific theory to be consistent with the rules of logic). However, we did not do this because this principle is too generic, general and obvious. Anyway, we will discuss the issue of logicality elsewhere (see § 2.5.1). Also see § 2.2.
[70] This is inline with our classification to them as "epistemological principles of science".

need to assume for building and justifying our scientific knowledge. In fact, the ontological reality (i.e. the ontological existence of reality) is a purely philosophical issue and is irrelevant to science.

4. It is also important to note that the principles of reality and truth should be regarded as an epistemological necessity (noting that they are an ontological choice in our view, i.e. we are free to accept or reject them from an ontological perspective). This is because no rational and consistent science (and knowledge in general) can be built without these principles (see for instance § 2.5.2).[71]

5. The principles of reality and truth do not mean that the world should look the same to every species and individual. In particular, the uniqueness of truth does not mean that the truth is absolutely definite and determined. So, we can say the truth in its exact details is not unique (although it is unique in its essence and within the condition of consistency). We may also say: the truth in its exact details is unique only for a given individual considering the entire set of conditions and considerations that determine the truth such as time, location, measuring equipment, ... etc. Also see § 2.4.3 as well as the subsections of § 1.3.

2.4.2 The Principle of Causality

This was investigated in § 1.7 within its philosophical, epistemological and scientific contexts and implications. In our view, this principle is an epistemological necessity (and possibly ontological necessity as well although this is not of concern to us here) for any rational theory (whether scientific or not) because without this principle no rational and consistent relations and explanations can be established. So, the (epistemological) demand for this principle is based on the (epistemological) demand for rationality in science (and indeed in any type of rational knowledge). Potential limitations (as well as other aspects) of this principle are inspected and assessed elsewhere (refer for instance to § 1.7 and 7.12).

In the following points we discuss briefly some issues related to the principle of causality (from an epistemological perspective):

1. We may suggest "rationality" as an epistemological principle and hence the principles of reality and truth (which essentially represent epistemological realism) and the principle of causality (and possibly other similar principles; see for instance § 2.4.5) become just instances of this "principle of rationality" (which is very general).

2. Since rationality is the basis and justification for causality, then a weak form of causality (i.e. with some indeterminism or uncertainty) may be acceptable (if we are forced to adopt such a weak form) as long as rationality is maintained.[72] In fact, this should also apply to realism as represented basically by the principles of reality and truth (i.e. a weak or partial form of realism may be acceptable as long as rationality is maintained).[73] However, it is important to note that this does not apply to logic (i.e. there is no acceptable weak form of logic) because any violation of logic is a violation of consistency and rationality (also see § 2.2 and § 2.5.1). These issues will be investigated in more details later on.

2.4.3 The Principle of Non-Uniqueness of Science

As indicated earlier and will be repeated later (see for instance § 1.10 and § 2.5.2), science is not unique and hence in principle any physical phenomenon can be described, quantified and predicted correctly (i.e.

---

[71] It is worth noting that being an ontological choice and an epistemological necessity (which allows ontological idealism with epistemological realism) may be seen as non-sensible. However, we may justify this by an analogy with dreams (which are real in themselves although they are not real in their content and what they depict). It should be obvious that the mere existence of dreams should make this (i.e. the existence of some realities whose contents are not real) "sensible". Yes, embracing ontological idealism with epistemological realism should mean that our "reality dreams" should be subject to more rigorous and strict rules to achieve optimal "adaptation" (i.e. to "enjoy these dreams maximally").

[72] We note that maintaining rationality with such a weak form may require some modifications and adaptations to our basic conceptual framework which basically rests on our classical intuition.

[73] It is worth noting that a "weak form of rationality" (whether by a weak form of realism or a weak form of causality) at the quantum level should not affect the (strong and total) rationality at the macroscopic level because at this level we deal with predictions that are deterministic and certain (at least in principle). In other words, at macroscopic level we do not deal with individual quantum events (which are supposedly indeterministic and uncertain) but with classical events which are determinate and certain (by the classical standards).

without violating the rules of logic or the principles of reality and truth) by more than one scientific theory. Hence, any scientific theory should be replaceable (at least in principle) by another scientific theory where both theories are empirically correct and practically equivalent although they may be epistemologically different (in addition to their difference in formalism as well as potential difference in merit and advantages).

Although this principle is not about the individual theories (i.e. each theory in itself) and hence it does not impose restrictions on the scientific theories independently and thus it may be seen irrelevant (i.e. as an epistemological principle for scientific theories), it is important for the construction and formation of science and its overall structure since it allows and legitimizes this sort of diversity in science (i.e. the existence of totally equivalent but different theories) which is not obvious and seems to be unrecognized (or at least controversial). In fact, it is important to propose (or at least highlight) this principle regardless of its status because we feel that there is a common undeclared belief that science is unique and hence new theories are legitimate to emerge and can be constructed only as corrections or improvements or as minor and superficial modifications of old theories (as long as old theories are working in general). This unspoken belief can be felt more strongly and particularly in the literature of quantum theory.

In fact, there are many examples in science for the validity of the principle of non-uniqueness of science. For example, we have different types and formulations of classical mechanics (e.g. Newtonian, Eulerian, Lagrangian and Hamiltonian). Such examples can also be found within the quantum theory itself such as the existence of two main valid formulations (i.e. wave mechanics and matrix mechanics; see 1.2 and § 5.3.2) as well as some hidden-variable variants of quantum mechanical formulations (which are claimed to be equivalent in prediction to quantum mechanics). Despite the fact that all these examples are rather simple and modest (in terms of the difference between the theories which is rather limited and mild), they legitimize this principle as a principle because if differences of these types can occur then differences of other types can also occur since we have no reason to believe that differences differ in this regard.

We should finally note that the principle of non-uniqueness of science is just an instance of the more general epistemological principle of non-uniqueness of knowledge. Hence, the specification to science here is because our book is about science (specifically quantum physics) and hence we are interested here in the epistemological principles of science; otherwise science is not special (as a form of knowledge) in its non-unique nature.

2.4.4 The Principle of Economy

The essence of this principle, which may also be labeled with other tags like "Occam's razor" or the "law of parsimony", is that the scientific theory should be as simple as possible, and hence if we have a set of theories (or formulations or interpretations, ... etc.) that are equivalent in their predictions and outcomes then we should choose the simplest one. It is worth noting that the principle of economy (in some of its instances and interpretations) should imply the validity of the principle of non-uniqueness of science.

It should be noted that the principle of economy does not necessarily require selection between different theories (as the first paragraph suggests) but it can be used in the creation or emergence of a single theory where economy considerations (e.g. in assumptions and postulates) are taken into account in its creation and formulation. It should also be noted that this principle does not represent a necessity or obligation and hence it can be violated. In fact, it is a kind of "recommendation" or "advice" rather than a principle. Yes, simplicity (and hence economy) is advantageous in general (e.g. in terms of saving time and effort and reducing the risk of mistakes and errors) although it is not necessarily so. For example, a simple theory may not be as good as a more complicated and elaborate theory for future development or unification with other theories and hence it is not the best from these perspectives. So, in our choice of a theory we should always consider all the theoretical and practical factors and (dis)advantages of all the available options, and one of these factors should be economy. Accordingly, as long as the other factors are the same (or irrelevant or taken into consideration) economy is an advantage.

In fact, the (old and modern) history of science is full of examples for the use of the principle of economy in the selection and creation of theories. For instance, the desertion of the Ptolemaic model of the solar system in favor of the Copernican model is based (at least with consideration of their general features rather than details and consequences) on this principle. There are also many examples for the adoption and employment of this principle outside science (e.g. in religion, philosophy, art, mathematics, etc.). For example, the theory of creation (of the Universe by a divine entity) could be an instance for the use of this principle.

It should be noted that this principle should not only be used for the selection or creation of theories but it should also be used for the application of theories. In fact, the latter use is a common practice (and hence it gives more legitimacy and extent to this principle). For example, we generally apply classical (i.e. non-Lorentzian) formulations rather than (the more rigorous but more complicated) Lorentzian formulations when the two formulations produce practically identical results (as it is the case in low speed systems where Lorentz mechanics converges to classical mechanics). In fact, the principle of economy is used (with and without declaration or awareness) in almost all sorts of our activities (mental and practical) and this may be explained by our "lazy" nature (although it may also be seen as a sign of wisdom). Other species of animals and plants also follow this principle in their voluntary and involuntary activities. Even Nature follows this principle in general where optimization is a common feature in the physical world.[74]

2.4.5 The Principle of Intuitivity

The essence of this principle is that certain theories and possibilities are intuitive (since they comply with our internal sense which is created by and based on our past experiences and observations) while others are not, and hence we can (and actually do) create or rule in or rule out certain theories and possibilities or make preferences according to this intuitivity criterion. In fact, the terms "intuitive" and "counter-intuitive" (or non-intuitive) are very common in the literature of science (and we can actually find a number of examples for the use of these terms in this book). This means that the use of intuitivity criterion for creating and selecting theories in science (and indeed in all types of intellectual and non-intellectual activities and products) is legitimate as a basis for making scientific judgments and preferences and is acceptable by the scientific community. It should be obvious that the justification of this principle is essentially based on the theoretical objective of science as an endeavor for understanding the world and making sense of it (see § 2.1) noting that intuition is one of the main sources or demonstrations of understanding.

The principle of intuitivity (like the principle of economy) is generally not compulsory and hence it is mostly used in making preferences and voluntary choices. Moreover, it is mostly related to the interpretation of theories (due to the link of this principle to the "understanding objective" of science which is the essence of interpretation) although it may also be used for other purposes and in other contexts. We should also note that the principle of intuitivity may be seen as an instance of the principle of economy (rather than being a principle on its own). Although this may be justified in some cases and circumstances, it is not justified in general (noting that intuitive theories may not be economic) and hence we regard it as an independent principle.[75] Also see § 2.7 and § 2.11.

2.5 Eligibility Criteria of Scientific Theory

These criteria (which will be investigated in the following subsections) represent the basic conditions that any theory is required to satisfy to be *eligible* (or *acceptable*) scientific theory. In other words, they are

---

[74] If we note that "optimization" includes "maximization" as well as "minimization" then we may say: Nature follows the principle of economy/extravagance. However, we should note that the distinction between economy and extravagance may be relative, i.e. it depends on our conceptualization and consideration and hence "extravagance" according to a given conceptualization or perspective can be "economy" according to another conceptualization or perspective (and vice versa). If so, then "optimization" (even in its "maximization" form) is essentially economy (at least according to certain conceptualizations and considerations).

[75] In fact, intuitive theories should be economic from the perspective of understanding (since they are easy to understand and appreciate) although they may not be so in general (e.g. from the perspective of formulation and application). It should also be noted that some instances of economy could be instances of intuitivity (and hence the principle of economy may be seen as an instance of the principle of intuitivity in these cases). Anyway, the relationship between these two principles (and their essence and content) is intimate. In fact, this should apply to the epistemological principles of science in general.

the necessary conditions for the entitlement of a theory to be classified as a "scientific theory" (or at least to be a good candidate for becoming a "scientific theory").[76] The validity conditions will be investigated later (see § 2.6).

2.5.1 Logicality

Logicality means being compliant with the rules of logic in its technical sense and value as the ultimate standard for the determination of truthfulness and falsehood from certain perspectives (which will be clarified later). As it should be known, logic represents the basic rules of thinking and reasoning and hence it should be followed in any rational intellectual process and product whose objective is to reflect and reveal the outside reality (whether this reality is physical or non-physical assuming the existence of non-physical reality such as the existence of divinity).[77] These rules simply determine what is true (or right or correct) and what is false (or wrong or incorrect) from formal and intrinsic perspectives (see the upcoming point 2). For example, the rules of syllogism dictate certain patterns that should be observed in the reasoning process (or arguments) to obtain correct (or truthful) conclusions. Accordingly, the following argument:

$$\text{All carnivores eat meat} \quad \& \quad \text{Lions are carnivores} \quad \rightarrow \quad \text{Lions eat meat}$$

is acceptable because it is compliant with the rules of syllogism, but the following argument:

$$\text{All carnivores eat meat} \quad \& \quad \text{Primates are not carnivores} \quad \rightarrow \quad \text{Primates do not eat meat}$$

is not acceptable because it is not compliant with the rules of syllogism. Also, the (rather generic and general) rule of consistency dictates the rejection of any contradictory (or inconsistent) statement and hence statements like:

$$\text{John exists (here and now) and John does not exist (here and now)}$$

or

$$\text{Lucy is taller than Sara and Lucy is shorter than Sara}$$

are simply rejected because they contain logical inconsistency.

More clarifications about logicality and its status as a basic eligibility criterion of scientific theories are given in the following points:

1. Despite our distinction in the above examples between the rules of syllogism and the rule of consistency, we may claim that all the rules of logic are essentially based on the principle of consistency (in the broadest sense of consistency) and hence even the rules of syllogism (and their alike) are actually no more than instances and derivatives of the fundamental principle of logic which is consistency.
2. Referring to our statement that the rules of logic determine what is true and what is false from formal and intrinsic perspectives, we note that "formal" refers to the form or shape of the propositions (rather than their content or substance) while "intrinsic" refers to the restriction of attention to the propositions on their own regardless of what they are supposed to reflect and represent of the outside world (which is another criterion for being true or not). Accordingly, a true proposition should satisfy the formal and intrinsic criterion of truthfulness (which is determined by the rules of logic) and the substantial and extrinsic criterion of truthfulness (which is determined by the outside reality). For example, "John

---

[76] Whether they are sufficient condition (i.e. collectively) as well should depend on the meaning of "scientific theory" and whether it means *potential* or *actual* scientific theory. However, it is obvious that these criteria (i.e. individually) are not sufficient conditions (due to the existence of other criteria).

[77] We may describe logic as the "grammar of thinking" and hence the validity of thinking depends on observing its rules (like ordinary language whose validity in expressing the meaning depends on observing the grammar of the language). We should also remind the reader that expressions like "rational intellectual process and product" mean intellectual activities and outcomes generated for the purpose (and with the condition) of being rational and they are supposed to be an honest reflection of reality (such as science and philosophy). This is unlike activities and outcomes related, for instance, to art or literature which are intellectual processes and products but without the purpose and condition of rationality and reality.

is tall" is logically consistent statement (and hence it is correct from formal and intrinsic perspective). However, its (unconditional) correctness is conditioned by its compliance with the reality (and hence it may be or may be not correct from substantial and extrinsic perspective). So, if John is tall then this statement is correct; otherwise it is incorrect (despite its logical consistency or correctness). On the other hand, the statement "John exists and John does not exist" is formally and intrinsically incorrect (and hence it is incorrect noting that in this case its compliance or non-compliance with the substantial and extrinsic criterion of truthfulness is meaningless).[78]

3. As indicated, the essence and objective of the rules of logic is to keep the rational intellectual process consistent and avoid contradiction. This can be simply justified from a theoretical perspective because one of the main objectives of any rational knowledge or theory (especially scientific theory) should be the rationalization of our observations (which is essential for understanding noting that understanding is the ultimate theoretical objective of any rational intellectual process; see § 2.1), and hence the compliance with logic should be an obvious requirement for the acceptance of any knowledge and theory.[79] However, even from a purely pragmatic perspective (which pragmatic individuals may find necessary to justify the embracement of logic and the payment of its costly price) the necessity of observing the rules of logic in any rational intellectual process can be easily justified because the product of any intellectual process that fails to observe the rules of logic will be useless (as a rational intellectual product). For example, if we accept the statement "John exists (here and now)" then we have a useful rational intellectual product (or a piece of information or knowledge) but if we accept the statement "John exists and John does not exist" or the statement "Lucy is taller and Lucy is shorter" then we do not have any useful piece of information (or indeed we will have a piece of confusion or misinformation or nonsense which could be harmful rather than being non-useful). So in brief, logical consistency (which is achieved by observing the rules of logic) is a practical, as well as theoretical, necessity. Accordingly, logicality is an epistemological necessity for any type of rational knowledge (since rational knowledge means sensibility, consistency, understandability, practicality, ... etc. and nothing of these can be achieved without logic).

4. Regarding the origin of the rules of logic, we can say that they are developed (or obtained) in the course of our evolution as an intelligent (or intellectual or thinking) species and hence they represent very general patterns which we extracted from the patterns of Nature. In other words, they can be seen as "imprints" or "patterns of Nature". In fact, a kind of "logic" can be observed even in the behavior of animals (which should reinforce the belief in the natural origin of logic).[80]

5. We should make a clear distinction between logic and common sense where some people seem to identify them as a single entity (or merge them). Our view is that common sense represents generally-accepted rules that are usually, but not necessarily, based on the rules of logic as well as common daily practices and observations. Accordingly, common sense depends on many non-logical (and potentially even illogical) factors such as culture (in its extensive sense), gender and past experiences of the individuals and societies. For example, religious education and ideological indoctrination (as well as most social practices) can create very strong common sense rules which are not logical at all (and some can even contradict logic). As a result, scientific theories (and knowledge in general) are not required to comply with the rules of common sense as long as they are compliant with the rules of logic. In fact, quantum theory can be a good example of a scientific theory that (potentially) contains elements that may not be consistent with some of our inherited or acquired common sense rules.

---

[78] This is justified by the fact that compliance with logic is a necessary but not sufficient condition for truthfulness (see point 9) which means that we have "logical truthfulness" and "actual truthfulness".

[79] We may also claim that the rules of logic represent a reflection of fundamental relations in the physical world (i.e. our internal logic is a reflection of a "physical" logic that governs the outside world). Hence, the theoretical need and necessity for logic does not arise only from the need and necessity for rationalization (which is essential for understanding) but from the necessity for the honest reflection of reality (see point 4).

[80] An issue of interest about logic is its basis, i.e. whether it is instinctive (and hence it is part of our biological blueprint) or it is acquired (either by the species or by its groups and individuals through interaction with the environment and for the purpose of optimal adaptation). In fact, this issue is not very relevant to our (quantum) investigation. After all, logic is an epistemological necessity (as indicated above) regardless of its origin. Anyway, if it is instinctive it should have been acquired by the species during evolution. On the other hand, if it is acquired it could become instinctive during evolution (noting that this is an evolutionary mechanism for obtaining and developing "instinct").

2.5.1 Logicality

6. As indicated earlier (see point 2), logic is related to the formal and intrinsic aspects of the propositions and hence it is independent of experiment and observation (which are related to the substantial and extrinsic aspects of the propositions). Accordingly, no experimental or observational evidence can defeat the rules of logic and no experimental or observational evidence can rehabilitate a logically inconsistent theory. For example, the defence of some special relativists (who try to refute the logical challenges to special relativity by the experimental and observational evidence in support of this theory) should be rejected on this basis (noting also that all the evidence are actually for the *formalism* of Lorentz mechanics and not for the *interpretation* of special relativity). So in brief, no experimental or observational evidence can challenge logic or degrade its value or diminish its authority. We may even claim that no correct experimental or observational evidence can (in principle) contradict logic because no "genuine" reality or truth can contradict logic whose patterns and rules represent a genuine reflection of reality. In fact, any experimental and observational evidence should be processed, rationalized and judged by logic not the other way around. Hence, we cannot reject or modify or ignore or dismiss the rules of logic by alleged experimental or observational evidence. Also, no experimental or observational evidence can rectify or justify the faulty logic of a logically inconsistent theory.
7. Compliance with logic is a strict and general condition and hence it should apply to all parts and aspects of any scientific (and even non-scientific) theory. In particular, logic should apply simultaneously and independently to the formalism and interpretation of any theory (that contains both these components). Therefore, we reject special relativity because of its logical inconsistencies although the formalism of the theory (i.e. Lorentz mechanics) is logically consistent as well as being generally supported by strong experimental and observational evidence (see point 6 and refer to our book "The Mechanics of Lorentz Transformations"). In other words, the logical inconsistency of the interpretation cannot be justified by the logical consistency of the formalism (also see § 2.8).[81]
8. A scientific theory that requires (as a necessity) a logically inconsistent interpretation should be a wrong (or illusory) theory because it should embed a logical inconsistency itself and hence all the observations that support the theory should be treated as empirical facts. Accordingly, a logically consistent new theory (i.e. new formalism that can sustain a logically consistent interpretation) should be sought to justify these empirical facts. In brief, any formalism that necessitates a logically inconsistent interpretation should be rejected (with its interpretation) and dismissed as an invalid theory.
9. As indicated earlier, compliance with logic is a necessary but not sufficient condition for scientific theories. In other words, any scientific theory that is not consistent with the rules of logic should be rejected but not every scientific theory that is consistent with the rules of logic should be accepted since its acceptance also requires the support of experimental and observational evidence (noting that logic ensures the formal and intrinsic aspects of truthfulness but not the substantial and extrinsic aspects).[82] For instance, classical mechanics is totally consistent with the rules of logic but it is not necessarily true as a scientific theory. In fact, the laws of classical mechanics are not consistent with our observations of the physical world at the quantum scale and hence classical mechanics is an incorrect theory for quantum systems despite its logical consistency (and even despite its correctness in classical systems).
10. "Quantum logic" is a term used in the literature by different authors (and sometimes even in different meanings). The main purpose of this logic is seemingly to avoid or/and explain the perplexities (and potential contradictions) of quantum physics which allegedly arise from applying the ordinary logic (which is supposedly valid only to the classical world) onto the quantum world. In fact, the invention (or discovery) of a new type of logic (which supposedly departs from the ordinary logic) may be seen to be as legitimate as the invention of the non-Euclidean geometries (which depart from the Euclidean geometry). However, the "conceptual legitimacy" of any type of non-Euclidean geometry is based on its self-consistency (which is the essence of the ordinary logic as indicated above) while its "practical legitimacy" is based on its agreement with physical observations (and its usefulness in this regard). Similarly, the "conceptual legitimacy" of any type of novel logic (e.g. quantum logic) should be based on its self-consistency (which originates from the ordinary logic) while its "practical legitimacy" is

---

[81] Similarly, the logical inconsistency of the interpretation does not affect the logical consistency of the formalism.
[82] It should be obvious that we are assuming here that the other eligibility criteria are satisfied.

based on its agreement with "observations" (and its usefulness in this regard). This means that the legitimacy of any novel logic should be (at least partly) acquired from the ordinary logic (and hence to a certain extent it should be consistent with the ordinary logic). After all, we are classical creatures (rather than quantum creatures) and hence even "quantum logic" (i.e. the logic that belongs to the quantum world assuming the existence of such a logic) should be subject to the rules of our "classical logic" (i.e. ordinary logic) to be "sensible" and "logical" to us.[83]

To sum up, as far as we are concerned and within our current investigation, there is no "quantum logic" as an alternative and substitute to the ordinary logic. Yes, there may exist some logical rules that are tailored specifically to the quantum world and they are consistent with the rules of ordinary logic. In this case, they are just instances of the rules of ordinary logic and hence they should be acceptable (although they are not expected to be of general validity like the rules of ordinary logic).

11. As we saw and will see (refer for instance to § 2.4.2 and § 8.6), we may accept a weak form of realism or causality (i.e. subject to certain conditions and limited to certain zones) but we cannot accept a "weak form of logic". In fact, there is no such "weak form of logic". So, the rules of logic should be respected strictly and thoroughly in any rational process and product.

Also see footnote [69] on page 35 about the possibility of suggesting the "principle of logicality" as one of the epistemological principles of science.

2.5.2 Compliance with the Principles of Reality and Truth

The principles of reality and truth were given and discussed in § 2.4.1. Important clarifications about the compliance with these principles and its status as a basic eligibility criterion of scientific theories are given in the following points:

1. As far as we are concerned exclusively with the epistemology of science, the principles of reality and truth are about knowledge and not about reality (i.e. in itself). In other words, these principles are of epistemological rather than ontological nature. So, what we are actually interested in here is "the existence of reality" in its epistemological value regardless of the "real" existence of reality in itself and beyond our knowledge (which is a purely philosophical issue). Accordingly, the "existence of reality" for example can be seen as a working principle or assumption that regulates and controls our knowledge even though from a philosophical perspective we may assume (or even accept) that reality does not exist. Therefore, the value and role of these principles are independent of any philosophical point of view or ontological choice and hence from this perspective the principles of reality and truth are safe from any suspicion or question mark. In fact, these philosophical issues (such as the "real" existence of reality) are outside the realm and reach of science and hence they should be excluded from any scientific investigation and debate. In simple terms, science is about knowledge (or epistemology) and not about existence (or ontology).

2. The compliance with the principles of reality and truth (to some extent) is a strict and general condition and hence it should apply (to some extent) to all parts and aspects of any scientific theory and to interpretation as well as to formalism.

3. In essence and from an epistemological perspective, the principles of reality and truth are no more than useful conventions[84] that are created and adopted (rather implicitly) for obvious pragmatic reasons. However, for any scientific theory to be qualified as such it should comply to some extent with these principles. This is because even if we do not find a solid theoretical justification to these principles or some of them (apart from this convention) any acquired knowledge in the absence of these principles becomes useless (and hence they are at least a pragmatic necessity). For instance, what is the value of "John is my brother" if reality (and hence John) does not exist, and what is the value of "John is my brother" if we have more than one reality where in one of these realities (at least) he may not exist or may not be my brother, and what is the value of "John is my brother" if we have more than one truth

---

[83] We may say: "logic" by nature is classical (because we are classical). By the way, none of the "quantum logic" systems proposed so far seem to be credible or sensible.

[84] Being conventions (from epistemological perspective) does not contradict the fact that they may have ontological roots and justifications as well as logical origins and links.

2.5.2 Compliance with the Principles of Reality and Truth

and hence we may have "John is my brother" according to one truth and "John is not my brother" according to another truth. As we see, the rejection of the principles of reality and truth usually leads to logical contradiction and inconsistency even though these rules (on their own) are not rules of logic and they may not be required by logic (i.e. directly).

4. The principles of reality and truth (as well as the rules of logic) do not rule out the possibility of having more than one correct theory about the same physical phenomenon. In fact, this should apply to interpretation as well as to formalism. This is because although differences that violate the principles of reality and truth (or the rules of logic) are unacceptable, differences in theoretical formulation and conceptualization, including those related to the philosophical and epistemological aspects of a physical theory, should be acceptable. So, although the reality and truth are unique, they can be described, expressed and theoretically conceptualized and structured in many different ways and that is why we see different branches of science that deal with the same or similar physical phenomena use different concepts and techniques in their theoretical approach to the physical reality. In brief, this sort of differences does not represent any contradiction or violate consistency since such differences are essentially in the shape and appearance and not in the content and essence.[85] Moreover, such differences do not violate the pragmatic considerations that we indicated earlier. Therefore, we can have more than one correct theory about the same phenomenon as long as they do not contradict each other since each theory can represent, describe, conceptualize and quantify the phenomenon from a certain perspective and hence it does not necessarily lead to a contradiction with the other theories. The justification of all this is the fact that science (and knowledge in general) is a mix of discovery and invention (where many observer-dependent factors contribute to the outcome). So, while we impose strict consistency requirements (as dictated by the rules of logic and the principles of reality and truth) on the "discovery" aspect which is supposed to provide an honest reflection of reality, we have a rather relaxed attitude towards the "invention" aspect which is a creative (or "artistic" or "impressionist") aspect. Accordingly, science is not unique and hence any physical phenomenon can be formulated in many different ways all of which are correct (which is a fact endorsed by the existence of many examples of multiple correct formulations related to a single physical phenomenon). It should be obvious that a highly-developed alien civilization (at the level of current human civilization) should have a different version of "science" from our earthly science although both versions of science describe the same physical phenomena "correctly" (at least from a practical perspective) as evidenced by the success of both versions to describe, predict and deal with the physical world. No rational person should expect this alien civilization to have Maxwell's equations or quantum mechanics for instance. Also see § 1.3, § 1.10 and § 2.4.3.

5. As indicated earlier, the compliance with the principles of reality and truth is a necessary but not sufficient condition for the eligibility of scientific theories. This should be obvious from the fact that there are other eligibility criteria (and this should apply to the other eligibility criteria).

6. Apart from the objective of consistency (which is primarily related to the individual thinkers), the uniqueness principles of reality and truth (as well as the rules of logic) should serve another important purpose which is what we may call "collective understanding" (or "collective consciousness") which ensures useful and effective communication between different individuals and thinkers because without uniqueness everyone will talk about his own reality and expresses his own truth and hence there will be no common "language" (and thus no common understanding and useful communication) in the absence of these principles. This should endorse the importance (and even epistemological necessity) of these principles of reality and truth.

To sum up, the compliance with the principles of reality and truth (at least to some extent) is an epistemological necessity (regardless of the ontological status of these principles). In fact, this is the essence of "epistemological realism" which no science (or rational knowledge in general) can exist without it. However, what is required is the compliance with realism in its moderate form and not in its extreme form (see for instance § 1.10 and point 4 of 8.6). Also see § 2.4.1.

---

[85] "Shape and appearance" should also include observer-dependent factors that we discussed earlier (see for instance § 1.3 and § 1.10).

2.5.3 Compliance with the Principle of Causality

Compliance with the principle of causality (at least to some extent) is an epistemological necessity to any scientific theory. This necessity is dictated by the rationality requirement that any rational knowledge (which includes science) should obey. However, as indicated elsewhere (see for instance § 1.7, § 2.4.2, § 7.12 and § 8.6) we may be forced to adopt a weak (or partial) form of causality at the quantum level (to overcome certain quantum mechanical difficulties) if the strong (or total) form of this principle cannot be maintained, and this should be acceptable (at least as a second choice or as a fix) as long as rationality is respected broadly. So in brief, causality should be respected as much as possible and where and when it is attainable.

2.5.4 Physicality

Science is about the physical world and hence any scientific theory should be about this world exclusively. This means that scientific theory should not only be about this world but it should not include any element from any alleged other world. More specifically, the theory should not involve any non-physical element, entity or concept in its framework and therefore the theory should not include any metaphysical or supernatural element or component. In fact, this criterion may be included in (or based on) the principles of reality and truth where "reality" means "physical reality" which is the only legitimate reality we have in science. If so, then the uniqueness of reality (where this reality is strictly physical) should also mean the denial of the existence of any other type of reality (at least from a scientific, and possibly epistemological, perspective).

As indicated, the sense of "physicality" (i.e. being physical) should be defined to exclude all (totally or partially) metaphysical theories even if they are formulated and presented as scientific theories and have physical content and essence. There are obvious examples of metaphysical theories such as those based on (or include elements of) spirituality and theology. In fact, these theories are the least dangerous to science from this perspective. The most dangerous of metaphysical theories are those which are fully fledged scientific theories in form and in essence but they embrace dodgy and disguised elements of metaphysics. One of the best examples of this category is the creation theories which are very common these days and have invaded many physical and non-physical branches of science (notably cosmology, physics and natural history). Another example is dark energy and dark matter (in one of its instances) which are used in cosmology and astronomy to address certain gaps and failures.

In the following points we investigate the criterion of physicality further and discuss some prominent examples and instances in science that may be questioned under this criterion:

1. It is important to note that when we describe something as being "physical" it should mean "having physical reality" (and hence "metaphysical" should mean "not having physical reality"). So, the essence of this is the concept of "reality" within the capability limits of detection by physical means, i.e. either by our senses directly or by their extensions in the form of devices and equipment that improve and extend our senses like lenses or microscopes or telescopes which improve and extend our vision or hearing aid devices or amplifiers or converters which improve and extend our hearing sense. So, it cannot be argued that "physical paradigms and patterns like electromagnetic fields and infra-sound waves are metaphysical because they are abstract models with no reality" because they have reality since we can observe them either directly (in the form of light or heat or audio waves) or through the sensory aids (i.e. devices that extend and improve our senses) since electromagnetic field in its directly-unobservable form of radio waves for example can be converted to observable forms (by exploiting its electric and magnetic effects)[86] and this similarly applies to infra-sound and ultra-sound waves.

2. Non-physical objects (like wavefunction; see § 1.2, § 3.2 and § 4.1) may be tolerated in scientific theories if they play the role of operational device that facilitates the formalism (without the need for the assumption of their reality and hence they are not treated as real objects), but they cannot be tolerated if they play a fundamental role in the theory (i.e. they cannot be considered as a purely

---

[86] In fact, some effects (like polarization) can be seen as virtually-direct evidence for the existence of electromagnetic waves and their physical reality.

operational device since the theory will collapse without them because the correctness of the theory is based on the assumption that they are real). An example of tolerable non-physical objects could be wavefunction (or the scalar and vector potentials in the theory of electromagnetism), while an example of intolerable non-physical objects could be dark matter and dark energy. This is because the former are needed as mere tools (to facilitate the formulation) while the latter are needed as objects that have physical reality.[87]

3. In our view, creation (as well as annihilation on the opposite side) is a nonsensical idea at least from a scientific perspective (i.e. even if we accept it from philosophical and theological perspectives). The emergence of existence from non-existence by creating something from nothing is (at least physically) unimaginable and even impossible although education and indoctrination make it look possible (and even unavoidable). This similarly applies to the annihilation of existence by reducing something to nothing. In fact, we are not aware of any known physical (and possibly even non-physical) processes or mechanisms that can realize and materialize creation and annihilation. So, in our view if creation and annihilation are possible at all then they should be metaphysical in nature and supernatural in essence and hence they should belong to other worlds and to different types of reality other than physical reality (and hence they are beyond the reach of science and physics). So, any "scientific" theory that is based on (or includes elements of) creation and annihilation should be classified as metaphysical or non-scientific and hence it should be rejected as a scientific theory.[88] Also see § 6.6.2.

2.5.5 Testability

Science is an evidence-based type of knowledge rather than a faith-based or trust-based type of knowledge (assuming this is a type of *real* knowledge). Hence, any scientific theory should have testable implications and consequences by which it can be proved to be right or wrong. Moreover, since science is about the physical world (according to the criterion of physicality which was investigated in § 2.5.4), the acceptable type of tests for accepting and rejecting any scientific theory should be physical, i.e. based on observation and experiment. Accordingly, any scientific theory should be physically-testable, i.e. verifiable and falsifiable by physical means and methods.[89] So in brief, any eligible scientific theory should not only be about the physical world (as required by physicality), but it should also be testable through physical means and techniques by being capable of producing physical results, predictions, consequences, implications, etc. that can be inspected and tested by observation and experiment to prove if the theory is right or wrong.

2.6 Validity Criteria of Scientific Theory

These criteria (which will be investigated in the following subsections) represent the basic conditions that any theory is required to satisfy to become *accepted* scientific theory. In other words, these criteria are the required conditions for the theory to be raised from the rank of "eligible scientific theory" to the rank of "valid scientific theory" or "correct scientific theory" (assuming the eligibility criteria of § 2.5 are already satisfied).

---

[87] Whether the "beables" of Bell belong to the first or second category is not very clear and may be case-dependent although they seem more appropriate to belong to the first category (and some beables may even be physical). We should also note that some physicists (like de Broglie and possibly Heisenberg) consider wavefunction as physical (like electromagnetic waves) and not just an operational device.

[88] In fact, we may even argue that creation and annihilation can be rejected logically by the logical rule of consistency or rejection of contradiction (with the aid of some physical premises) because any physical process should occur in a finite time, so during the transition between existence and non-existence we should have a state between existence and non-existence which is contradictory (and hence it is impossible and not only unimaginable). To put it in a different form, if creation and annihilation occur instantaneously (i.e. in no time) then no physical process can occur in no time (where in our case something suddenly becomes nothing or vice versa) while if they occur in a finite period of time then we should have an inconsistent mix of existence and non-existence. However, this argument is not sufficiently rigorous and may be questioned from different perspectives.

[89] We would like to emphasize that being physically-testable means being physically *verifiable* and *falsifiable* (i.e. it can be proved right or wrong by observation or experiment based on the theory). Therefore, if a theory cannot be falsified then it cannot be accepted even though it was *seemingly* valid and correct. We may claim that a theory that cannot be falsified cannot be *genuinely* verified because verifiability and falsifiability are correlated and interdependent.

2.6.1 Scientific Evidence

In simple terms, any valid (or correct) scientific theory should be supported by scientific evidence (and hence physical-testability is materialized through a physical test that yields an observational or experimental evidence in support of the theory). In fact, "scientific evidence" should mean that the theory is not contradicted by any observational or experimental evidence (which is a passive requirement), and the theory is proactively and specifically endorsed by observational or experimental evidence (which is an assertive requirement). More clarifications about this criterion and its role in validating scientific theories are given in the following points:

1. Being "correct" or "incorrect" (or "right" or "wrong" or any similar adjectives) in the context of validating scientific theories by scientific evidence should be based on practical (rather than theoretical or idealistic) criteria related to the compliance with experiment and observation regardless of being right or wrong theoretically (assuming such theoretical criteria are meaningful and viable). So, being "correct" (or "right" or "true") in this context should be interpreted practically and pragmatically (rather than idealistically) as providing correct physical predictions in the cases and circumstances to which the theory supposedly applies. Accordingly, being correct or right is no more than a practical measure of acceptance. In fact, because we adopt a practical measure, expressions like "valid theory" should be more appropriate than expressions like "correct theory" (since "valid" is more suggestive of practical value while "correct" is more suggestive of theoretical or idealistic value).[90]

2. Correct predictions from wrong theories is logically consistent and historically documented where it is well known from the history of science that many wrong theories provide correct predictions and results. This should raise an important question about the value of any scientific evidence in support of a given theory because the theory could be wrong (even though it is endorsed by certain evidence) since further evidence can emerge in the future and invalidate the theory. However, since we are adopting practical criteria for correctness (as explained in point 1), any scientific theory supported by evidence (and not contradicted by evidence) should be valid but conditionally (i.e. until the emergence of invalidating evidence) and hence it should be followed and treated as a valid (or "correct") theory. However, we should always be cautious about the validity (or "correctness") of an already-endorsed theory (or a conditionally-valid theory) when the theory is used to investigate and probe new areas and fields where the theory could fail. In fact, this should mean that all our scientific knowledge is conditionally-valid especially when it is used in new areas of research and investigation or applied under new conditions and circumstances.

3. When an invalidating evidence emerges against a previously-endorsed (and hence valid) theory then the theory becomes invalid and hence a new theory should be sought or the original theory is amended (e.g. by generalization or specialization or conditioning) to comply with the new evidence. Yes, the theory may be restricted in its domain of validity and applicability and hence it is kept as a valid theory but only within its new (restricted) domain. The obvious example of such a restricted theory is classical mechanics which is restricted in validity to macroscopic and slow systems to cope with the challenging evidence that emerged from quantum and Lorentzian systems (where quantum mechanics and Lorentz mechanics replaced classical mechanics).

4. If a theory has validating and invalidating evidence at the same time then the deliberation of the previous points should apply.

5. When evidence emerges in support of a certain aspect (or a specific part or a conditioned instance) of a scientific theory then the validity of the theory (which the theory acquires from the emerging evidence) should be restricted to the endorsed aspect (or part or instance). For example, if evidence emerged in support of time dilation aspect but not length contraction then this should endorse and validate this aspect of Lorentz mechanics. Yes, if some other aspects (or parts or instances) are logically or physically correlated to the endorsed aspect then they should get similar endorsement (as much as the correlation implies). Moreover, such limited endorsement should even increase the likelihood of the validity of the other aspects of the theory (which are not endorsed directly or indirectly) although this should depend on the nature of the theory (e.g. if the theory is based on a single theoretical structure

---

[90] In fact, we may refer to this type of correctness using terms like "empirical correctness" or "empirical success".

or not) and how its aspects and parts are related to each other.
6. Evidences vary in their significance and strength and hence the value of any evidence should be assessed to evaluate the validity acquired by a given scientific theory from that evidence. The reader is referred to our book "General Relativity Simplified & Assessed" where we discussed some aspects and criteria (as well as specific examples) for evaluating the significance and strength of scientific evidence.
7. When a given evidence endorses two theories then in principle it should be accepted as an endorsement to both theories. Yes, if the two theories contradict each other or they have contradictory aspects then detailed assessment (which depends on many case-specific factors) is required. However, if the two theories are competing (i.e. they have the same domain of validity and they are supposed to be representing the same physical systems and phenomena) then the value of the evidence in endorsing one or both theories (assuming it has endorsing value) should diminish (noting that the two theories supposedly cannot be correct simultaneously and hence the implication of the evidence should be damaged).
8. If two conflicting theories are supported by different evidences then in principle the stronger evidence should be followed (but cautiously and conditionally) although this should depend on many case-specific factors and considerations.
9. Scientific evidence is usually obtained and established for the technical aspect (or formalism) of the theory rather than its interpretation. The evidence for the interpretation (as well as the relationship between the evidence for the formalism and the evidence for the interpretation) will be investigated later on (see for instance § 2.8, § 2.9.5 and § 2.10).
10. The general consensus is that a theory is correct *iff* all its predictions are correct and incorrect *iff* (at least) some of its predictions are incorrect. However, this should be regarded as a logical convention rather than a logical rule and hence we may define partial correctness/incorrectness (where a theory can be seen as correct conditionally and within the limits of its validity). Anyway, since we adopt practical criteria for correctness (see point 1) this has no practical significance and hence we follow the rules that we outlined in the previous points regardless of what label (i.e. "correct" or "incorrect" or "partially-correct") we use to classify the theory. As seen earlier, classical mechanics may be seen as an example of a partially- or conditionally-correct theory (since it is valid in certain systems and regimes and not valid in other systems and regimes).[91]

2.6.2 Compliance with the Established Theories

This validity criterion means that any newly-proposed theory should be compliant (i.e. non-contradicting) with other well-established theories whose domain of validity overlaps with the domain of validity of the newly-proposed theory. In fact, "well-established theories" should include well-established facts (even though these facts are not in the form of technically-formulated scientific theories). More clarifications about this criterion and its role in validating scientific theories are given in the following points:

1. This criterion is obviously based on the eligibility criterion of compliance with the principles of reality and truth as a requirement for the uniqueness of truth (see § 2.5.2). However, we classified this as a validity criterion (rather than an eligibility criterion or as an application or instance of the criterion of § 2.5.2) considering the actual validity of the proposed theory within the overall scientific body (or knowledge to be more general) as well as its specific nature (opposite to the general nature of the criterion of § 2.5.2).
2. This criterion may also be seen as an instance or application of the previous criterion of validity (see § 2.6.1) especially as a passive requirement. However, the two criteria are different although they are implicitly related (since "established theories" should be supported by "scientific evidence"). Anyway, the distinction between the two criteria should provide more clarity and hence it is justified regardless of whether these criteria are distinct or not (noting that this is a trivial matter and has no practical consequences).

---

[91] In fact, if we embed the conditions and restrictions of validity within the theory then the theory should become "unconditionally correct" (i.e. within its restricted and conditioned domain)

3. The correspondence principle (see § 4.7.2 and § 6.7) is based on this criterion because classical physics is a well-established theory in its domain of validity and hence quantum physics (whose domain of validity supposedly extends *possibly partly* to the domain of validity of classical physics)[92] should not contradict the predictions of classical physics in this domain. In brief, quantum physics should not contradict the correct predictions of classical physics (regardless of where this common area of validity occurs).

2.7 Optimality Criteria of Scientific Theory

We discuss in the following subsections the basic criteria for assessing the optimality of scientific theories so that we have certain measures by which we can determine if a given scientific theory is ideal or non-ideal (considering the current state of development of science). It is noteworthy that for some theories some of the optimality criteria may not be applicable or realizable or achievable due to the special nature of the theory as well as to potential conflict with other criteria and requirements (see for instance point 3 of § 7.2).

2.7.1 Simplicity

This means that the theory should be as simple and clear as possible, i.e. it does not include unnecessary complications and twists. The benefit of simplicity and clarity (over complexity and ambiguity) should be very obvious (e.g. avoiding potential errors, ease of use and application, ease of further development, etc.) and hence we do not need to argue in favor of this. For example, if we create a theory that is highly abstract theoretically and complicated mathematically (and hence it is complex and difficult to understand) then, apart from the risk of being too abstract to represent reality, the theory could consume considerable resources (e.g. in its digestion, integration with the existing theories, using in practical and theoretical applications, further development, etc.) and become very costly.

This criterion also implies that the theory should be as close as possible to common sense and classical heritage and based on natural intuition. For example, a theory that is based on bizarre concepts and methods (and hence it is against common sense, heritage and intuition) will be costly and difficult to integrate with the preexisting theories and can lead to illusions and fantasies (as well as other definite and potential lesions and troubles). On the opposite side, a theory that is based on common sense and natural intuition and inline with the heritage can be easily digested, integrated, used, developed, etc. and hence it will not only be safer (i.e. in representing reality) but also cheaper in cost.

In fact, the very nature of science (as an attempt to understand the physical world; see § 2.1) should impose simplicity as (at least) an optimality criterion. Accordingly, only the minimum required complexity and ambiguity should be tolerated in the development of science and formulating its theories. Unfortunately, this does not seem to represent the trend (or rather the fashion) in modern science where many unnecessary complications are introduced deliberately for the sake of prestige and esteem. Finally, it should be obvious that this criterion can partly be based on the principles of economy and intuitivity (see § 2.4.4 and § 2.4.5).

2.7.2 Empiricism

This means that the theory should emerge from observation and experiment rather than from pure theoretical speculations and mathematical modeling. The merit and relevance of this criterion should be obvious because science is about the physical world and hence we should learn science from this world (by observing it and experimenting on it) rather than by dictating our ready-made models and patterns (and even illusions and fantasies) on it. Unfortunately, modern scientists are obsessed with excessive

---

[92] In fact, no one should expect quantum mechanics to be a *practically* valid theory for classical systems in general (as indicated by *possibly partly*). Yes, there are some "microscopic" systems with large quantum numbers (which are on the verge of the validity domain of quantum physics) that can be seen as an example for the application of the correspondence principle. Also, quantum physics can be at the foundation of some classical branches and applications of physics. Also see § 3.7, § 4.7.2 and § 6.7.

theorization and over-mathematization (where theoretical sophistication and abstraction are admired and valued regardless of their empirical justification and physical value). This criterion can partly be based on the principles of economy and intuitivity (see § 2.4.4 and § 2.4.5).

2.7.3 Realisticity

This means that the theory must be realistic and practical and hence it should not include non-realistic elements and non-practical aspects like theoretical fantasies and illusions and mathematical artifacts (such as singularities).[93] For example, if we create a theory about gravity that leads to illusions and fantasies (like white holes and travel in time) then this theory will be a burden and liability to science and can lead to many disastrous consequences (e.g. wasting resources on investigating illusions and fantasies, misleading science to wrong directions in research and development, leading to traps and wrong conclusions, etc.). It should be noted that some types and extents of fantasies and illusions can make the theory non-scientific (rather than non-ideal). Hence, this criterion is about scientific theories that (marginally) have non-realistic aspects and non-practical features without losing their scientific status. Modern science is full of examples of theories (e.g. general relativity) which have strong non-realistic aspects and illusive implications. This criterion can partly be based on the principles of economy and intuitivity (see § 2.4.4 and § 2.4.5).

2.7.4 Harmony

This means that the theory should be compatible and harmonious (in its concepts, formalism, methodology, style, etc.) with other well established scientific theories (that are related in subject or/and domain to it) so that it can be easily integrated and merged with other (preexisting) theories and branches of science. For example, if we create a theory about the transmission of electromagnetic waves in material media that is totally different (in concepts, symbolism, formulation, etc.) from the preexisting theory of electromagnetic transmission in vacuum then this will create many problematic issues (e.g. about compatibility and integrability) and will be unnecessarily costly. So, an ideal theory in this case should be as harmonious as possible with the existing theory of electromagnetism and should ideally be a generalization or extension or instantiation of the existing theory rather than being a theory from scratch and alien to the existing theories in the field of electromagnetism. In fact, any theorist should look (before attempting to create and develop a theory) to the existing knowledge and theories in that field (as well as similar fields) and try to make as much use as possible of the existing reliable knowledge and theories (noting that good science is based on integration and progression). This criterion can partly be based on the principle of economy (see § 2.4.4).

2.7.5 Objectivity

This means that the theory is objective in its description of the phenomenon that it supposedly represents and hence it is free of subjectivity or involvement of the state or attitude of the observer for instance. This is not only to ensure that the theory is an honest depiction of the reality as it is but also to eliminate any external dependency (i.e. on factors that do not belong to the phenomenon such as the attitude or emotions of the experimenter) that could diminish the value of the theory. In fact, the root of this criterion originates from the consideration of inertness (or passivity) of observation which we investigated in § 1.3.3.

It should be noted that a minimum amount of objectivity is required for any scientific theory to be truly scientific and hence when we talk here about objectivity as an optimality criterion we mean scientific theories that are generally-objective but they are potentially smeared with elements of subjectivity (without affecting their scientific and generally-objective status). It should also be noted that the objectivity here is different from the ontological objectivity (whose essence is the existence of independent and determinate

---

[93] We use "realisticity" rather than "realism" because "realism" is already used in a different meaning (see for instance § 1.10).

reality) which is a philosophical issue and hence irrelevant to science. We also note that the objectivity here does not contradict the (inherent) observer-dependent aspects and features that characterize any type of observation and knowledge (see for instance point 5 of § 1.3.3). Also see § 7.9.

2.7.6 Quantitativity

This means that an optimal scientific theory should produce a quantitative type of knowledge represented, for instance, by equations, inequalities, numbers, tables, charts, etc. Moreover, the more quantitative the theory is (and hence the more precise and accurate its quantitative predictions) the better it should be. The justification of this optimality criterion is the presumption that science is based on accuracy and precision and hence any optimal scientific theory should produce this (accurate and precise) form of knowledge.

However, quantitativity should not be considered an optimality criterion for a scientific theory unless the expected type of knowledge from such a theory is supposed to be quantitative. For example, if we have to develop a theory about how to measure temperature then ideally we are required to produce a quantitative theory (and its quantitative aspect should be as precise as possible), but if we are supposed to develop a theory about how to recognize the fertility by color (e.g. in some species of animals) then we do not (or may not) need to produce a quantitative theory. Anyway, we should recognize the value of quantification in science (due to the measurability, objectivity, accuracy and certainty that quantification provides) and hence quantitative theories are generally superior (in comparison to the corresponding non-quantitative or less-quantitative theories).

2.7.7 Predictivity

This means that an optimal scientific theory should be able to provide predictions and expectations about non-existing phenomena rather than being restricted to describing what is already existing. This criterion may originate from the desire to have a more reliable form of knowledge whose validity can be tested not only by what is already existing but by what is expected to exist. It may also originate from pragmatic considerations and values since predictive science can foresee the future and hence it provides more useful knowledge and better potential for further development. However, these justifications do not mean that this criterion is required (as a necessity) for a theory to be truly scientific because reliability can be (conditionally) established by what is existing. Moreover, purely-descriptive science is also useful and should have useful applications as well. Therefore, we classify this criterion as an optimality (rather than eligibility) criterion.

2.7.8 Interpretativity

This means that an optimal scientific theory should have an acceptable interpretation (the more intuitive and rational the interpretation is the better the theory should be). The demand for this optimality criterion should be obvious because science is essentially an attempt to understand the physical world (see § 2.1), and without an acceptable and sensible interpretation no understanding (or at least no profound understanding) of the physical content of the theory will be achieved (noting that the formalism usually provides somewhat blind and abstract rules whose essence and significance are difficult or impossible to understand and appreciate without a sensible interpretation). This criterion can partly be based on the principle of intuitivity (see § 2.4.5).

It is important to note that classifying interpretativity as an optimality criterion (rather than an eligibility criterion) is inline with our view that some scientific theories may not be interpretable and hence interpretativity is not a necessary condition for scientific theory. In fact, quantum mechanics may belong to this category of non-interpretable scientific theories (as will be investigated later on; see for instance § 2.8, § 7.15 and § 8.6).

2.8 Relationship between Formalism and Interpretation

Before investigating the relationship between formalism and interpretation we should discuss the technical meaning of "interpretation" of scientific theories (noting that the technical definition of interpretation should shed light on this relationship).[94] In brief, the "interpretation" of a scientific theory may be defined generically as the qualitative equivalence of the technical formalism (which is usually quantitative or/and symbolic) of the theory. However, in our view the interpretation is more than a qualitative reproduction of the technical formalism because the role of interpretation is not just to provide a more comprehensible and simple version of the theory (as represented primarily by the formalism) but it extends beyond this to areas and aspects like providing justifications, indicating causes and consequences, and relating the formalism to observable physical entities and quantities.[95] Accordingly, the interpretation provides an added value to the original theory and vitalizes it, rationalizes it and breathes life into it by describing its contents and aspects using familiar concepts and easily-understood terminology (e.g. from daily life and classical physics).

Returning to the relationship between the scientific theory (as represented primarily by its formalism) and its interpretation, we may outline this in the following points:

1. If we consider the formalism and interpretation as the two main components of a typical scientific theory then we may say (while referring to § 2.1): the essential role of the formalism is to provide a set of rules for correct empirical prediction of the behavior of the physical world, while the essential role of the interpretation is to provide a comprehensible picture (or description) of the physical world. In more simple and loose terms, formalism is mainly about conquering the world while interpretation is mainly about understanding the world.

2. The validity and invalidity of the formalism and interpretation are independent of each other in general. Hence, we may have correct/incorrect formalism with correct/incorrect interpretation (considering all the four possibilities).[96] For example, we reject special relativity as an incorrect interpretation (due to its axiomatic and logical weaknesses) although the formalism of the theory (i.e. Lorentz mechanics) is generally supported by strong experimental and observational evidence and hence it is correct (i.e. in a practical, and possibly limited, sense).

3. We may have more than one acceptable or valid interpretation to a given theory (as well as may have more than one theory or more than one formalism that correctly describe a given physical phenomenon). Also see § 2.13.

4. In principle, we may not have an interpretation (or may not have an acceptable or valid interpretation) of a particular scientific theory even though the theory (as represented primarily by its formalism) is well established. A potential example of this case (i.e. non-existence of an acceptable or valid interpretation of a well established theory) is quantum theory whose all existing interpretations are seemingly questionable and short of evidence (see for instance chapters 6 and 8). In fact, even Lorentz mechanics may be included (tentatively) in this category considering its most common interpretations (with special relativity being the most notable example). So in brief, although it is nice to have an interpretation (and it may be nicer to have a *single* interpretation), there is no necessity for the existence of interpretation (or its uniqueness if it exists). In other words, a valid scientific theory is valid regardless of having and not having an acceptable or valid interpretation and regardless of having one or more acceptable or valid interpretation.[97] In fact, the essence of a scientific theory and its real scientific content is in its formalism (noting that the interpretation, despite its scientific character, can be seen as an epistemological component or accessory attached to the theory although in an ideal

---

[94] We draw the attention that an initial definition of "interpretation" (as well as "formalism") was given in § 1.1.

[95] In fact, one of the main roles and objectives of any interpretation (according to the explicit meaning and suggestion of the word "interpretation") should be providing a convincing explanation and justification to the formalism. From this perspective we may say: the formalism is about "how" while the interpretation is about "why" (although this should be understood broadly rather than literally).

[96] Yes, if the formalism is incorrect and the interpretation is correct then this interpretation should not be really an interpretation of the theory represented by that formalism.

[97] We may even claim that some scientific theories may not be interpretable, i.e. they are in such a messy state (epistemologically) that no satisfactory interpretation can fit within their structure. In fact, quantum mechanics is a potential candidate for this type of non-interpretable theories (see for instance § 2.13 and § 8.6).

2.8 Relationship between Formalism and Interpretation

situation it should emerge naturally from the formalism). Also see § 2.13.

5. From a broad epistemological perspective, the formalism may be seen as the test bed for the interpretation (i.e. in general not in its specific details and individual forms and instances) where the legitimacy (or "correctness") of the latter is based on the legitimacy (or "correctness") of the former. In other words, the empirical correctness of the formalism legitimizes the theoretical (or conceptual) correctness of the interpretation (assuming the interpretation meets the required conditions and criteria).[98] Yes, specific details of the interpretation, as well as choosing between contradicting interpretations or favoring certain interpretations over others, should require specific evidence.

6. From the perspective of the relationship between formalism and interpretation, ideal scientific theory is a theory whose formalism embeds its own interpretation (i.e. the formalism intuitively suggests its interpretation and hence the theory can be interpreted naturally and effortlessly). This is a characteristic feature of classical physics (in general and over most of its branches) but not of modern physics (especially quantum theory).

7. In certain branches of physics, the difference between the formalism and interpretation of scientific theory is rather delicate and subtle. This is especially notable in those branches of modern physics in which observation and measurement take a central role in the theory itself (and not only in its application) which usually affects issues like inertness and objectivity (see for instance § 1.3.3 and § 2.7.5) and can lead to muddle and confusion between what belongs to formalism and what belongs to interpretation. The obvious examples of this type of branches are Lorentz mechanics and quantum mechanics where the former is centered on the idea of "frame of reference" (or "observer") while the latter is centered on the idea of "measurement" (or "observation"). As a result, the border line between formalism and interpretation becomes rather blurred (see for instance our book "The Mechanics of Lorentz Transformations" as well as § 7.10 of the present book).

Anyway, it is very important to make the distinction between formalism and interpretation clear and transparent to avoid traps and pitfalls. For example, the confusion between formalism and interpretation in Lorentz mechanics (where the interpretation of special relativity is identified with the formalism of Lorentz mechanics) led to many troubles and fallacies such as giving the credit of the (empirical) success of Lorentz mechanics to special relativity while putting the blame for the (epistemological) failure of special relativity on Lorentz mechanics (and hence we see the majority of physicists accept special relativity, despite its epistemological inconsistencies, because of the empirical success of Lorentz mechanics, while we see some physicists reject Lorentz mechanics, despite its empirical success, because of the epistemological failure of special relativity).[99] Another example is from the quantum theory itself (which is the subject of this book) where the status of the issue of measurement (and wavefunction collapse) was confused, i.e. whether it belongs to the formalism (as suggested by its inclusion in the axiomatic framework of quantum mechanics according to its commonly-accepted version) or it belongs to the interpretation (as suggested by its epistemological nature and content). As we will see, the issue of quantum measurement is the source of many (if not most) troubles and confusions in and around the quantum theory (particularly in its interpretation) noting that one of the reasons for its problematic nature is the lack of distinction, i.e. if it belongs to formalism or interpretation (see for instance § 5.7, § 6.11 and § 7.10).[100]

In fact, there are other aspects and issues about the relationship between formalism and interpretation (e.g. the relationship between the evidence of formalism and the evidence of interpretation). These aspects and issues will be investigated in the rest of this chapter (see for instance § 2.9.5 and § 2.10).

---

[98] This could explain why having an incorrect formalism with a correct interpretation may not be sensible (see footnote [96]).
[99] The reader is referred to our book "The Mechanics of Lorentz Transformations". Also see footnote [6] on page 9.
[100] As we will see later (refer for instance to chapter 8), the postulate of measurement is controversial where some schools of interpretation consider it part of the axiomatic framework while other schools do not.

2.9 Eligibility Criteria of Scientific Interpretation

These criteria (which will be investigated in the following subsections) represent the basic conditions that any interpretation is required to satisfy to be *eligible* scientific interpretation. So, they are the necessary conditions for the entitlement of an interpretation to be classified as a "scientific interpretation" (noting that the validity criteria will be investigated later; see § 2.10).[101] However, before we go through the details we draw the attention to the following remarks:

1. The first four eligibility criteria of scientific theory (i.e. logicality, compliance with the principles of reality and truth, compliance with the principle of causality, and physicality) that we investigated earlier in § 2.5.1, § 2.5.2, § 2.5.3 and § 2.5.4 apply to the interpretation as to the formalism and hence we do not repeat here although we will include them in the following subsections (for the sake of completeness and structure) with a rather brief investigation to outline and highlight some of their specific features with regard to interpretation. So, the main focus in the present section is the investigation of the additional eligibility criteria of interpretation.
2. Regarding the testability criterion, the previous remark may also apply (noting that testability was investigated in § 2.5.5). However, because testability seems to be a controversial issue it requires additional investigation in this section (as well as later in the book).

2.9.1 Logicality

As indicated in the preamble of this section, this criterion was investigated earlier and hence we refer the reader to § 2.5.1 for details. However, we should remark that the logicality here (i.e. with regard to interpretation) should be more pertinent and demanded since the interpretation is supposed to provide a more comprehensible and rational version of the theory with the use of more common concepts and ordinary language and hence it can be seen as the primary and familiar domain for the applicability of logic and logicality.

2.9.2 Compliance with the Principles of Reality and Truth

As indicated earlier, this criterion was investigated earlier (see § 2.5.2). However, we remark again that the compliance with the principles of reality and truth here (i.e. with regard to interpretation) should also be more pertinent and required for the same reason as for logicality.

2.9.3 Compliance with the Principle of Causality

Again, this criterion was investigated earlier (see § 2.5.3) noting that the compliance with the principle of causality here should also be more pertinent and required for the same reason as for logicality.

2.9.4 Physicality

As indicated in the preamble of this section, this criterion was investigated earlier and hence the discussion of § 2.5.4 should generally apply here. Accordingly, any eligible scientific interpretation should be physical in essence and spirit because it is supposed to be an interpretation of a scientific theory. Therefore, no interpretation of theological or ideological or metaphysical or philosophical nature is eligible to be a scientific interpretation. Accordingly, no religious interpretation based, for instance, on divinity should be accepted (i.e. from a scientific perspective). Similarly, ideological interpretations (such as those proposed by some communist philosophers to interpret certain scientific theories) should also be rejected

---

[101] Whether they are sufficient conditions as well should depend on the meaning of "scientific interpretation" and whether it means *potential* or *actual* scientific interpretation. However, it is obvious that these criteria (i.e. individually) are not sufficient conditions (due to the existence of other criteria).

scientifically.[102] Also, pseudo-physical interpretations based, for instance, on dark matter or dark energy (which are essentially metaphysical in nature) should also be rejected.[103]

We should also classify the interpretations that are based on (or include) the paradigm of creation in this category (i.e. non-scientific) because creation is a philosophical and theological paradigm (rather than scientific) in nature. In fact, no creation-based interpretation can be genuinely scientific because it is metaphysical or at best it stands on the border between the physical world and the non-physical world and hence it cannot be purely physical. As we explained elsewhere (in this book as well as in some of our previous books) the rejection of the involvement of metaphysical elements in the formulation and interpretation of scientific theories is justified by the nature of science and metaphysics which are different and have different domains of application and validity. Moreover, metaphysical premises and propositions do not have scientific meaning and cannot be proved or disproved (at least by scientific means) and hence they have no scientific substance or value. In other words, they are scientifically nonsensical and meaningless and non-testable (see § 2.9.5).

2.9.5 Testability

We may also say: this criterion was investigated earlier and hence we refer the reader to § 2.5.5 for details. However, the testability of interpretation seems to be different from the testability of the theory (as represented primarily by its formalism) and hence it requires more investigation and attention. Accordingly, we should investigate what "testability of interpretation" means. We should also investigate if the testability of interpretation is achievable or not.

Regarding the meaning of "testability of interpretation", we can say that the testability of formalism is generally obvious since this testability (or rather this type of tests) can be achieved (or realized) by observations and experiments that produce certain outcomes which can be compared to the predictions of the formalism (whether these predictions are qualitative or quantitative). For example, if an astrophysical theory correlates the color of stars to their age then we can (in principle at least) collect data (by observation) about stars of known ages and known colors and hence we can test (qualitatively) if this correlation is correct or not. Similarly, if a physical theory predicts the amount of heat generated by certain thermodynamic systems under certain conditions then we can (in principle at least) experiment on these systems to test (quantitatively) if this prediction is correct or not. So, in both cases the testability (of formalism) is sensible and achievable, i.e. the tests can be conducted according to the requirement of the testability criterion.

However, when it comes to the interpretation we may feel a difficulty about proposing a sensible meaning for testability because if "testability of interpretation" should have any sensible meaning as such then the interpretation tests should be independent of the formalism tests as if we actually have two independent theories (i.e. one about formalism and one about interpretation) rather than a single theory that have formalism and interpretation. In fact, this issue is related to the issue of the meaning of "interpretation" and its technical definition which we investigated (rather briefly) in § 2.8. Accordingly, we can say if the interpretation is just a qualitative equivalence of the technical formalism of the theory then it is difficult to propose a sensible meaning for testability that is different from the testability of formalism. However, as we indicated in § 2.8 the interpretation is more than a qualitative reproduction of the technicality of formalism because it extends to areas and aspects like providing justifications, indicating causes and consequences, and so on. So, in reality what interpretation (or rather some types of interpretation) actually represents is a theory (of epistemological nature) that is parallel to the technical (or formal) theory and hence the testability of interpretation is completely sensible and (in principle) independent of the testability of formalism although it could depend on the nature and type of the interpretation and its content.

---

[102] Interestingly, during the era of the Soviet Union some communist theorists interpreted quantum theory in an "idealistic way" and hence it is seen as inconsistent with the "dialectical materialism" (which was the official "religion" of the state at that time). Accordingly, it was recommended to ban quantum theory (but "realism" proved to be stronger than ideological dogmatism and religious devotion).

[103] We note that things like dark matter and dark energy usually enter even in the formalism. This may also apply to creation which will be discussed next (see for instance § 6.6.2).

2.9.5 Testability

Regarding the achievability of testability (or test) of interpretation, it is obvious (from our discussion about the meaning of testability) that the testability of interpretation is achievable although there is no guarantee that every type of interpretation of any type of scientific theories can be tested and be testable (which is inline with the fact that testability is an eligibility criterion since some interpretations may not have any testable content or implication or consequence and hence they are not eligible to be classified as scientific interpretations). So in brief, the testability of interpretation is achievable if we can find a type of tests that can target interpretations specifically and hence these tests can distinguish between one interpretation and another interpretation (where our ability to find this type of tests depends primarily on the nature of the interpretation which also depends on the nature of the formalism). Accordingly, for the testability of interpretation to be sensible and achievable we should, in principle, have (or can have) certain contents or implications or consequences of the theory that depend specifically on the particular interpretation and hence we can distinguish between different interpretations and whether they are right or wrong. In other words, we need an interpretation-dependent (but possibly formalism-independent) tests that can distinguish between correct interpretation and incorrect interpretation.[104]

For example, the interpretations of time dilation phenomenon (which is implied by the formalism of Lorentz mechanics) as real or apparent and the interpretations of Lorentz mechanics itself that hypothesize the existence or non-existence of an absolute frame of reference (where all these interpretations rest on the same formalism) can be tested (at least in principle) by proposing or designing specific observations and experiments that can distinguish between correct and incorrect interpretations (where all these tests are consistent with the formalism) and hence these interpretations are testable.[105] On the other hand, metaphysical interpretations in general (e.g. creation theories or dark energy theories) are not testable (at least within the current scientific horizon) and hence they should be excluded from being scientific interpretations not only by the physicality criterion but also by the testability criterion.[106] So, to conclude this subsection we can say categorically and unequivocally: "testability of interpretation" is meaningful and achievable and hence it should be considered as a definite eligibility criterion for scientific interpretations (and accordingly only testable scientific interpretations are eligible scientific interpretations while non-testable "scientific" interpretations are not eligible although they could be eligible to be classified as philosophical or theological interpretations for instance).[107]

It is noteworthy that testability in its negative sense (i.e. having a test against a particular interpretation rather than in support of it) is not only acceptable but it is a reality recognized and accepted by the mainstream quantum physicists. An example of testability in this sense is represented by the essence and implications of Bell's theorem (see § 9.3.2). As we will see, many experiments (such as Aspect's experiment) have been designed and conducted in the last decades for rebutting the local hidden-variable interpretations (see § 8.2) where these experiments are mostly based on Bell's theorem and inequality (or variants of them). In fact, testability even in its positive sense have been claimed in support of particular interpretations. For example, it is claimed that the slit-grid experiment (see § 6.8.1) endorses the transactional interpretation (see § 8.4) and refutes some other interpretations (e.g. the Copenhagen interpretation; see § 8.1). If so, then this experiment should be a realistic example for positive, as well as negative, testability. Overall, these examples should indicate that the achievability (let alone sensibility) of testability (or test) of interpretation is generally accepted by quantum physicists.

We should finally note that the discussion in the last paragraph (as far as quantum mechanics is concerned) is largely about the interpretations of quantum mechanics as a whole (as represented by the

---

[104] As we will see next, the experiments related to Bell's theorem (see § 9.3.2 and § 9.3.3) like Aspect's experiment (see § 10.4) may be considered as an example of experiments that test interpretations (mainly local hidden-variable interpretations). Also, the slit-grid experiment (see § 6.8.1) may be classified (by some) in this category. In fact, many variants of the double-slit experiment could be classified in this category (as we will see later).

[105] See for instance our book "The Mechanics of Lorentz Transformations".

[106] In fact, they could be excluded by the testability criterion even if we suspect that they can be tested in the future because they are not definitely testable. Yes, it may be argued (as in footnote [107]) that they should in principle be testable (due to their scientific content) although we currently cannot see how they can be tested.

[107] In fact, this should lead to the exclusion of some (if not most or all) of the existing interpretations of quantum mechanics from being scientific (in a strict sense) at least for the present time where their testability seems unlikely (if not impossible). However, on the other hand we may also argue that these interpretations should in principle be testable (due to their scientific content) although we currently cannot see how they can be tested.

schools of interpretation). Otherwise, the testability of specific interpretative aspects and features of quantum mechanics should be obvious noting that there are many types of experiments and categories of tests (e.g. experiments of double-slit type) that are specifically designed and conduced for testing certain quantum mechanical aspects and features (of interpretative nature). In fact, the slit-grid experiment (as an experiment for testing the complementarity principle which is essentially a feature or at most a marginal principle that is not part of the formalism; see for instance § 4.7.3 and § 5.9) is an obvious example of this type of experiments and tests. We may even include in this category the "tests" of the statistical interpretation (of Born) since this (as part of the basic or standard interpretation of quantum mechanics and wavefunction in particular) is an interpretative aspect although it is of rather different nature (due to its intimate relation to the formalism and its inclusion in the axiomatic framework of the theory; see § 4.2).

2.9.6 Comprehensibility

Any eligible interpretation should provide (or represent) a comprehensible version (including explanations and justifications) of the formalism using the ordinary language (rather than the technical or symbolic language).[108] For example, we can verbalize Newton's third law comprehensibly as: forces occur in Nature in equal and opposite couples. In fact, comprehensibility should also mean that the interpretation must be easier to understand than the interpreted theory because the role of interpretation is mainly to explain and clarify the theory and justify its formal and technical content. As we will see later (refer to chapter 8), some schools of interpretation of quantum mechanics are not less difficult to comprehend than the theory itself (and some can even be more difficult). Accordingly, these interpretations can be eliminated (as eligible scientific interpretations) by this criterion.

2.9.7 Thoroughness

Any eligible interpretation should be thorough and hence it should provide an explanation and justification to the theory as a whole and not to just part of it.[109] This is because what is required is an interpretation to the theory in its entirety and not to just selected elements and aspects of it.[110] In fact, partial interpretations usually lead to contradictions and inconsistencies where they may succeed in explaining and justifying certain elements and aspects but they can clash with other facts. Accordingly, we may claim that partial interpretations should be discarded because their partial and normally *ad hoc* nature is usually based on their lack of correct and sufficiently general framework to explain and justify the theory as a whole; otherwise they should be thorough and general.

In fact, partial interpretations are usually arbitrary fixes and hence they should not be considered legitimate interpretations because they fail to provide consistent and thorough explanation and justification to the essence of the theory. Yes, partial interpretations can be accepted provisionally as a first step in building and constructing thorough and legitimate interpretations. They may also be accepted as interpretations to specific parts and aspects of the theory (but not as interpretations to the entire theory) if they satisfied the conditions of acceptable interpretations and did not lead to contradictions and inconsistencies.

As we will see later (refer to chapter 8), some schools of interpretation of quantum mechanics seem to be partial in their explanation and justification to the theory. An example of this may be the many-worlds interpretation which apparently cannot explain and justify the quantitative aspect of the probabilistic nature of quantum observations (see § 8.3). We may include in this category even those schools of

---

[108] As indicated, providing (or representing) a comprehensible version of the formalism is just part of the interpretation (related to the qualitative aspect which we pointed out in § 2.8).

[109] It should be noted that "thoroughness" should also be understood as thoroughness in acceptability and satisfaction and hence a thorough interpretation that is partially satisfactory is not thorough.

[110] It is worth noting that we are considering here interpretation of scientific theories in their entirety; otherwise if the objective was to interpret a certain part or aspect of a theory then in principle the proposed interpretation could be acceptable (as an interpretation to that part or aspect) as long as it satisfies the criterion of thoroughness within that part or aspect (although it is obviously not acceptable as an interpretation to the theory as a whole and may not be acceptable even to that part or aspect if it leads, for instance, to inconsistencies and contradictions).

interpretation that do not offer rational physical mechanisms to some aspects of the interpretation (e.g. the Copenhagen school in its failure to offer a comprehensible physical mechanism for the collapse of wavefunction). Accordingly, these interpretations can be eliminated (as eligible scientific interpretations) by this criterion.[111]

2.9.8 Rationalization

Any eligible interpretation should provide a reasonable explanation and justification to the theory and its various aspects (and hence to the observed phenomena that the theory is supposed to describe and formalize). As indicated earlier, rationalization should be seen as one of the main roles and objectives of interpretation and what in reality makes an interpretation interpretation. In fact, eligible interpretation should rationalize all the major aspects of the theory (and hence it should be thorough in this regard). This is not the case in some interpretations (particularly of quantum mechanics) and hence they should not be accepted as eligible interpretations. An example of this type of interpretations is the aforementioned schools of interpretation that do not offer rational physical mechanisms to some aspects of the theory (such as the Copenhagen and many-worlds schools in their failure to propose meaningful physical mechanisms for the collapse of wavefunction and branching of worlds; see § 8.1 and § 8.3). Accordingly, these interpretations may be eliminated (as eligible scientific interpretations) by this criterion.

2.10 Validity Criteria of Scientific Interpretation

The validity criteria of scientific interpretation are essentially the same as the validity criteria of scientific theory (as represented primarily by its formalism) which we investigated in § 2.6. So in essence, what is required to validate a given interpretation is scientific evidence (in support of the interpretation) and compliance with the well-established theories and interpretations (as well as other known facts to be more general). This should be justified by the fact that the interpretation itself is virtually a scientific theory (of epistemological nature) even though it is seen as an attachment to a (formal) scientific theory, and hence the interpretation should be validated by the same criteria that validate scientific theories. In the following points we provide more clarifications about the validity criteria of scientific interpretation:

1. Despite some opposite claims, we believe that experimental evidence can belong to interpretation (and not only to the formalism of scientific theory). In other words, a given interpretation of a scientific theory can be endorsed or refuted by a given experimental evidence. Also see § 2.9.5 and § 7.8.
2. Although the evidence for (or against) the interpretation of a theory is (in principle) independent of the evidence for (or against) the theory itself, some types of evidence may belong to both. Accordingly, we can have scientific evidence that support a theory and at the same time support or oppose its interpretation.[112]
3. If an interpretation of a given theory is the only possible interpretation of the formalism of the theory, then any scientific evidence in support of the formalism may be regarded as evidence in support of the interpretation and hence this interpretation may get the legitimacy and endorsement of the scientific evidence like the formalism itself (see point 5 of § 2.8). However, it is almost impossible to rule out other potential interpretations that can emerge in the future and can be similarly valid in explaining and interpreting the formalism even though we currently have only one possible interpretation. Hence, it is difficult to regard the scientific evidence for the formalism as an endorsement to this interpretation

---

[111] We should note in this regard that there are many examples of "partial" interpretations in the sense of being related to specific phenomena (like photoelectric effect or Compton scattering). Although most of these interpretations are supposedly based on more general interpretations and models (e.g. particle-wave duality or wavepacket), the scope and background of many of these specific interpretations is limited (i.e. they are not intended or designed to offer more general interpretation than that of the specific phenomena) and hence they can be accepted (if they are supported by evidence and are consistent with other facts) as such because although they are specific and limited (and hence they are "partial" in this sense) they are complete within their limited domain and scope (and hence they are not partial from this perspective and by this consideration).

[112] If we have evidence against a theory then it should not be very meaningful or useful to talk about supporting or opposing its interpretation.

unless the evidence actually belong to both (i.e. formalism and interpretation) as explained in the previous point.

2.11 Optimality Criteria of Scientific Interpretation

These criteria are generally similar to those of scientific theory (see § 2.7) and hence they do not require further investigation. Yes, some of the criteria of § 2.7 (e.g. interpretativity; see § 2.7.8) are not applicable here while some of the rest require some modifications and amendments to be inline with the nature of interpretation. We may also add some useful remarks in this regard:
1. An ideal interpretation should be based on analyzing the formalism and hence it reflects the essence and spirit of the formalism (as much as possible).
2. For some theories, some of the optimality criteria of interpretation may not be applicable or realizable or achievable due to the special nature of the theory as well as to potential conflict with other criteria and requirements (see for instance point 3 of § 7.2 noting that our concern here is interpretation). In fact, some theories may not be interpretable at all (let alone being optimally interpretable or not, or meeting certain optimality criteria or not).

2.12 Role and Importance of Interpretation

As indicated earlier, the role of scientific interpretation (essentially and broadly) is to provide explanation and justification to the formalism and this should include things like providing a more comprehensible version of the formalism, indicating causes and consequences, relating the formalism to observable physical entities and quantities, linking our scientific knowledge to other types of knowledge (such as philosophy), and so on. So in brief, the interpretation of a scientific theory is an *interpretation* in the broadest and most extensive sense of this word (with its scientific and epistemological significance) and that is what actually determines its role and contribution to science (as well as other potential contributions).

Regarding the importance of scientific interpretation and its value to science, it may be claimed that all we need in science is formalism. So, as long as we get the right formalism we should not bother about its interpretation and hence we can leave this to the epistemologists and philosophers of science. However, we can claim that scientific interpretation can be as important to science as the formalism itself for the following reasons (among other reasons):
1. Science is not only an attempt to deal with or exploit the physical world and benefit from its resources and harness its powers and forces (which the formalism may be sufficient to achieve), but it is basically and primarily an attempt to understand this world (see § 2.1). In fact, even these practical factors and benefits cannot be achieved ideally without deep understanding to the physical world (i.e. better understanding leads to better utilization and more benefits). So, interpretation (through providing explanation, justification, rationalization, linking, etc.) is essential for providing deep understanding and insight into the physical world and hence it serves the objectives of science and contributes to its progress and advance. In brief, science is an attempt to understand (as well as to exploit) the physical world and this understanding will not be achieved (in a complete and ideal manner) unless we get the right interpretation for the formalism of our scientific theories (noting as well the direct and indirect practical benefits that we get from understanding).
2. The interpretation provides a sense of direction and a compass for the future development and hence it offers guidance for the future investigation and progress and can play an important role in steering the scientific research. So, a good interpretation can lead to positive progress in the right direction while a bad interpretation (or lack of interpretation) can lead to failure and disorientation.

The important role of interpretation in motivating research and indicating the direction of future development is highlighted by the recent enhancement of interest in the interpretation of quantum mechanics as a cause or effect (in part) of the research on quantum mechanical applications in new fields like quantum computing and quantum cryptography.[113] In fact, this shows the nature of the

---
[113] We note that the mutual influence between formalism and interpretation is so strong in these novel fields of quantum mechanical applications that some formal theories are built on certain interpretations (and similarly certain interpretations

interactive relation between the theory and its interpretation where each one can drive, motivate and direct the other in a manner that is beneficial to both.
3. One of the main roles of interpretation is to give a qualitative and descriptive shape to the reality behind the formalism so that we incorporate this "image" of reality into the bigger (preexisting) picture of the reality (or outside world) that we already have. As such, the interpretation is fundamentally and intrinsically based on our presumption and notion of the existence of a unique, external and independent reality and hence it demonstrates our deep feeling (or need) of this external reality. In fact, this is deeply rooted in our need and desire for optimal adaptation with our environment which is the ultimate objective of any living being and the driving force of evolution.[114]

We should finally note (with reference to the above claim that interpretation is entirely irrelevant to science and hence science is all about formalism and therefore scientists should not waste their time on interpreting their theories) that interpretation in its technical sense and form may be so (arguably) but interpretation in some sense and form is adherent to any scientific theory and scientific effort. The formulation of any scientific theory is initiated by a sort of vague generic "interpretation" of the observed physical phenomenon and hence no theory can be detached entirely from interpretation (in some shape or form). In fact, the formulation of a scientific theory starts with the use of some familiar concepts, patterns, models, etc. which are usually of classical macroscopic nature and origin and hence a sort of (implicit) interpretation is already embedded in the formalism and is part of the scientific process and content which scientists do and generate (noting that this sort of "primitive interpretation" does not necessarily meet the technical criteria of eligible scientific interpretation).

Anyway, if we note that interpretation is important like formalism (because understanding, apart from its practical benefits, is a need for us as intellectual species; see § 2.1) and consider the fact that scientists are the most (if not the only) capable individuals of interpreting science then we should accept that interpretation (despite its epistemological character) should be an essential part of science (in its broad sense) or at least an important accessory to science and hence it should be one of the duties of scientists.

2.13 Existence and Uniqueness of Scientific Interpretation

As indicated earlier (and will be repeated later), there is no necessity that every scientific theory must have an interpretation and its interpretation (if it exists) is unique. So, neither the existence nor the uniqueness of interpretation is guaranteed. In fact, even the interpretability of scientific theory is not guaranteed and hence a scientific theory may not be interpretable at all (i.e. it cannot have an interpretation).[115]

We can argue in support of these claims that the existence and uniqueness (and even interpretability) are not obvious at all and hence they require proof. As far as we can see, there is no such proof and hence we can legitimately make these claims. The onus of proving the opposite should then be on the other side. In fact, we can do more than this by offering some reasonable justifications for our claims (as we will see next).

---

are inspired or boosted by the development of certain formal theories in these fields). In fact, this sort of formalism-interpretation interaction is not novel or unusual in quantum mechanics (noting that the formalism and interpretation of quantum theory have no well-defined border; see for instance § 7.10). For example, certain formal hidden-variable theories (e.g. Bohm's formulation of quantum mechanics) are built on and inspired by the hidden-variable interpretation (see § 8.2). The interaction can also be seen for instance in the strong relation between the aforementioned novel quantum fields (e.g. quantum information, computation and cryptography) and the theoretical and experimental research related to Bell's theorem (which is largely about interpretation). In fact, Bell's work is seen as the foundation or cornerstone for the science of quantum information. The importance of these novel and thriving branches of science (which are intimately-linked to the interpretation and epistemology of quantum mechanics) is not restricted to the extension and expansion of classical potentials and possibilities but they open the gate for quantum potentials and possibilities that have no classical parallel.

[114] In our view, optimal adaptation to the environment is the ultimate objective of any form of knowledge of human species throughout his history of evolution (where optimal adaptation means the best available or possible satisfaction of his physical and mental needs whether as individual or as group or as species). In fact, this should apply to any living individual and group and species (with some obvious extension to "knowledge").

[115] It is important to note that if "existence" means the potential existence (i.e. in itself as in mathematics) rather than the actual existence then it should be equivalent to interpretability. However, we decided to give it a rather loose and more general meaning and hence we generally treat it as different to interpretability to stress on this distinct possibility.

In the following points we investigate the issues of existence and uniqueness of scientific interpretation (as well as the issue of interpretability):

1. It is important to note that "interpretation" here is used in its (rather strict) technical sense according to the eligibility criteria of § 2.9. As indicated in § 2.12, scientific theories usually have generic and implicit interpretations to their formalism (since the formalism is usually initiated by and associated with an implicit and generic interpretation).

2. It may seem strange that scientific theories (especially if they are as "perfect" and accurate as quantum mechanics) may not have interpretation and may not even be interpretable despite their empirical success which should indicate their success of capturing and representing reality. However, as indicated in point 1 "interpretation" here is technical and restricted by rather stringent criteria and conditions and hence there is nothing strange about this. Also see the upcoming point 4.

3. It may also seem strange that scientific theories (despite their rigorous and specific nature) can have more than one interpretation. However, this can be justified by the non-uniqueness of science (in a rather broad sense that should include all types of knowledge) since interpretation (like formalism and like any other type of knowledge) is affected by many factors (e.g. personal, cultural, social, tempo-spatial, etc.) that make it interpreter-dependent and hence non-unique.[116]

4. Regarding the issue of interpretability,[117] we believe that a scientific theory (as represented primarily by its formalism) can be interpretable only if it satisfies certain "conceptuality" conditions, and these conditions are not guaranteed to be met by every scientific theory. For instance, if the formalism is based on capturing and depicting associations and correlations without a proper underlying rational conceptual framework then the theory may not be interpretable due to this conceptual deficiency (even if the theory was based on correct and realistic associations and correlations and hence it is empirically correct).

5. It seems to us (from inspecting the scientific literature especially in the field of quantum theory) that there is a rather implicit consensus among the scientists and philosophers of science (or at least among the majority of them) about the necessity of the existence and uniqueness of interpretation of scientific theory (and quantum theory in particular). This consensus can be understood or felt from the lack of attention to this issue where all the discussions and debates about quantum mechanics, for instance, and its interpretation take place in the absence of any indication to the possibility of non-existence or non-uniqueness of interpretation as if the existence and uniqueness of interpretation are guaranteed and obvious.

2.14 Evolution and Achievements of Science

It is important to discuss briefly the development of science and its evolution and achievements along its history. Although this seems unrelated or irrelevant to the scope of this book (which is the epistemology of quantum physics) and to the issues discussed in this chapter (considering their predominantly epistemological nature), we think it is an important factor in assessing the value of scientific theories and appreciating science in general as well as recognizing the contributions of science fairly and squarely. In the following points we make some remarks about the evolution and achievements of science (noting that we will try to be more specific by linking our remarks to quantum theory):

1. We would like first to stress on the indeterministic nature of the history of science, i.e. the evolution of science is affected by historical factors as well as by factual factors.[118] This is inline with our proposal

---

[116] As explained earlier, despite the rules of logic and the uniqueness principles of reality and truth, science is not unique due to the fact that it is not a perfect and exact image (or mirror reflection) of reality but it is a mix of reflection (or discovery) and invention and hence each scientist "reflects" reality in his own mirror (which is shaped by his genes, past experiences, environment, etc.) and from his position in space-time. These justifications of the non-uniqueness nature of science should apply to interpretation as well as to formalism (and in fact they apply to all types of knowledge as indicated above).

[117] As indicated earlier (see point 1), "interpretability" can have two meanings (or "flavors"): being interpretable in a general sense, and being interpretable in a technical sense (i.e. in accord with the eligibility criteria of § 2.9). Although our attention in the present point is mainly on the former, the latter is not entirely excluded or out of sight. In fact, the former (at least here) should include the latter as a special case.

[118] "Factual" here should mean something related directly to science and of scientific nature.

about the non-unique nature of science (see for instance § 2.4.3). The evolution of science is not subject to an inevitable and deterministic process but rather to an uncertain and probabilistic process in which many factors determine its outcome. In fact, most of these factors are seemingly unrelated to science such as social and political factors, historical accidents, characteristics of individuals involved in this process, and so on. Accordingly, quantum theory is a product of its history and it bears all the hallmarks of this history. In other words, quantum theory is not inevitable event and development in the history of science and hence we could have a different "quantum theory" if history took a different path or direction. Also see § 7.1 and § 7.15.

2. Although the general trend in the evolution of science is progress and advance, it may fluctuate and even go backwards from time to time (in fact we can find many examples of this in the medieval science and even in modern science). So, it is not necessarily that the science of today is better than the science of yesterday. Similarly, it is not necessarily that correct and good science leads to correct and good science (neither incorrect and bad science leads inevitably to incorrect and bad science) although the general trend is so. We also note that there are always cross roads for the evolution of science where science can take (probabilistically) one route or another where the routes differ in their virtues and even in their correctness. Moreover, a theory or an idea or a development in the history of science may contribute positively in some aspects and negatively in other aspects. Accordingly, quantum theory (despite its undeniable empirical success and achievements) is not necessarily a progress and success in its entirety. For example, quantum theory could be a setback (or retreat or failure) to science from the epistemological perspective. In fact, this applies to modern physics in general (although we will not pursue this issue here since it is out of scope).

3. As indicated earlier, the evolution of science is affected directly and indirectly by many human and non-human elements and factors (and similarly science affects many other elements and factors). In fact, science is a living being that affects and is affected by all the elements and factors of its environment (in the broadest meaning of "environment"). Therefore, the interrelation between science and its environment is very involved and intricate. However, there is one element or factor of special significance in its relation to science, that is technology. Although technology in its broad and generic sense is about making tools and techniques to benefit from resources, overcome difficulties and solve problems it is essentially and more specifically a knowledge-based activity. This is because knowledge (which science represents its best example) is essential to technology because these objectives (e.g. solving problems) cannot be achieved without knowledge (or at least they are optimally achieved with knowledge). This intimate relationship between science and technology makes the distinction between them delicate and subtle (at least occasionally and in some zones). In fact, there is no precise border between science and technology especially in modern times where technology is heavily systematized and sciencetized. This ambiguity and lack of clear distinction between science and technology is very common these days (thanks to the aforementioned intimate relationship between modern science and modern technology).

Although the blur between science and technology is harmless in general (and may even be useful sometimes), it is harmful sometimes. For example, many technological successes are attributed to the supposedly underlying scientific theories (which gives these theories more credit than they deserve). For example, many recent advances and achievements in the fields of computing and communication are attributed to quantum physics. Although this is true in general it is not true in many details, because many of these details are not worked out according to the strict theoretical models and scientific principles of quantum physics but they are achieved through "empirical methods" such as trial and error, using semi-classical and artificial models, using approximations and imitations, employing engineering tricks, and so on. So, a significant part of the quantum-related modern advancements and achievements do not belong to the success (or "correctness") of quantum theory and hence the credit for this part should not be attributed to quantum physics.

To make the idea clearer we say: crediting these successes to quantum physics is like crediting the technological achievements of ballistic missiles and artificial satellites to Newton's laws of motion and gravity. Certainly Newton's laws are at the foundation of these technological achievements but they are just part of the theory underlying these technologies and hence Newton's laws should take only part of the credit for these achievements. Accordingly, we should not exaggerate the success of Newton's laws by attributing the entirety of the missiles and satellites technologies and their credit to these laws. Similarly, we should not exaggerate the success of quantum mechanics by attributing the entirety of the computing and communication technologies and their credit to this mechanics because many details are worked out according to (and thus they belong to) techniques and methods (and even theories) unrelated to quantum mechanics.

To sum up, many of the modern achievements (whose credit is usually attributed to quantum theory) are mostly technological achievements rather than scientific (or quantum mechanical) achievements and hence they should not be credited (at least in their entirety) to quantum theory and its "correctness".

4. When we talk about "correct" and "wrong" scientific theories (or theories in general) we should put this into its proper perspective. First, there is no such absolutely correct or absolutely wrong theories since scientific theories (and human knowledge) in general contain correct elements and aspects and wrong (or bad) elements and aspects. Second, being correct or wrong is not the only factor that determines the value of a theory (especially from a historical perspective and by practical standards). In fact, even "wrong" (or "bad") theories can have a positive role in the evolution of science. In general, we can find many examples of scientific theories (or "theorems") that made positive impact and contribution to the progress of science but they were rejected later because they were found to be "wrong" or because they were replaced by "better" theories (and this is likely to be the fate of the current quantum theory). The least positive contribution of a "wrong" scientific theory is that it provides an initial proposal (or initiative) and a working hypothesis to start with and hence it helps to exclude (subsequently) certain wrong possibilities. For example, the "wrong" atomic model of Thompson (i.e. plum pudding model) provided an initial proposal whose examination led to the development of the more realistic model of Rutherford. So in general, many (if not most) of the "wrong" theories (i.e. the respected ones not the junk) have positive contributions (as well as negative contributions) and hence they participate in the progress and development of science (possibly as much as the "correct" theories do). This should reflect the nature of science (and human knowledge in general) which is an ever-lasting and generally-improving process of trial and error, and this should justify the claim that science is a history of mistakes (noting that the biggest mistake of science, or scientists, could be the belief that science has an ultimate end and a final destination where it becomes complete and perfect).

We should also draw the attention to the casual discussion in § 2.15.1 about the level of advancement of physics (and science in general) and the prospect of its future development.

2.15 Freewill and Science

The issue of freewill (and its nature and demonstrations) is one of the most challenging perplexities of this world. In fact, freewill is as bewildering and puzzling as life itself which is the source of freewill. In this section we will try to investigate very briefly this issue in a scientific context considering its significance to quantum theory in particular. However, before we start this investigation in the following subsections we note that many (and possibly most) of the issues and question marks about freewill in this context are related to the issue of biology (and life ultimately) and if it is part of physics or not, i.e. whether biological phenomena and life obey the physical laws (possibly in a very complex form) and hence they belong to physics or not (in which case they should be subject to a different type of laws and hence they are beyond the reach of physics). This issue, as well as its consequences and implications, is not clear cut and could be a matter of choice and convention. Therefore, most of the aforementioned issues and question marks have no definite answers or treatments, and hence large part of our investigation is contemplative in nature and reflects our personal views.

2.15.1 Freewill and Physicality

We start this investigation by asking: does freewill contradict physicality (see § 2.5.4 and § 2.9.4)? Is freewill something that can be explained by physical factors and may also be a physical factor that affects physically other physical factors and phenomena? If no, can we classify freewill as metaphysical (like spirit or soul) or something neither physical nor metaphysical? If yes, does the study of freewill belong to physics (and possibly to quantum physics in particular)?

It should be obvious that everything that we observe in this world belongs to the physical world (i.e. it is not metaphysical) and freewill (like many other observable things related to life, biology, society, etc.) is observable and hence it is physical in this sense. However, it should also be obvious that not everything that is physical (i.e. belonging to the physical world) belongs to the science of physics. In fact, physics should be restricted by definition and convention to phenomena that do not involve things like life (as well as many other things). Hence, biological and social sciences for instance (as well as many other branches of science and knowledge) should not be seen as branches of physics (at least in its current state) despite being physical in the aforementioned sense.

Yes, if physics becomes so powerful and broad to the limit that it can formulate laws and principles capable of describing and predicting biological and sociological phenomena (as well as similar phenomena) then these branches of science could be included as branches of physics in its extended sense. However, this seems to be a non-achievable (and possibly impossible) goal. If our current physics cannot find (exact) solutions even to some of the simplest systems (like helium atom) then there is no hope (at least for the foreseeable future) to extend physics to this ambitious extent.[119] In fact, physics in its current state (despite its complexity and elaboration) is a very simple and primitive science that effectively deals only with the very simple systems using very simple models, patterns, techniques, methodologies ... etc. most of which are no more than imitations and approximations. So, we should not be very ambitious at this stage of scientific development to hope that we can develop physical theories about life or freewill or society or economy for instance. This should explain why branches and topics like these are shrouded with many ambiguities and uncertainties and are virtually impossible to tackle by the current physics. This should also highlight another important issue about the scope and prospect of the future physics and the level of advancement that we reached so far in our understanding and conquering of the physical world.

Opposite to the common belief that we reached a very high level of advancement in these regards and we may even be close to finalize physics (as was once thought that the 19th century version of physics is virtually the complete and final version of physics) it seems to me that we are still at the very first stages in our exploration to the Universe and cracking its secrets and mysteries. Hence, we should not be content with our physics or quantum theory (or indeed with any other theory). We must always look to more improvements and new challenges some of which could be on the verge of science fiction although we should always be realistic about our capabilities and our current state of development so that we do not jump carelessly here and there and waste our resources in illusions and fantasies. We should also be driven in our search and investigation by observation and experiment rather than by illusory models and theories so that we make advances without losing touch with reality.

To conclude, the investigation of freewill within physics (or quantum theory specifically as done by some) is entirely inappropriate. Physics, and quantum physics in particular, is too simple to deal with such complex phenomena like freewill even though freewill is physical and could belong ultimately to the quantum world and it is of quantum origin and has quantum demonstrations.

2.15.2 Freewill and Indeterminism

We start this investigation by asking: does freewill implies indeterminism (which is a characteristic feature of quantum mechanics; see for instance § 5.2 and § 6.4)? If so, does freewill contradict or violate causality? In fact, this sort of questions and the issues behind them may explain the alleged intimate link between quantum theory and freewill since both are associated with things like indeterminacy and uncertainty as well as possible violation of causality. In general, freewill should imply, to some extent, indeterminacy and uncertainty, and this in itself should have no harm. Yes, the possibility of freewill being violating causality may be harmful since causality is a pillar for any rational thinking and knowledge (see for instance § 1.7,

---

[119] It is worth noting that the interesting physical problems that can be solved exactly by quantum mechanics are few. This applies to science in general and not restricted to quantum mechanics.

§ 2.4.2 and § 2.5.3). So, the issue that we need to focus on in this investigation is causality and this is what we will do in the next subsection.

2.15.3 Freewill and Causality

We start this investigation by asking: does freewill contradict or violate causality (at least in its strict classical sense)? At **basic level** (i.e. the level of making specific decisions at specific occasions and events) it should not because at this level the cause is not the "freewill" in its generality but the specific will at each individual occurrence and event. For example, if John has a freewill to eat apple or orange and he decided to eat apple on Monday and orange on Tuesday then the cause in each event is not the "freewill" in its generality but the cause for eating apple on Monday is the will to eat apple (on Monday) while the cause for eating orange on Tuesday is the will to eat orange (on Tuesday). So, at this level the cause is specific and definite (and hence there is no indeterminacy or uncertainty) and thus at this level causality is certainly safe.

Yes, at **higher level** (i.e. the level of having the ability to make different decisions in *seemingly* similar or identical circumstances) we should need to justify the indeterminacy and uncertainty by identifying a cause for making these different decisions. In other words (using the example of John), why he has freewill and hence he can choose to eat apple on Monday and orange on Tuesday? Anyway, this issue is similar to the issue of possible violation of causality by quantum physics which we investigated elsewhere (see for instance § 7.12). But we should note that some of our treatments to this issue within the context of quantum physics (e.g. claiming the possibility of having a different quantum theory) do not apply here. So, what we can say is that some of the treatments within the context of quantum physics applies here and this should be sufficient. Moreover, the issue of freewill is of interest to us as far as and as much as quantum physics is concerned and hence our treatment within the context of quantum physics is sufficient (and thus any related issue of non-quantum nature is out of scope and should be pursued elsewhere).

We should finally note that freewill is essentially classical (even if it is assumed to be of quantum origin) and hence this may require addressing the issue of potential violation of causality (as well as the issue of indeterminism and uncertainty) at the classical level which we generally assumed (over the entirety of this book) to be safe from such quantum mechanical type of shocks. Yes, if freewill is assumed to be of quantum origin then we may be able to evade classical-type violation of causality since the classical violation in this case is of quantum origin and cause and hence a convincing and sufficient treatment at the quantum level (which we assume we have) should be sufficient to address this issue (since in this case it becomes essentially a quantum rather than a classical issue).

In fact, this fix should apply to all classical aspects that originate from a problematic or perplexing quantum mechanical aspects where we can simply refer the matter to the quantum level and hence we can keep our classical science and epistemology safe from such quantum mechanical type of shocks. For example, an uncertainty at the classical level originating from an uncertainty at the quantum level does not make the uncertainty principle (of quantum mechanics) applicable at the classical level, and hence we should still be able to negate the existence of intrinsic uncertainty (of the uncertainty principle type) at the classical level. Also see point 4 of § 5.2.

Chapter 3
Quantum versus Classical

In this chapter we make thorough comparisons between what is described as "classical" and what is described as "quantum" (e.g. phenomena, waves, etc.) and investigate some of the relationships between them. These comparisons and investigations should clarify the difference between the two attributes in their different instances and contexts and provide background knowledge that is useful and necessary for the upcoming investigations.

3.1 Comparison between Classical and Quantum Phenomena

In this section we briefly compare the properties of classical phenomena (including objects) to their quantum counterparts. As we will see, many of the distinctions between classical theory and quantum theory originate from these properties. We outline this comparison in the following points:

1. **Size**: the difference between the macroscopic size of the classical world and the microscopic size of the quantum world is very obvious and it has direct and profound impact on our physics about these worlds. Due to the scale factor (see § 1.3.2) many of our direct experiences and patterns that we use in describing classical phenomena do not apply (at least strictly and exactly) to quantum phenomena due to the difference in size and our inability to have direct experience of the quantum world.

2. **Distinguishability**: due to their big size and elaborate structure, classical objects of the same type (e.g. two cars or two planets) are distinguishable (at least in principle) because it is improbable that two macroscopic objects (considering their big size and complex structure) can be entirely identical. On the other hand, quantum objects of the same type (e.g. two electrons or two neutrons) have rather simple structure (at least as we perceive them and from our perspective) since they are "elementary" objects and hence they are generally indistinguishable. Accordingly, indistinguishability (which is generally unrecognized in classical physics) introduces some features on the physics of quantum phenomena that are not found in classical physics, e.g. effects on the statistics of quantum particles and their wavefunction and spatial distribution which lead for instance to the concept of exchange symmetry (or exchange force or exchange energy) and classification of quantum particles to fermions and bosons. It is useful to be aware of the following remarks about distinguishability:
   • Indistinguishability in the quantum world is not a necessity but a possibility supported by experimental facts (or so claimed).
   • Indistinguishability is conditioned by certain requirements and hence not any two quantum objects of the same type (like two electrons which are intrinsically identical) are indistinguishable. For example, if two electrons are too far apart then they are distinguishable (by the extrinsic factor of position). So, indistinguishability (intrinsically and extrinsically) can be realized in certain situations such as in scattering phenomena and atomic orbitals where the wavefunctions of the scattered particles or the orbiting electrons overlap and hence the particles become totally indistinguishable (i.e. extrinsically as well as intrinsically).[120]
   • Indistinguishability may be found (exceptionally and accidentally) in classical physics but it does not cause the physical effects that it causes in quantum phenomena.
   • Also see point 10 of § 7.11.

3. **Continuity and Discreteness**: while classical objects are generally continuous (e.g. they can be divided and have continuous properties), quantum objects are discrete and come in the form of quantized units and discrete (or countable) entities.

---
[120] We note that "indistinguishability" my be replaced by "identicality" to refer to the intrinsic indistinguishability. Accordingly, all quantum objects of the same type (such as electrons) are (always) identical although they may be distinguishable (extrinsically) or indistinguishable (when they cannot be distinguished by an extrinsic factor like position).

4. **Real and Complex**: while classical phenomena are described *primarily* by real physical parameters (i.e. real numbers and variables), quantum phenomena are described *primarily* by complex physical parameters (i.e. the wavefunctions which are complex).[121]
5. **Particle/Wave Nature**: classical objects have definite nature from the perspective of being waves or particles,[122] while quantum objects do not have definite nature as they generally behave partly as waves and partly as particles.

We should finally note that although the direct effect of quantum phenomena is generally negligible (and may not even be sensible) on a large macroscopic scale they can still have a significant (indirect) effect on a macroscopic scale. This is because quantum phenomena can (and generally do) underlie certain macroscopic phenomena. For example, the size of certain stars (e.g. white dwarfs and neutron stars) which is a macroscopic phenomenon is determined by the degeneracy pressure (of electrons and neutrons) which is a quantum phenomenon. Similarly, the direct effect of the uncertainty principle is negligible on a macroscopic scale but it could have a tangible indirect effect (supposedly) on the scale of the entire Universe (according to certain theories and models). In fact, macroscopic phenomena should in general be ultimately based on quantum phenomena (although this does not necessarily mean that the formalism of classical mechanics can emerge from the formalism of quantum mechanics; see for instance § 3.7 and § 7.3).

3.2 Comparison between Classical and Quantum Waves

It is important to distinguish between two main types of waves: classical waves (or physical waves) like sound or electromagnetic waves, and quantum waves (or matter waves or probability waves or non-classical waves) like the waves represented by the wavefunctions of quantum systems. Although both types of wave demonstrate similar characteristic wave properties (like interference) they have other dissimilar properties and characteristics which we investigate in the following points:

1. **Physicality**: classical waves are physical while quantum waves are not, i.e. they are essentially abstract mathematical apparatus (see point 2 of § 2.5.4).[123] Accordingly, classical waves propagate in the ordinary (or physical) 3D space while quantum waves "propagate" in the so-called "configuration space" (which can have more than 3 dimensions and hence it is not physical).[124] Also, classical waves are described by real variables (which indicate their physical nature) while quantum waves are described by complex variables (which indicate their non-physical nature).

   There are many indications and justifications for the non-physical nature of quantum waves. For example:
   - As pointed out, quantum waves are "complex" and "in configuration space" which should indicate their non-physical nature.
   - The predictions of quantum mechanics (which are usually obtained from the wave mechanics) can also be obtained from the matrix mechanics[125] which is a non-wave type of mechanics (see § 1.2 and § 5.3.2) and this should also indicate that matter waves (as typically represented by the wavefunction) are just mathematical apparatus for providing probabilistic predictions.
   - As we will see, quantum waves lack certain physical properties of physical waves (at least in some

---

[121] The reader should note the significance of *"primarily"* because classical phenomena may be described by complex parameters but conveniently (rather than out of intrinsic necessity) where the complex numbers are used as a convenient mathematical device with the physical parameters being represented by the real or imaginary parts (or both parts but independently) noting that the real and imaginary parts are real. This is unlike quantum phenomena which are described by complex parameters out of necessity (not out of convenience). However, it is important to note that all the observable (or measurable) aspects of quantum phenomena (such as position or momentum) are represented and quantified by real parameters.
[122] "Particle" here should mean "not wave" and hence it should include extensive material objects.
[123] So, while classical waves represent physical entities on their own (e.g. as a form of energy), quantum waves do not (noting that quantum waves represent essentially spatial probability amplitudes and are related primarily to our observation and knowledge and hence they are more of mathematical devices than physical objects).
[124] It is noteworthy that the "configuration space" is different from the Hilbert space which we mentioned earlier (see § 1.2).
[125] In fact, these predictions can also be obtained from a more general non-wave formulation that includes both types of mechanics.

3.2 Comparison between Classical and Quantum Waves

instances) such as carrying energy, and this should reveal their non-physical nature in general.
- Some of the properties of quantum waves (such as having phase and group velocities in opposite directions) are counter-intuitive from a physical perspective and this should indicate their non-physical nature.
- The statistical interpretation of quantum waves (as probability amplitudes) is different from the interpretation of classical waves, which should also indicate their non-physicality.
- The association of matter waves with ensembles of quantum objects (instead of individual quantum objects) could also indicate their non-physical nature (see point 3 of § 6.1).
- The dependency of matter waves on the frame of reference should also indicate their non-physical nature (see point 5).
- Also see point 3.

2. **Energy**: classical waves carry energy while quantum waves do not. For example, the energy-frequency relation $E = h\nu$ applies only to classical waves (noting that quantum waves have wavelength but their frequency seems ambiguous or undefined).[126]

   This similarly applies to **other physical properties** of classical waves (where these properties do not generally characterize quantum waves). For example, while it is common to associate wavelength with matter waves (since it is required for identifying interference pattern, for instance, and can be obtained from the de Broglie equation $\lambda = h/p$; see Eq. 16), no one seems to be interested in their frequency or speed or other similar physical properties (which are normally associated with classical waves). In fact, these properties are generally ignored as if they are irrelevant or not defined although they may appear casually in the justification and derivation of some propositions and relations (such as the de Broglie hypothesis and equation).[127]

   In fact, the confusion about the real physical properties of some types of quantum waves (where some physical properties may be attributed to them) originates from the confusion between real and matter waves in some quantum objects (like light) which possess physical wave characteristics where some of the physical attributes of their physical wave nature are attributed to their matter wave nature (thanks to this confusion); otherwise we can easily see that "wavefunction" and other potential forms of matter wave do not possess real physical properties like energy or frequency or speed (at least in well-defined form).[128] Also see point 4 of § 7.11.

3. **Detection**: classical waves can be detected (or observed or measured) themselves while quantum waves cannot (although they can be "detected" through their statistical effects as probability amplitudes as indicated earlier). This is an obvious consequence of the physicality of classical waves and the non-physicality of quantum waves. We should also refer the reader to § 2.5.4 where we discussed certain aspects of detection.

4. **Dispersion**: classical and quantum waves show different dispersion behavior (e.g. quantum waves show dispersion in vacuum unlike classical waves which do not).[129]

5. **Frame dependency**: according to the de Broglie equation $\lambda = h/p$ (see Eq. 16), matter wave is frame-dependent (since it depends on the magnitude of momentum which depends on the speed

---

[126] We note that relations like $E = h\nu$ may be used (in some formulations) with quantum waves. However, they are used essentially to facilitate the formulation rather than to characterize the quantum waves and attribute real physical properties to them. We should also note that the energy that is supposedly carried by wavepacket is due to its "particle nature" rater than to its "matter wave nature". This also applies (perhaps more obviously) to the "pilot wave" model.

[127] We note that the collapse of wavefunction (instantaneously and globally as will be seen later on; see for instance § 4.1 and § 5.7) should suggest that the "propagation" of matter waves (if "propagation" is defined and meaningful at all) occurs with infinite speed or at least immeasurable superluminal speed (unlike classical waves in their commonly-recognized forms and according to the common consensus about their behavior). This unusual nature of matter waves should put a question mark on the "wave" model in quantum physics and should indicate that matter "wave" is no more than imitation and approximation to the "true" nature of quantum objects rather than an accurate and "exact" model.

[128] A typical example in this regard is the standard wave equation which correlates speed to wavelength and frequency, i.e. $v = \lambda\nu$. It is obvious that this equation does not apply to matter waves because they do not have well-defined speed (and hence frequency).

[129] We note that the dispersion shown by quantum waves is essentially in their models (considering their non-physical nature). It should be known that dispersion occurs when the speed of wave is a function of its frequency and hence phase and group velocities differ.

which is frame-dependent). So, a given particle can have different waves at the same time (relative to different frames), and it may even have no wave at all (since if the particle is at rest, e.g. in its own frame, then its momentum is zero and hence from Eq. 16 we get $\lambda = \infty$ which effectively means it has no wave).[130] This is unlike classical waves whose wave nature is intrinsic and frame-independent although the wavelength may also depend on the observer to some extent. Accordingly, the wave property of quantum objects is not the same as the wave property of classical waves (e.g. electromagnetic waves) whose existence and characterization are generally independent of the observer and observation although some of their properties may be dependent on the observer and observation.

6. **Scale**: quantum waves effectively and practically exist only at microscopic level while classical waves are (or seem to be) more general.[131] The reason is that quantum waves are about the very nature of quantum objects[132] while classical waves can be seen as a form of disturbance of objects or fields. In fact, this may explain why the fuzziness of quantum objects (due to their "wavy" or "undulating" nature) reduces as their size increases and vice versa.

As we see, the above comparison should indicate that quantum and classical waves are very different in nature and hence the concept of "quantum wave" is a sort of imitation or approximation (based on the familiar classical concept of wave) rather than an exact model. Accordingly, we should be cautious about using and interpreting the concept of "quantum waves" and treating them as real waves (i.e. as we understand "waves" classically) which may lead to incorrect implications and consequences especially with regard to the interpretation of quantum mechanics. In fact, this could be one reason for the difficulty of interpreting quantum mechanics consistently and rationally as will be investigated later (see for instance § 7.11 and § 7.15). Also see § 6.13.1.

3.3 Comparison between Classical and Quantum Particles

We outline in the following points the fundamental differences between classical particles and quantum particles:

1. **Localization**: classical particle is sharply-localized entity, while quantum particle is diffusely-localized entity, i.e. its location is subject to probabilisticity and uncertainty. This is reflected in the concept of wavepacket (which is commonly used in quantum mechanics to model quantum particles) as well as probabilisticity and uncertainty in determining the spatial position of quantum particles (see § 4.7.1 and § 5.2.2).
2. **Definiteness**: classical particle has definite (even if unknown) physical properties (like momentum) while quantum particle may not necessarily have such properties (since its properties can be subject to intrinsic indeterminism and uncertainty which may touch even its reality; see § 4.7.1 and § 5.2).
3. **Distinguishability**: as seen earlier (refer to § 3.1), classical particles are distinguishable in principle while quantum particles may not be so.
4. **Divisibility**: classical particles are divisible in principle[133] while quantum particles may not be so (e.g. electrons cannot be cut in two pieces).
5. **Physical properties**: quantum particles possess certain unusual physical properties, such as exchange symmetry and spin (in its quantum sense; see § 4.6), which are unknown in classical particles. On the other hand, classical particles possess certain physical properties, such as size and shape, which may not be recognized or defined in quantum particles.[134]

---

[130] In fact, the restriction of the wave property to moving objects and the variability of the wavelength (relative and according to different frames) indicate that the (matter) wave property is not an intrinsic attribute of quantum objects but it is rather extrinsic and frame-dependent (unlike, for example, mass or charge which are intrinsic attributes).

[131] In fact, the paradigm of classical waves is basically macroscopic (noting the quantization at the microscopic level) and hence the above-indicated generality may be rejected. If so, then quantum waves are microscopic while classical waves are macroscopic. In this regard, we should emphasize that "scale" here refers to the size of physical reality to which the wave is attributed rather than the "size" of the wave itself (as represented mainly by its wavelength).

[132] The use of "wavepacket" and "pilot wave" as quantum wave models may indicate this.

[133] We note that although classical particle may be modeled (approximately) as a geometric point for the purpose of simplicity (considering for instance its center of gravity) it is still physical in nature (e.g. has finite size and mass) and therefore it should be divisible in principle (e.g. where its mass is divided between its parts).

[134] In fact, properties like size may be recognized and defined (although vaguely and diffusely) in certain types of quantum

6. **Dualism**: quantum particles have wave nature as well as particle nature, while classical particles do not have wave nature.

3.4 Comparison between Classical and Quantum Measurement

The difference between classical and quantum measurement is a big subject in the literature of quantum theory charged with debates and controversies (especially from its epistemological and interpretative perspectives). In fact, this difference can go to such extreme extent that makes quantum measurement a novel concept that has very little in common with classical measurement. As we will see (refer for instance to § 5.7 and § 6.11), quantum measurement is one of the most troubling issues in quantum physics (unlike classical measurement which is generally straightforward and basically focused on practical and operational issues). Moreover, the issue of measurement is one of the main features that distinguish quantum physics from classical physics (see chapter 5).

In the following points we outline some of the most important differences between these two types of measurement (i.e. classical and quantum):

1. **Nature**: classical and quantum measurements differ in nature and not only in practical and operational aspects. In classical measurement the measured phenomenon and the measuring equipment are compatible and comparable in size and nature since both are macroscopic and are subject to the same principles (of classical physics) and hence we can assume that in general the interaction between the two takes place directly and by the same physical rules. On the contrary, quantum measurement is realized through an interface between the macroscopic world (represented by the measuring equipment) which is subject to classical physics and the microscopic world (represented by the observed phenomenon) which is subject to quantum physics. This difference in nature should require formulating two different theories for measurement: one classical and one quantum.

2. **Function**: according to some theories of quantum measurement based on well-accepted interpretations of quantum mechanics (which sometimes permeate even through the "formalism"), the function of quantum measurement is not to reveal (an unknown but already-existing and determined) reality but to create this reality or at least give it a definite identity (assuming such a reality exists and is recognized by these interpretations). This is in complete contrast to classical measurement whose function is (supposedly) just to reveal (an already-existing and completely-determined) reality. Accordingly, the difference between classical measurement and quantum measurement is not really between two types of measurement that differ in nature but between measurement (i.e. classical) and "creation" or "materialization" of reality (i.e. quantum "measurement" which is not really measurement in its known and traditional sense).

3. **Disturbance**: classical measurement usually (but not necessarily) does not disturb the observed (macroscopic) phenomenon. Moreover, in principle such disturbance can be eliminated actually or virtually (although it may be difficult to do so). However, due to their extreme tininess (and possibly other reasons) the disturbance to quantum systems by quantum measurement seems inevitable (or almost) and hence it is likely that even in principle we cannot eliminate such disturbance (at least in some cases and circumstances or according to certain interpretations and scenarios). So, the inertness of observation (see § 1.3.3) may be theoretically or practically unachievable in quantum physics. In brief, the disturbance of the observed phenomenon by measurement is a more serious factor in quantum measurement than in classical measurement and it may even be a theoretical (not only practical) issue in the physics of quantum phenomena.[135]

4. **Role in the theory**: measurement does not play any role in the classical theory (i.e. it is not part of the theory itself although it plays a role in its application) or at most it plays a marginal role. On the

---

particles like atom and nucleus but not in other types such as electron (although some may talk rather laxly about its size and may also be described oppositely as point or point-like). Anyway, most classical properties (including those belonging to particles such as size and shape) should not be attributed literally and exactly to quantum objects. In fact, this should include even the property of being "particle" (which is the subject of this section) since quantum particles are not really (or "classically") particles (due for instance to their dual particle-wave nature).

[135] It is important to note that even consciousness is suggested as a possible element or factor or agent of disturbance in quantum measurement (see point 5).

contrary, measurement plays a central role in the quantum theory itself (not just in its application) and it is part of its axiomatic framework and formal structure (noting that in general the physical state of quantum system changes by measurement; see for instance § 4.2 and § 4.8).[136] In fact, in this respect quantum theory is unique among all scientific theories (whether old or modern) since no other theory gives measurement this central role. This should be understood and appreciated in the light of our investigation of the differences between quantum and classical (or non-quantum) theories which is partly presented in the other points.

5. **Role of consciousness**: a role of consciousness in quantum measurement is suggested (and even accepted by some physicists and philosophers of science) which is unimaginable in classical measurement. For example, according to some respected physicists (such as Wigner) what constitutes a measurement in quantum physics is the intervention of the consciousness of observer. In fact, this should negate any meaning or possibility of objectivity (i.e. in its classical sense and significance) in quantum physics. Also see § 1.11 and § 9.1.3.

Also see § 5.7 and § 6.11.

3.5 Comparison between Classical and Quantum Theories

A fair and thorough comparison between the two theories requires detailed and thorough inspection (which is distributed throughout this book). However, in the following points we outline and compare some of the most prominent characteristic features of the two theories:

1. **Domain of validity**: the domain of validity of classical theory is the macroscopic world while the domain of validity of quantum theory is the microscopic world. However, we should be aware of the claim that the domain of validity of quantum theory extends even to the macroscopic world (although we deny this at least from practical, if not theoretical, perspectives and reasons).
   So, both theories are valid (i.e. each in its primary domain of validity) and they are supported by massive experimental and observational evidence. However, classical theory is definitely invalid in the domain of quantum theory, while the validity of quantum theory in the domain of classical theory is questionable (at least practically) despite the opposite claims. More thorough discussion about this will follow (see for instance § 3.7 and § 7.3).

2. **Epistemology**: classical theory is based on a clear and "common sense" epistemology, while quantum theory struggles in this regard, i.e. its epistemology is shrouded with haziness, confusion and controversy and hence there are many contradicting schools about its interpretation (see for instance chapter 8). In fact, this feature (i.e. clarity of classical epistemology and confusion in quantum epistemology) should explain why classical epistemology and interpretation are rarely indicated or discussed while their quantum counterparts attract a great deal of attention and interest.

3. **Determinism**: classical theory is deterministic while quantum theory is indeterministic (see § 5.2). However, it is important to note that probabilisticity (as a form of indeterminism; see § 5.2.2) also exists in some classical theories such as statistical mechanics. Nevertheless, an important difference between the classical and quantum probabilisticity is that the classical probabilisticity underlies (or/and is underlied by) the classical *deterministic* physical phenomenon,[137] while the quantum probabilisticity is in the essence and heart of the quantum physical phenomenon and hence the quantum phenomenon itself is (supposedly) probabilistic and indeterministic. It is also important to note that deterministic aspects also exist in quantum theory and hence it is not entirely indeterministic (see for instance § 5.2).

Similarly, uncertainty (as another form of indeterminism; see § 4.7.1 and § 5.2.3) also exists in classical theories but classical uncertainty is different in nature from quantum uncertainty (see for instance § 4.7.1 and § 6.6). In brief, some forms of uncertainty (i.e. casual uncertainty due for instance to limitations on the accuracy of the measuring equipment) are common to both classical and quantum theories, and one form of uncertainty (i.e. intrinsic uncertainty due to the uncertainty principle) is

---

[136] Although not all quantum physicists accept this view, it is a dominant view.

[137] For example, the rules of probability in statistical mechanics are generally based on definite deterministic classical laws and principles and are used to derive definite deterministic classical laws and principles.

specific to quantum theory and does not exist in classical theory (or at least it is not claimed to exist in classical theory in the same sense). Also see point 7 of § 6.6.
4. **Causality**: classical theory (thanks, in part, to its deterministic nature according to point 3) is based on the principle of causality (see § 1.7) while quantum theory may not be so (according to certain implications and interpretations of quantum mechanics as well as the principle of causality). So in brief, classical theory is based on strict causality while quantum theory may not be so. It is worth noting that the classical causal determinism (as discussed in this point and point 3) is typically represented by the so-called "Laplace's demon" which summarizes the causal deterministic nature of classical theory.
5. **Realism**: classical theory is definitely realistic (i.e. belonging to realism; see for instance § 1.10) while quantum theory may not be so (or at least it may include some non-realistic aspects and features).

Also see § 3.7.

3.6 Relationship between Our Views to Classical and Quantum Worlds

Before we discuss the relationship between our views to the classical and quantum worlds, we discuss (very briefly) the relationship between these worlds which is one of the most important and problematic issues in the quantum theory and the philosophy of science.[138] It should be understood that the central issue in this relationship is how the classical and quantum worlds interact to produce the classical and quantum realities (or rather our conception and vision of these realities assuming the existence of these realities in some form). So, the central issue here is the relationship between our view to the classical and quantum worlds rather than the relationship between the two worlds in themselves. In fact, the existence of two worlds (i.e. classical and quantum) irrespective of our view is nonsensical since if there is any outside world it should (at least for the sake of simplicity and intuitivity) be single and unique although it may be divided justifiably to classical and quantum by our views and theories due to our humanist nature and scale factor (see § 1.3.1 and § 1.3.2).[139]

It is obvious that the macroscopic classical world is observed directly (since it is commensurate to our cognitive scale), while the microscopic quantum world is observed indirectly (i.e. through macroscopic means and apparatuses) and hence our views to these worlds (as shaped by our observations) should be different although they should be linked and related. As a consequence, it is natural to ask: what is the relationship between our classical view (as represented typically and partly by classical mechanics) and our quantum view (as represented typically and partly by quantum mechanics)? In fact, we can address this issue from two perspectives: ontological and epistemological.

From the ontological perspective, it may seem logical to assume that the classical behavior of the world (as represented by classical phenomena) is shaped and formed by the quantum behavior of the world (as represented by quantum phenomena) and hence our view from this perspective is shaped and formed by the underlying quantum phenomena. In other words, our classical view can be seen as a sort of scaling up or averaging the underlying quantum behavior which is at the foundation of the classical phenomena. In fact, this opinion is essentially based on the assumption that the classical reality (as seen by us) is a sort of scaling or averaging of the quantum reality. However, this view may be challenged noting that the effect of the scale factor may be more than just scaling and averaging. In other words, the quantitative difference (brought in by the scale factor) may lead to a qualitative difference and hence the classical reality is not necessarily a scaling or averaging of the quantum reality. In fact, this should depend on many factors and hence some classical realities are scaling and averaging of corresponding quantum realities while others are not.[140] This, actually, is reflected in our physics (which typically represents

---

[138] One of the prominent demonstrations of this subtle and bewildering relationship (as well as its problematic aspects) is seen in the issue (or theory) of quantum measurement (see for instance § 3.4, § 5.7 and § 6.11).

[139] It should be understood that this relationship (whether between the two worlds or the two views) is essentially of philosophical and epistemological nature (although it has obvious scientific consequences and implications) and hence it is strongly related to the interpretation of quantum mechanics. This should explain (at least partly) why the issue of quantum measurement is at the heart of almost all schools of interpretation where these schools struggle to rationalize quantum measurement through this relationship (see for instance chapter 8).

[140] For example, Brownian motion and temperature or pressure (in their kinetic interpretation) may be seen as examples of scaling and averaging while color or ductility may not be so.

our view in its classical and quantum aspects) where some of our classical formulations can be obtained from our quantum formulations while others cannot (at least not intuitively and practically).

From the epistemological perspective, it should be logical to take the opposite direction in this argument and hence we can say: our quantum view is actually derived (or based on or affected by) our classical view. This is because we are, after all, classical creatures and hence we cannot penetrate (cognitively and conceptually) to the quantum world except through our classical machinery (whether theoretical or practical). In fact, this is just a repetition of what we indicated earlier of the effect of factors like humanism and scale (see § 1.3.1 and § 1.3.2) in shaping our view to the world, both in our meticulously-organized and highly-abstract knowledge (in the form of science for instance) and in our rather primitive form of knowledge in the daily life.

So to sum up, the relationship between the two worlds is effectively a relationship between two views. Moreover, these views are shaped ontologically by scaling and averaging (as well as other types of interaction and influence) and shaped epistemologically by the extension of the classical view to the quantum domain to produce (or derive) the quantum view. In fact, this extension is at the root of many problematic and perplexing issues in quantum theory since the classical view is generally inappropriate to represent and reflect the quantum reality.

3.7 Relationship between Classical and Quantum Physics

Based on the fact that quantum physics has emerged as a correction to classical physics (when applied to microscopic systems), it may be thought that quantum physics is a more general theory and hence classical physics (in its domain of validity) is a special or limiting case of quantum physics. This thought may be supported by the correspondence principle which suggests that classical physics is a limiting case for quantum physics (see § 4.7.2 and § 6.7).[141] However, the relationship between classical and quantum physics is more complicated as will be explained, rather briefly and partially, next (also see § 3.8).

It should be obvious that classical physics is not valid in quantum systems such as atoms and electrons (and that is why quantum physics was developed to replace classical physics in these systems). It should also be obvious that, in general, quantum physics in its quantum mechanical formalism is not applicable (at least from a practical perspective) to macroscopic systems due for instance to the non-practicality of employing quantum mechanical paradigms and models, like matter waves and wavefunctions, to macroscopic objects.[142] So, in general the two theories may be seen as disjoint in their domain of validity. However, there are a number of areas and zones where the two theories converge and meet. In fact, these areas and zones are what justify the correspondence principle and its alike. For example, it can be shown that certain quantum mechanical formulations converge for large quantum numbers to corresponding formulations that were obtained earlier from classical mechanics.[143] It can also be shown (using for instance the Ehrenfest theorem) that certain classical laws and principles (like the laws of motion and the conservation principles) can be obtained from quantum mechanical formulations representing expectation values and average behavior of physical systems.[144]

---

[141] In fact, this thought may also find its roots in the fact that Lorentz mechanics (which is a correction to classical mechanics) is a more general theory than classical mechanics.

[142] The reader should be aware of views like: "quantum mechanics is believed to be a universal theory, capable of describing macroscopic as well as microscopic objects" (see Rae and Napolitano in the References) about the general validity of quantum mechanics. Another useful quote in this context is: "We have experimental evidence that quantum theory is successful in the range from 10^{-10} to 10^{15} atomic radii; we have no evidence that it is universally valid. Yet, it is legitimate to attempt to extrapolate the theory beyond its present range, for instance, when we probe particle interactions at superhigh energies, or in astrophysical systems, including the entire universe" (see "Quantum Theory Needs No 'Interpretation'" by C.A. Fuchs and A. Peres). We can also find in the literature declarations like "Classical mechanics can be considered as a special case of quantum mechanics".

[143] Quantum numbers are discrete numbers (integers or half-integers) used to label and characterize certain (discrete) values of physical quantities (such as energy) of quantum systems in certain quantum states.

[144] The essence of the Ehrenfest theorem is that the classical equations of motion are satisfied by the expectation values of the corresponding quantum operators. However, it should be noted that the Ehrenfest theorem and its alike cannot produce the classical conceptualization of these classical formulations even if they succeed in producing the formalism. For instance, there are no well-defined (or classical) trajectories in quantum theory (at least according to the mainstream understanding) and hence the Ehrenfest theorem cannot produce them conceptually.

It should be obvious that there are many examples in the quantum theory for the convergence of quantum mechanics to classical mechanics in certain cases and circumstances although this does not qualify quantum mechanics to be such a general theory that incorporates classical mechanics as a special or limiting case of quantum mechanics. This is unlike the situation in Lorentz mechanics and its relation to classical mechanics where the latter can be seen as a special or limiting case to the former (see for instance our book "The Mechanics of Lorentz Transformations"). It should also be obvious that there are certain areas and zones in classical physics (representing macroscopic systems) that cannot be dealt with by the formalism of quantum mechanics (at least due to practical reasons) and hence classical mechanics is still needed to deal with these cases. So, neither classical mechanics is a special or limiting case to quantum mechanics nor quantum mechanics is of such general validity that includes the macroscopic world.

We should note (with reference to our earlier claim that classical physics is not valid in quantum systems) that some correct quantum results can be obtained from classical (or semi-classical) formulations. In fact, we have many examples of such cases from the early days of the development of quantum mechanics where quantum phenomena were investigated using classical physics (and even later including some continuous attempts throughout the history of quantum physics by fans of classical physics to obtain all the quantum results and predictions from classical physics). It is common to explain such successes by fortunate coincidence, but we do not think fortune can explain all these many "coincidences". Our belief (which is inline with our opinion which we expressed earlier that science is not unique; see for instance § 2.4.3) is that to a certain extent even classical physics has a share of the truth in the quantum world although it is not thorough or ideal. In fact, this issue should also be linked to the issue of the relationship between our views to the classical and quantum worlds (which was investigated in § 3.6) and may be seen as an indication to the intimate relation between the two worlds which should be reflected in our views to them. So, if we consider these partial successes of classical physics in the quantum world alongside the partial successes of the quantum physics in the classical world (as indicated earlier within our discussion to the extent of the validity of quantum physics and the correspondence principle) then we can conclude that the relationship between classical physics and quantum physics is more complicated than it might be thought initially, and this relationship is more delicate and intricate to be realistically represented by the correspondence principle alone or the claim that quantum physics is a more general theory. In fact, each of these physics seems to have a shared validity zone in the domain of validity of the other physics (apart from the validity zones that distinguish each and are specific to each).[145] The reader is also referred to § 5.3.3 and § 7.2 for further investigation to this relation.

We should also note that historically (and even recently) there were many attempts to formulate or style quantum mechanics classically, e.g. by deriving the quantum mechanical rules (or at least some of them) or their equivalents from classical rules or by formulating quantum mechanics on the style of classical theories like continuum mechanics.[146] The common factor of all these attempts is to extend the validity of classical mechanics to the quantum world (although not necessarily exactly and thoroughly) and create a harmonious theory of "classical-quantum mechanics" which supposedly mixes the old rules and methods of classical physics with the new rules and methods of quantum physics (or derive the new rules from the old ones or harmonize and reconcile the two sets of rules or ... etc.). In fact, these theories (or some of them at least) may be seen as a sort of unification theories despite their limited success and lack of popularity.

To conclude, classical mechanics in its domain of validity is not less reliable and authentic than quantum mechanics in its domain of validity (considering the allowed level of errors and approximations in each domain at the appropriate scale and size to that domain and noting the general independence of their formulations). Moreover, the claim that quantum mechanics has general validity (and hence it applies, in principle at least, even to the domain of classical mechanics) is baseless because we cannot find a single

---

[145] Another example for the intricate (and even intimate) relation between classical and quantum physics will be met later (see for instance § 4.2 and § 4.4) where we will see how the operators of quantum mechanics are obtained from classical expressions representing their corresponding observables.

[146] In fact, quantum mechanics itself can be seen as one of these attempts (and is actually a successful one!) because it is a "wave mechanics" (in its most prominent and dominant formulations) which finds its roots in the classical theory of waves (noting that Schrodinger's equation is a modified version of the classical wave equation).

3.7 Relationship between Classical and Quantum Physics

credible example for the (direct) application of quantum mechanics in classical systems. For instance, we cannot find any application of Schrodinger's equation or wavefunction or uncertainty principle or particle-wave duality or ... etc. in macroscopic systems. Yes, there are allegations of such extensions and generalizations but all these allegations are hypothetical in nature and lack credibility and authenticity. Interestingly, most of these allegations are found within the context of discussing the interpretation of quantum mechanics and they usually belong to dodgy schools of interpretation, and this should expose their true nature.[147]

In fact, we may even claim that classical mechanics seems to have a larger domain of validity (or rather potential validity as an approximation) than quantum mechanics. For example, we can find many classically-based quantum theories (new as well as old) and this should include the old quantum theory (which is the genesis of the current quantum mechanics) which despite its failures and shortcomings has also some genuine successes and merits (and that is why it is the "mother" of the modern quantum theory). Moreover, we can find many classically-based models and techniques in the current quantum mechanics.[148] Although these models and techniques are normally used for the purpose of approximation and imitation (or they are alleged to be so) they are as useful as any "exact and genuinely-quantum" counterpart and they are, after all, part of the modern quantum kit and machinery. So, the alleged generality and supremacy of quantum mechanics over classical mechanics is unfounded (or at least questionable). Accordingly, each one of these theories (i.e. classical and quantum mechanics) should be used in its domain of validity, and in this regard they should be treated and "respected" equally. The failure of classical mechanics (i.e. when applied to quantum phenomena) is because it was the "elder" theory and hence it failed when it was tested outside its domain. If we were quantum creatures and quantum mechanics was (supposedly) our prime theory then we will equally find quantum mechanics fails when we try to step outside our quantum world and test it outside its domain of validity (i.e. in the "macroscopic" world).[149] So, there is no superiority or inferiority and no generality or specificity relation between these theories.[150]

We should finally draw the attention to the following useful remarks:

1. As indicated earlier, macroscopic phenomena should in general be ultimately based on quantum phenomena. However, this should not necessarily mean (as some might think) that the formalism of classical mechanics can emerge from the formalism of quantum mechanics.[151] This is because the formalism of any theory is the creation of human and hence it contains an element of invention and therefore a theory about a phenomenon that underlies another phenomenon does not necessarily underlies the theory of the other phenomenon. Also see § 3.6.

---

[147] In fact, some even proposed (or expressed their desire to find) a wavefunction for the entire Universe and hence through the use of quantum mechanics they can find solutions to chronic cosmological problems and get answers to perplexing questions related to the alleged Big Bang and the expansion of the Universe. This sort of daydreaming or wishful thinking should be classified as science fiction (at least for the current stage of scientific development).

[148] It is worth noting that the emergence of statistical mechanics (which is a successful classical theory) from classical laws and models for the underlying quantum (or microscopic) phenomena (where atoms and molecules are modeled and treated as classical particles that have deterministic trajectories, speeds, positions, etc.) may be seen as an example for the conditional or limited validity of classical physics in the quantum domain. However, this does not demonstrate the validity of classical physics at the quantum level for the description of quantum phenomena but for the validity of classical physics at the quantum level for the description of (the emerging macroscopic) classical phenomena and hence this is not an example of the validity of classical physics in the quantum domain as such since no quantum laws are derived from this application of classical physics.

[149] In fact, this rationale should put question marks on (at least some of) our theories in fields like astronomy, astrophysics and cosmology which deals with scales much larger (e.g. in space and time) than the "macroscopic" scale that we are familiar with in our daily life (and even in classical physics although this physics is allegedly valid to all "macroscopic" scales). Also see point 7 of § 7.3.

[150] Despite this, there are certain advantages for each theory in its domain. For example, classical mechanics provides an insight into the physical world (within its domain) that quantum mechanics seemingly fails to provide (within its domain). On the other hand, quantum mechanics seems to provide a higher level of precision (within the accepted rules and principles) in its calculations and results (although this should be appreciated within the consideration of the relative size of the two worlds, i.e. macroscopic and microscopic).

[151] As we will see, even if we assume the emergence of classical mechanics from quantum mechanics it does not mean that classical mechanics (as a theory) is a special or limiting case to quantum mechanics (as a theory) since each theory is needed (as it is) in its domain.

2. The general validity of quantum mechanics may be challenged from a practicality perspective by arguing that quantum mechanics is too complicated to be applicable to typical macroscopic systems (which are in the domain of classical mechanics) regardless of the theoretical validity or invalidity of quantum mechanics in principle at the macroscopic scale. This may be supported, for instance, by noting that quantum mechanics is too complicated even in its applications to some of the simplest quantum systems (as can be seen for example by considering the huge complexity of the physics of atoms, molecules and elementary particles). So, the applicability of quantum mechanics to the much more complicated macroscopic systems (with a huge number of degrees of freedom) is untenable.[152]
3. Another aspect of the relationship between classical and quantum physics is the epistemology of reality. As indicated earlier, a weak form of rationality (whether by weak form of realism or weak form of causality) at the quantum level (if we are forced to adopt such form at this level) should not affect the (strong and total) rationality at the macroscopic level because at this level we usually deal with averages and expectation values which are deterministic and certain (and governed by definite causal relationships). In other words, at macroscopic level we do not usually deal with individual quantum events (which are supposedly indeterministic and uncertain) but with the result of a large number of these quantum events and hence their statistical outcome is determinate and certain (by the classical standards). So, there should be no inconsistency or difficulty in the epistemology of reality at these levels.
4. As indicated earlier, quantum phenomena generally underlie classical phenomena and hence it is important (especially for interpretation) to understand this relation. However, primarily this is not about interpreting or understanding the quantum phenomena but about interpreting and understanding the classical phenomena which supposedly stem from the quantum phenomena (i.e. this is in the domain and responsibility of the interpretation of classical physics which is outside the scope of this book). Yes, interpreting the quantum phenomenon correctly is likely to help in interpreting the classical phenomenon correctly (and the reverse may also be true). So in general, interpreting and understanding such quantum phenomena is actually not only about interpreting and understanding quantum phenomena but about interpreting and understanding classical phenomena (which stem from the quantum phenomena). Also see § 3.6.
5. The non-applicability of quantum mechanics to macroscopic systems should include applying the theory to macroscopic objects directly and as they are and applying the theory to macroscopic objects as ensembles of quantum objects. However, the reason for the non-applicability is different in these two cases where it is usually conceptual or theoretical in the former case and practical in the latter case. We should also note that the application in the second type (i.e. as ensemble) is not really an application of quantum mechanics at the macroscopic level, but rather an application of quantum mechanics to macroscopic objects but at the microscopic level (i.e. by treating these objects as ensembles of quantum objects to which quantum mechanics applies).

Also see § 3.5, § 4.7.2, § 6.7 and § 7.3.

3.8 Border between Classical and Quantum Domains

The classical and quantum worlds may be identified (rather grossly) as the domains of validity of classical and quantum theories. However, this identification has two main problems: the first is the lack of obvious and objective distinction and the second is the potential overlap where the quantum domain (to which quantum theory supposedly applies) may extend to include even the classical domain (at least partly). In fact, the second problem was effectively addressed in § 3.7 where we concluded (somewhat broadly and indirectly) that the validity domain of quantum theory does not extend to the domain of classical physics and hence the two domains should not overlap (at least extensively). Accordingly, the two domains are broadly distinct and hence we can address the first problem by trying to identify the border between the two domains so that the distinction between the two domains becomes obvious and objective.

---

[152] Referring to the upcoming point 5, we are mainly considering here the case of applying quantum mechanics o macroscopic objects as ensembles of microscopic objects.

3.8 Border between Classical and Quantum Domains

Let agree first that any border between classical and quantum domains is actually artificial and is not really between the two realities (i.e. classical and quantum) "as they are", but between the two realities "as we perceive them". In other words, such a border is based on the difference between the two realities because of factors like humanism and scale (see § 1.3.1 and § 1.3.2). This is because the two domains are actually created by our theories (i.e. classical and quantum) rather than being based on the existence of two different realities. In fact, there is no basis or reason (neither ontologically nor epistemologically) to believe that we have two different realities that can be identified and distinguished independently of our theories and conceptions. So, this border should more appropriately be regarded as a border between our theory (or conception) about the classical world and our theory (or conception) about the quantum world.[153]

Let also agree that any border between the two domains is not sharply defined and hence such a border is a "zone" rather than a "line".[154] In fact, we may even claim that not only the "width" of this zone is not sharply defined but even its "location" or "position" could be dependent on cases, circumstances, contexts, applications, etc. In other words, the border is blur in location as in extension and this depends on various factors and considerations. We should also consider that there might be "isles" of validity of each theory in the domain of the other theory. So in brief, the border and validity domains with regard to these theories are complicated and subject to various factors and considerations although there are certain areas and zones of obvious and unique applicability of each one of these theories (e.g. solar orbits or ballistic trajectories for classical theory and atomic "orbits" or electron spin for quantum theory).

---

[153] As we will see, this border (in extent and location) may even depend on the particular application of our physical theories.

[154] In fact, this "zone" is what provides (at least primarily) the domain of applicability of the correspondence principle. This "zone" may also provide the interface between the two physics (where they meet and interact) with regard to the issue of measurement.

Chapter 4
Formalism of Quantum Theory

In this chapter we briefly investigate the formalism of quantum theory (i.e quantum mechanics). In fact, this investigation focuses on the generalities of the formalism of this theory rather than on the delicate and highly-technical details of this formalism. Moreover, the investigation is not comprehensive at all and hence the contents of this chapter are no more than selected parts and aspects of the formalism that we need in our discussions and investigations in this book (noting that some of these parts and aspects may not strictly belong to the formalism according to some interpretations and opinions). We will also try (for the sake of clarity and common benefit) to minimize the use of symbols and equations and hence our presentation of the formalism is mostly descriptive in nature.

It should be noted that the current formalism of quantum theory is commonly cast in a language and style that generally reflect the attitude and spirit of a specific interpretation (i.e. the Copenhagen interpretation) and hence some interpretative aspects (belonging to the Copenhagen school) are associated with the formalism and may be seen as part of the formalism. In fact, even the standard (or basic) interpretation of quantum mechanics (as represented mainly by Born's statistical interpretation of the wavefunction as a probability amplitude) may be questioned as a formal aspect of the theory (i.e. if it belongs to the formalism of the theory or it could be an interpretative aspect and hence other interpretations may be possible in principle).

Anyway, we will follow in our presentation of the formalism of this theory the mainstream literature of quantum theory irrespective of what should be classified as formal or interpretative. We will also follow the mainstream literature in the formulation of quantum mechanics and hence we will ignore in this regard other quantum mechanical formulations (e.g. hidden-variable formulations) that are supposedly equivalent in their predictions to the common (or official or mainstream) formulation of quantum mechanics.[155]

As indicated earlier (see § 1.2), there are two main types of formalism: wave mechanics and matrix mechanics. However, since the investigation of the formalism is not the primary objective of this book, we will not go through the details of these types of formalism but we will take just what we need (which is generally taken from wave mechanics because it is more common in use and more intuitive and hence it is more sensible epistemologically).[156] As the scope of the book is the epistemology of quantum physics, we will generally focus in this chapter on the epistemologically-significant aspects of the formalism and what we need in our epistemological investigation in other parts of the book.

4.1 Wavefunction

The wavefunction (which may also be called the state function) of a given quantum system in a given physical state is a complex-valued function of position and time that contains all the physically accessible information about the system in that state. The wavefunction is commonly symbolized by the Greek letter ψ and since it is in general a function of position and time it is usually written as $\psi(\mathbf{r}, t)$ or $\psi(x, y, z, t)$ where $\mathbf{r} = (x, y, z)$ represents position (which is a 3D vector) and t is time. However, it may also be written as $\psi(\mathbf{r})$ or $\psi(x, y, z)$ when the wavefunction is time-independent and can even be written without argument (i.e. just ψ) when the dependency is obvious or irrelevant. As we will see, the wavefunction represents a probability amplitude function of space and time, and hence the probabilities of the possible

---

[155] As we will see in § 8.2, there are two main types of hidden-variable theories: interpretative type (which does not suggest any modification to the current formalism of quantum mechanics), and formal type (which proposes modification to the current formalism of quantum mechanics).

[156] Matrix mechanics is generally used to describe spin (see for instance the appended subsection § 10.3.1) noting that spin is not included in the Schrodinger equation (which is the heart of the *basic* wave mechanics formulation of quantum physics).

4.1 Wavefunction

outcomes of measurements made on the system (as functions of space and time) are obtained from it. In the following points we outline some of the general properties of wavefunction:

1. The wavefunction of a quantum system represents the physical state of the system and hence it describes its evolution in space and time.[157] From this perspective the wavefunction may be likened to the "trajectory" (e.g. of a particle) in classical mechanics or (rather more appropriately) to the "world line" in Lorentz mechanics.
2. "*Wave*function" may suggest that it undulates like an ordinary wave (by going up and down, e.g. in a sinusoidal wave fashion). However, it is not necessarily so since in some situations and circumstances it may decay exponentially, for instance, or have other non-undulating forms.[158]
3. The wavefunction of a quantum system is generally the solution(s) of the Schrodinger equation (although certain extensions and exceptions may apply).
4. The space and time dependencies are the main dependencies of the wavefunction (noting that they are the only ones in the Schrodinger equation). However, the wavefunction (or rather "state function") may have dependencies on other variables such as spin. But we should note that if "wavefunction" is conceptualized (considering its association with the Schrodinger equation) to have only space-time dependencies then other dependencies should be conceptualized differently (e.g. as attachment or extension to the wavefunction or something else noting that matrix mechanics is generally used to describe spin for instance). Anyway, as far as the wave mechanics of the Schrodinger equation (which is the basic quantum mechanics that we are considering as the scope of this book) is concerned, other dependencies should not be included in our investigation although they may be investigated in this book casually or out of necessity due to their epistemological value and significance (see for instance § 4.6). Also, "wavefunction" may be used occasionally to refer to "state functions" that include such non-tempo-spatial dependencies (even if such dependencies are not considered to belong to the wavefunction in its technical and conventional conceptualization, as indicated already). In fact, the use of "state function" should be more appropriate in such cases and circumstances (as indicated already).
5. As indicated, there are two types of wavefunction: time-dependent wavefunction $\psi(x,y,z,t)$ which has temporal dependency as well as spatial dependency, and time-independent (or stationary) wavefunction $\psi(x,y,z)$ which has only spatial dependency.[159] The time-dependent wavefunction may be expressed (in convenient circumstances) as a product of the time-independent wavefunction times a function of time $f(t)$, i.e. $\psi(x,y,z,t) = \psi(x,y,z) f(t)$.[160] It should be noted that the time-independent and time-dependent wavefunctions may be distinguished by using lower case and upper case letters (respectively), i.e. $\psi \equiv \psi(x,y,z)$ and $\Psi \equiv \Psi(x,y,z,t)$. Other conventions and variations can also be found in the literature (so the reader should be aware of each to avoid confusion and misunderstanding).
6. In the absence of measurement, the wavefunction (as a representative of the quantum state) evolves *deterministically* according to the time-dependent Schrodinger equation (see § 4.3). This means that if the wavefunction in known at a given time then the Schrodinger equation can be used (in the absence of measurement) to determine its form at any time in the future.[161] So, from this perspective quantum physics is deterministic.
7. According to the dominant (or traditional) interpretation of quantum mechanics, the wavefunction collapses by measurement (or observation). This means that the process of measurement changes the wavefunction irreversibly from being a combination (or superposition) of a number of possible states to a specific state which is the obtained (or "observed") state and is one of the possible states.

---

[157] Accordingly, the wavefunction is commonly described as "state function" or "quantum state".

[158] In fact, this property of wavefunction (i.e. the possibility of having non-undulating forms) may be added to the list of § 3.2 as a feature that could distinguish quantum waves from classical waves (which are generally undulating).

[159] In fact, labeling the time-independent as "wavefunction" is not rigorous (at least in some cases) and hence time dependency should be considered as part of the (complete or true) wavefunction. It should also be noted that "time-independent wavefunction" may be used in reference to time-dependent wavefunction whose temporal dependency is not observable, i.e. its squared magnitude (which represents probability density) is time-independent since its time dependency is represented by a complex phase factor of unity magnitude such as $e^{-i\omega t}$.

[160] A time-dependent wavefunction whose spatial part is an eigenfunction should have no time-dependent observable consequences (i.e. its temporal part has no physical significance).

[161] By time reversal (with some technicalities) it can be determined even in the past.

4.1 Wavefunction

Accordingly, the measurement replaces the (old) wavefunction of the quantum system by a (new) wavefunction representing a new quantum state that matches the outcome of the measurement where the new wavefunction represents (in general) a new combination of possible states that are allowed by the outcome of the measurement.[162] This cycle of "measurement → collapse → replacement" continues indefinitely as long as the system exists and the measurements are going on.

8. From a mathematical perspective, the wavefunction should be a well-defined and single-valued function (considering its dependency on position and possibly time).[163] It should also be continuous and has a continuous first order derivative (with some exceptions regarding the continuity of derivative at infinite discontinuity in the potential). Moreover, it must be neither zero everywhere nor infinite somewhere. The purpose of all these conditions is to ensure that the wavefunction is physically meaningful (i.e. in its attributes and implications) and practically usable and to avoid ambiguity and lack of definiteness.

9. From another mathematical perspective, the wavefunction may be seen as a vector in a Hilbert space representing all the possible quantum states (where the points in this space correspond to quantum states).[164] In fact, "wavefunction" and "state vector" are commonly used interchangeably. However, it is important to note that "vector" should be understood in an abstract sense (considering mainly the rules of formation and manipulation in vector spaces) rather than visual sense.[165]

10. Referring to § 3.2, the wavefunction represents a quantum (or matter) wave and not a classical (or physical) wave like electromagnetic wave.

11. As indicated above, the wavefunction of a quantum object represents a probability amplitude. This means that the square of the wavefunction modulus (i.e. $|\psi(\mathbf{r},t)|^2$) represents the probability density function for the presence of the object in a tiny volume of space around a given point $\mathbf{r}$ in space at a given instant t in time.

12. Although the wavefunction is complex (i.e. not real) and non-observable (since it is non-physical; see point 10), the tempo-spatial probability density $|\psi|^2$ is real and observable (or rather has observable consequences). This means that the wavefunction provides specific (but usually probabilistic) predictions for the outcome of measurements conducted on the system. So in brief, although the wavefunction itself is not observable, all the predictions of the observable quantities and attributes of quantum systems are obtained (directly or indirectly) from the wavefunction (usually through complicated formulations and procedures that involve other physical and mathematical entities like operators; see for instance § 4.5).

13. In general, the wavefunction should be normalizable (i.e. the integral of its modulus squared over the entire space is finite)[166] to be usable as a probability distribution function (i.e. to become physically

---

[162] Whether the wavefunction (following measurement) is an eigenstate or a superposition of eigenstates depends on the observable under consideration. So, the wavefunction is an eigenstate with regard to the already-measured observable but it is a superposition with regard to other (incompatible) observables (noting that the next measurement will determine which observable, and hence which wavefunction, should be considered). In fact, this should indicate the non-physical nature of the wavefunction (i.e. being a mathematical device) since the wavefunction is a matter of consideration and conceptualization of the physical situation.

[163] In fact, being single-valued essentially applies with respect to its spatial dependency.

[164] When the wavefunction represents a combination of attributes (such as position and spin) then the Hilbert space is a *direct product space* of the Hilbert spaces that correspond to these attributes.

[165] In fact, functions (such as wavefunctions) may be seen as vectors of an infinite-dimensional vector space. In other words, their continuous (or non-countable) argument (which represents infinite dimensionality) corresponds to the discrete (or countable) index (which represents finite dimensionality) of ordinary vectors. Also see footnote [12] on page 11.

[166] This type of function may be called square-integrable or quadratically-integrable or L^2 function. It should be noted that some wavefunctions (i.e. solutions of Schrodinger's equation) are not normalizable in this sense (because $\psi = 0$ identically or because the normalization integral is infinite) and hence they are generally rejected (unless alternative normalization schemes apply, as will be indicated in footnote [167]).

admissible). Accordingly, the normalization is expressed mathematically as:[167]

$$\int_{\text{all space}} |\psi|^2 \, dV = \int_{\text{all space}} \psi^* \psi \, dV = 1 \qquad (2)$$

where dV is an infinitesimal volume element (of space) and where 1 represents the probability of finding the object somewhere in space (which is certainty).[168]

14. The normalization of wavefunction is a direct consequence of the probabilistic nature of the quantum theory (as reflected by its probabilistic interpretation which regards the wavefunction as a probability amplitude; see points 11 and 12).
15. The normalization of wavefunction affects only its modulus which is the part that is physically significant from this perspective (noting its relation to probability). This should be physically sensible since the argument has no physical role to play in this regard.
16. The normalization of wavefunction (i.e. by a constant and when normalization is relevant) should always be possible due to the linearity of Schrodinger's equation (see § 4.3).[169]
17. As indicated earlier, "time-independent wavefunction" (as well as similar expressions) may be used for the spatial part of the wavefunction (which has no temporal dependency at all) and may be used for the (time-dependent) wavefunction whose time dependency is not observable since it does not contribute to its probability density. However, it should be noted that both uses are rather loose (since in the first the use of "wavefunction" is lax because actual wavefunction should have tempo-spacial dependency, while in the second the time-independence actually belongs to the probability density rather than the wavefunction). So, attention is required (when reading the literature) to avoid confusion and misunderstanding.
18. The wavefunction is a single and simple entity, i.e. it is not made of parts and components even if the system represented by the wavefunction is made of parts and components. Hence, when the wavefunction collapses it should collapse globally and instantaneously.[170] In other words, the wavefunction either exists (as a whole) or does not exist (as a whole). Hence, it is meaningless for the wavefunction to collapse in one part but not in another part (since it has no parts to make the scenario of partial collapse sensible). In fact, wavefunction (according to the common understanding of quantum theory) is like a living being which is either alive as a whole or dead as a whole, and hence it is meaningless to talk about a human who is alive in his right side and dead in his left side.[171]

Also see § 6.1.

4.2 The Axiomatic Framework of Quantum Mechanics

Quantum mechanics is based on a number of postulates or axioms whose number, order, content and form may differ between authors although they generally follow the five postulates of von Neumann. In fact, these postulates also depend on the adopted interpretation of quantum mechanics and hence these postulates may differ (in number and content) between different schools of interpretation. Anyway, these postulates (which form the axiomatic framework of quantum theory) represent the basic principles and

---

[167] In fact, this equation (and its alike in this chapter) essentially applies to very simple quantum systems consisting of a single object (or particle). More elaborate equations and formulations usually apply to more complex systems. However, from the epistemological perspective (which is the subject of this book) the restricted formulations (like this equation) are sufficient for fulfilling our objectives. We should also note that some wavefunctions are not normalizable in this way (e.g. wavefunctions of free particles). More specifically, the above normalization scheme applies to square-integrable functions; otherwise other normalization schemes (such as using the delta function or enclosing the system in a large box) should be used when possible.

[168] It is common in the literature to use $d^3\mathbf{r}$ for dV.

[169] It can be easily shown that if the wavefunction is initially normalized then it would stay so, and hence normalizing by a constant (i.e. with no time dependency) is possible.

[170] In fact, collapse is just an example. Hence, the essence of this discussion applies to any change (or development or evolution) of the wavefunction which should apply to it as a single and simple entity that responds globally and instantly to that change.

[171] We are not talking here about certain technical medical criteria and conventions that may allow this.

4.2 The Axiomatic Framework of Quantum Mechanics

rules that should be followed in the formalism of quantum physics and hence they actually represent the essence of quantum mechanics, i.e. all the rest of quantum mechanics are just technical details to the application of these principles and rules (see for instance § 4.9). These postulates (or rather their contents which represent the axiomatic framework) are briefly investigated in the following points:[172]

1. Any physically-realizable state of a given quantum system is quantum mechanically described by a wavefunction $\psi(\mathbf{r},t)$ that contains all the accessible information on the system in that state. This wavefunction should have the properties given in § 4.1 (i.e. continuous, normalizable, etc.).[173]

2. If $\psi_1(\mathbf{r},t)$ and $\psi_2(\mathbf{r},t)$ are two physically-realizable states of the system, then the linear combination (or superposition):

$$\psi(\mathbf{r},t) = c_1\psi_1(\mathbf{r},t) + c_2\psi_2(\mathbf{r},t) \tag{3}$$

is another physically-realizable state of the system (where c_1 and c_2 are complex constants). This can be easily extended to the linear combination of n states ψ_i, that is:[174]

$$\psi(\mathbf{r},t) = \sum_{i=1}^{n} c_i \psi_i(\mathbf{r},t) \tag{4}$$

3. If a quantum particle is in a state represented by the wavefunction $\psi(\mathbf{r},t)$ then $|\psi|^2 \, dV$ represents the probability of finding the particle in the infinitesimal volume dV around the position $\mathbf{r}$ at time t. Accordingly, the integrated probability density $|\psi|^2$ over a patch of space represents the probability of finding the particle in that patch at time t (and hence the integral should equal unity if the patch represents the entire space assuming ψ is normalized).

4. Every observable O of a quantum system (described by a given wavefunction ψ) is represented by a mathematical (Hermitian) operator $\hat{O}$ acting on ψ.[175] The eigenvalues of this operator represent the possible outcomes of measurement of the value of that observable (noting that the eigenvalues are real as they represent observables).

5. In general, the wavefunction before measurement is a superposition of states (which correspond to the possible outcomes of the measurement). The wavefunction immediately after the measurement is the eigenfunction corresponding to the eigenvalue obtained by the measurement. This change of the wavefunction (i.e. from being a superposition to becoming an eigenfunction of the obtained eigenvalue) by the measurement is commonly known as the "collapse" or "reduction" of the wavefunction (and the postulate that describes and formalizes this change is called the "measurement postulate" or "reduction postulate" or "projection postulate" of quantum mechanics). Also see point 13.

6. Following the collapse (by a measurement) to a given eigenstate ψ_i the system will remain in that eigenstate as long as it remains in existence and away from an external influence that changes this eigenstate (including a measurement of an incompatible observable; also see point 12).

7. The Hermitian operator $\hat{O}$ (as identified in point 4) can also be used to extract the required physical information about the system from the wavefunction (see for instance § 4.4 and § 4.5). In particular,

---

[172] We note that we are presenting the contents of these postulates and hence the points generally do not represent separate postulates. Moreover, we may go a little beyond these postulates for the sake of clarity and completeness and this is why we used in the title of this section "Axiomatic Framework" rather than "Postulates". Also, our relaxed approach prioritizes clarity over formality since we are not much interested in the formalism or in building a mathematically-rigorous theoretical structure. We also consider in this regard the fact that these postulates differ between authors and schools.

[173] As indicated earlier, the wavefunctions are commonly seen (from a certain mathematical perspective) as points or vectors in a Hilbert space that corresponds to the quantum system and represents its states.

[174] In fact, here we are essentially considering the case of discrete eigenstates (which should be sufficient for outlining the rationale). The summation should be replaced by an integral in the continuous case (with some delicate mathematical technicalities). The formulation can also be extended to represent the mixed case. The details should be sought in the literature.

[175] "Observable" means physically observable or measurable quantity (and may also be labeled as "dynamical variable") such as position or momentum or energy. Also, "operator" may be defined generically as a mathematical entity that transforms a given function into another function. We also note that observables may also be represented in quantum mechanics by matrices instead of ordinary operators (which are usually of differential type noting that these matrices can be seen as operators). Also, quantum operators associated with observable quantities are Hermitian (and hence when we use "operator" in this section it should mean Hermitian with no need for explicit declaration; see § 4.4).

the average (or mean) value of an observable is equal to the expectation value of its operator (see point 5 of § 4.5).

8. The operators that represent the position **r** and momentum **p** of a particle are **r** and $-i\hbar\nabla$ respectively (where **r** is the position vector, i is the imaginary unit, $\hbar$ is the reduced Planck constant and ∇ is the nabla differential operator).[176] The operators representing other observables take the same functional relation to these operators as do the corresponding classical quantities to the classical position and momentum.

9. For an observable that is represented classically by a function $f(O_1, \cdots, O_n)$ the corresponding operator is $f(\hat{O}_1, \cdots, \hat{O}_n)$.

10. Commuting operators represent compatible (or *simultaneously-measurable*) observables while non-commuting operators represent incompatible (or *non-simultaneously-measurable*) observables.[177]

11. In the absence of measurement, the tempo-spatial evolution of the wavefunction ψ of a given quantum system is determined (*deterministically*) by the time-dependent Schrodinger equation $\hat{H}\psi = i\hbar\frac{\partial \psi}{\partial t}$ where $\hat{H} = \hat{T} + \hat{V}$ is the Hamiltonian operator of the system (noting that H, T, V are the total, kinetic and potential energy; see § 4.3).[178]

12. The outcome of a measurement of an observable O of a system in an eigenstate ψ_i is given by the eigenvalue o_i of ψ_i corresponding to the operator $\hat{O}$, that is:

$$\hat{O}\psi_i = o_i\psi_i \tag{5}$$

In particular, the outcome of an energy measurement is certain to be E when the wavefunction is an eigenfunction of the Hamiltonian operator with eigenvalue E (and hence an eigenfunction of the Hamiltonian operator always represents a state of definite energy).

13. When a measurement of an observable O represented by an operator $\hat{O}$ is performed on a given quantum system represented by a wavefunction $\psi = \sum_i c_i \psi_i$ (with c_i being complex constants), the probability of the outcome of the measurement being equal to a certain eigenvalue o_i is $|c_i|^2$ (where ψ_i is the eigenfunction of $\hat{O}$ corresponding to the eigenvalue o_i).[179] As indicated earlier, the development of the wavefunction during the time between measurements is governed (*deterministically*) by the aforementioned time-dependent Schrodinger equation (noting that this development is commonly labeled as "unitary evolution").

We should finally note that the phrasing or/and content of some of the above axiomatic statements are inline with certain interpretations (and in particular the Copenhagen interpretation which historically is the dominant interpretation). Also see § 6.2.

---

[176] In fact, the operator **r** is commonly distinguished from the position vector **r** by a hat, i.e. $\hat{\mathbf{r}}$.

[177] Two observables are described as "compatible" if the operators that represent them have a complete set of common eigenstates (and hence a measurement of one of these observables, following a measurement of the other observable, should have a predictable outcome since the first measurement will lead to the reduction to a specific state which is common to both). Also, "simultaneously-measurable" means without an intrinsic uncertainty in the outcome.

[178] Classically, the Hamiltonian H of a mechanical system represents its total energy (i.e. kinetic plus potential) usually as a function of momentum and position (noting that "potential energy" in this book is generally restricted to be a sole function of position). It is noteworthy that Schrodinger's equation itself (in its above-given form) is a "postulate" (i.e. it is hypothesized rather than being derived mathematically or obtained logically from more fundamental principles) and hence it is part of the axiomatic framework of quantum mechanics.

[179] So in brief, if the wavefunction before the measurement is represented by the superposition $\psi = \sum_i c_i \psi_i$ then the probability of obtaining the eigenvalue o_i (as an outcome of the measurement) is $|c_i|^2$. Moreover, if the outcome of the measurement is o_i then the wavefunction after the measurement is ψ_i. It is noteworthy that c_i is given by:

$$c_i = \int_{\text{all space}} \psi_i^* \psi \, dV$$

where the integral is commonly called the overlap integral (connecting ψ and ψ_i).

4.3 Schrodinger's Equation

The wavefunctions of quantum systems are described and quantified by the Schrodinger equation which is a differential equation given (in its generic and general form) by:

$$\hat{H}\psi = i\hbar \frac{\partial \psi}{\partial t} \qquad (6)$$

where $\hat{H}$ ($=\hat{T}+\hat{V}$) is the Hamiltonian operator (or energy operator) of the quantum system, $\psi(x,y,z,t)$ is the time-dependent wavefunction of the system, i is the imaginary unit, $\hbar$ is the reduced Planck constant and t is time. In fact, this is the time-dependent form of the Schrodinger equation and hence we have another form (i.e. time-independent form) which is given by:[180]

$$\hat{H}\psi = E\psi \qquad (7)$$

where E ($=T+V$) is the total energy of the system (with T and V representing kinetic and potential energy).

More clarifications about the Schrodinger equation and its role in quantum mechanics are given in the following points:

1. The Schrodinger equation is the fundamental equation of quantum mechanics (in its wave mechanics formulation) and hence most of the work in this mechanics is focused on finding solutions to this equation. As indicated before, the objective of solving the Schrodinger equation is to obtain the wavefunction (which is the solution that contains all the required information about the quantum system).
2. Mathematically, the Schrodinger equation is a linear homogeneous partial differential equation of first order in time and of second order in space. Physically, it is a wave equation (noting that it is about *wave*function and it is formulated on the style of the classical wave equation). This is also reflected in its name "Schrodinger's wave equation" which is used occasionally.
3. Neither the existence nor the uniqueness[181] of the solution of Schrodinger's equation is guaranteed. Hence, physical considerations and restrictions (as well as approximations, modifications, extensions, etc.) may be needed to find a (unique) solution to the Schrodinger equation of a given quantum system.
4. Exact (analytical) solutions to the Schrodinger equation are rare and hence even if a "solution" is found it is usually an approximation obtained by numerical methods or semi-classical models or perturbation theory or ... etc.[182] In fact, among the *real* quantum systems (i.e. excluding the hypothetical and pedagogical systems like "particle in a square well") the one-electron atoms (i.e. hydrogen atom, singly-ionized helium, doubly-ionized lithium, etc.) seem to be the only systems that have exact solutions from the Schrodinger equation (at least within the domain of atomic and molecular physics).
5. It is important to note that obtaining an exact (and unique) solution to the Schrodinger equation (assuming we were lucky and smart enough to find such a solution) of *real* quantum systems is not the end of the story because (even the simplest of) real systems are too complicated to be fully described and quantified by the Schrodinger equation. For example, getting an exact solution for the hydrogen atom is just the beginning (which may be sufficient for describing the basic physics of this atom in itself) and hence we need many extensions and corrections (e.g. for fine structure, hyperfine structure, Zeeman effect, and Lamb shift) to account for various physical effects and influences

---

[180] The significance of the time-independent Schrodinger equation (which required a special attention) is that many (if not most) of the commonly investigated quantum systems are stable and hence they are in a stationary state (with no time dependency) and hence only the time-independent equation is needed in their investigation. It is worth noting that the time-independent Schrodinger equation is distinguished by the assumption that the quantum system has definite energy (and hence it is the energy eigenvalue equation).

[181] "Existence" and "uniqueness" here should be understood in their technical mathematical significance (and hence they are not about practical considerations). It is important to note that the type of the potential in the Schrodinger formulation (corresponding to the physical setting) is pivotal in determining the existence of a unique solution or not.

[182] It is useful to note that approximate solutions in this context can be classified into three main types: those based on exact model with approximate method of solution (e.g. by using numerical methods), those based on approximate model (e.g. by using a semi-classical model) with exact method of solution, and those based on approximate model with approximate method of solution (e.g. by using an approximate potential with numerical methods).

4.3 Schrodinger's Equation

that are not included or considered in the Schrodinger equation. So in brief, getting exact solution (to the Schrodinger equation) does not mean getting exact physics or reaching the final destination. In fact, this should shed some light on the rather primitive and basic nature of this equation and its many limitations and deficiencies. This should not only impose restrictions and limitations on the science behind this equation and its limited value but should impose restrictions and limitations on the epistemology and interpretation of quantum physics (noting that this equation represents the heart of the formalism of quantum theory and considering the intimate relation between formalism and interpretation and hence limitations in formalism should normally lead to limitations in interpretation; see for instance § 2.8). Also see point 8 as well as § 6.3.

6. The Hamiltonian operator $\hat{H}$ is the quantum mechanical operator that corresponds to the classical Hamiltonian H of a mechanical system which represents its total energy (i.e. kinetic energy T plus potential energy V) and is commonly given (for a particle of mass m, speed v and momentum of magnitude p) by:[183]

$$H = T + V = \frac{1}{2}mv^2 + V = \frac{p^2}{2m} + V \tag{8}$$

From this perspective, the Schrodinger equation may be seen as the quantum mechanical analogue of the classical energy conservation principle (and this may be proposed as a reason for its non-derivable nature; see point 11).

7. Since the Schrodinger equation is linear and homogeneous (in the wavefunction), any linear combination of solutions of the (time-dependent) Schrodinger equation is also a solution. However, this does not apply to the time-independent Schrodinger equation unless the solutions are degenerate, i.e. they have the same energy.

8. The above versions of the Schrodinger equation (see Eqs. 6 and 7) represent its "classic" and basic form (as opposite to its Lorentzian or "relativistic" form and possibly other forms). Accordingly, there are other non-classical forms of this equation in which Lorentzian effects (and possibly other effects) are included in the formulation. In fact, the Lorentzian form of the Schrodinger equation (or rather the Lorentzian equivalent of the Schrodinger equation) is called the Dirac equation which provides the basis for the quantum field theory (which is out of scope).

9. As indicated earlier, the Schrodinger equation is deterministic from the perspective of describing the evolution of the wavefunction between measurements (although it cannot describe measurement or predict its outcome noting that indeterminacy is largely related to measurement).[184]

10. If the spatial form of the wavefunction is known at a given instant in time then it is possible in principle to know its spatial form at any instant in the future by using the Schrodinger equation. This is because Schrodinger's equation is a differential equation of first order in time that can be integrated step-wise in time (possibly by using numerical methods) to obtain the future change during an infinitesimal time interval.[185] This means that if the spatial form of the wavefunction is known at a given time

---

[183] If we follow the prescription of points 8 and 9 of § 4.2 (also see § 4.4) then the Hamiltonian operator corresponding to the above classical Hamiltonian is:

$$\hat{H} = \frac{(-i\hbar\nabla)^2}{2m} + \hat{V} = -\frac{\hbar^2}{2m}\nabla^2 + \hat{V}$$

where ∇^2 is the Laplacian (scalar) operator noting that $p^2 = \mathbf{p}\cdot\mathbf{p}$ where $\mathbf{p}$ is the momentum and $-i\hbar\nabla$ is the momentum (vector) operator.

[184] As seen in § 4.2, the collapse of wavefunction (as a consequence of measurement) is not subject to the Schrodinger equation but to the measurement postulate.

[185] This may be expressed formally by the following equation:

$$\psi(\tau + \delta t) = \psi(\tau) + \left.\frac{\partial \psi}{\partial t}\right|_\tau \delta t$$

where τ is the time at which the spatial wavefunction in known and δt is an infinitesimal time interval while the other symbols are already known (noting that we suppressed the spatial dependency of the wavefunction and its derivative for clarity as well as being obvious). In fact, the above phrasing is somewhat lax because what we are actually using is the time derivative (noting that this type of temporal evolution, as given by the first order time derivative, is legitimized and formalized by the Schrodinger equation, and is actually obtained from it, and hence this formulation is ultimately based on the Schrodinger equation).

then the entire future evolution of the wavefunction can be determined (as long as it is uninterrupted by measurement).[186] In fact, this is a demonstration of the deterministic nature of Schrodinger's equation which is indicated in the previous point.

11. Because the Schrodinger equation is a postulate (rather than being derived from more fundamental principles and laws; see for instance § 4.2 and § 6.3), the validity of this equation (as well as the validity of the quantum theory which is based on it) is entirely based on the experimental evidence and the ability of this equation to provide a formal framework for the description and quantification of the quantum phenomena in a correct and predictable way. So far, this equation and the formalism that is based on it (taking account of its limitations or taking into account many extensions and improvements that are required to address its limitations) passed all the experimental tests and hence it is still regarded (after about a century from its first appearance) as the solid basis for the quantum theory.

12. The role of Schrodinger's equation in quantum mechanics may be likened to the role of Newton's second law in classical mechanics. But apart from this basic analogy, the two are totally different. For example, "force" is central to Newton's second law but it does not appear in Schrodinger's equation[187] (and in fact it does not appear in the formalism of quantum mechanics primarily although it may appear occasionally in classical analogies and semi-classical formulations or in certain extensions, applications and instantiations, e.g. originating from electromagnetism or related to Ehrenfest theorem and its applications).

13. The essence of the physics of quantum theory is represented by its postulates (see § 4.2 and § 6.2 as well as § 4.9) plus the Schrodinger equation (noting that this equation is referred to in these postulates and hence it is part of the axiomatic framework of quantum mechanics; see § 4.2 and § 4.9). So, all the rest (excluding interpretation and specific applications) is essentially mathematics, methods and techniques (i.e. details about how to apply and make use of this essence).[188] Accordingly, if the postulates are put aside (ignoring the reference to the Schrodinger equation) then all that remains in the formalism of the theory is Schrodinger's equation.

14. As indicated earlier, the time-independent Schrodinger equation is an eigenvalue equation (i.e. for energy).[189] This does not apply to the time-dependent Schrodinger equation.

15. As we will see (refer for instance to § 4.6), the dependencies in the Schrodinger equation (which are only tempo-spatial) are not sufficient to characterize and describe all types of quantum behavior and properties. Accordingly, additional dependencies (like spin; see § 4.6) should be added to complete the characterization and description. This should reveal one of the limitations of Schrodinger's equation and its need for supplements.

Also see § 6.3.

4.4 Operators in Quantum Theory

Here, we investigate the operators in quantum theory and their relationship to the observable physical quantities. It is one of the postulates of quantum mechanics (see § 4.2) that observables (such as position and energy) are represented by operators (like $\hat{r}$ which is the position operator and $\hat{H}$ which is the energy operator). Operators play typical and essential roles in the formalism of quantum theory. For example,

---

[186] It is worth noting that knowledge of the tempo-spatial dependency of the potential may be needed in this determination. We also note that we should also exclude possible influence by external factors during this evolution (although this may be included in "measurement"). Also, past (or back-in-time) evolution may be determined similarly.

[187] We should note, however, that (conservative) force is present in Schrodinger's equation implicitly through the potential energy but we should also take into account that the relation between (conservative) force and potential energy is essentially classical. Anyway, the difference in nature and essence between Schrodinger's equation and Newton's second law is obvious.

[188] In fact, we may need to make some exceptions such as the treatment of spin (although spin, in one of its formulations, belongs to a Lorentzian treatment which is beyond the elementary quantum mechanics that we are talking about). We should also note that we are essentially talking about the wave mechanics version of quantum physics (see § 1.2 and § 5.3.2). Anyway, these issues will be discussed in more detail in § 4.9.

[189] So, an eigenfunction of the Hamiltonian operator corresponding to an eigenvalue E represents a quantum state with sharply defined energy.

4.4 Operators in Quantum Theory

the eigenstates and eigenvalues are linked through the corresponding operators (e.g. $\hat{H}\psi = E\psi$). Also, the expectation value of an observable is obtained by the so-called sandwich integral which involves the operator of that observable (see § 4.5). The behavior of operators (i.e. whether they commute or not) also plays a role in determining the compatibility of observables and hence the uncertainty in their measurements (see for instance § 4.2). More clarifications about operators and their roles in quantum mechanics are given in the following points:

1. As indicated earlier, "operator" may be defined as a mathematical entity (or abstract object) that transforms a given function into another function. It may also be seen as a rule or recipe (represented by a mathematical symbol) that requires a specific action on its "argument" or "operand" (i.e. what the operator operates on). However, "operator" may also be used to refer to the mathematical symbol (rather than the entity or rule or recipe or action).

2. Every observable physical quantity (which may also be called dynamical variable) is represented by a specific quantum operator (where in the eigenvalue equation involving that operator the quantity plays the role of eigenvalue of that operator corresponding to its eigenstate). However, it should be noted that not all operators in quantum mechanics represent observables (i.e. some operators used in quantum mechanics do not represent observables).

3. It is common in quantum mechanics to distinguish an operator from the variable that it represents by putting a hat on the symbol of that variable, i.e. the symbol without hat represents the variable while the symbol with hat represents the operator. For example, $\mathbf{p}$ stands for momentum while $\hat{\mathbf{p}}$ stands for momentum operator. In this book we follow this convention.

4. The operators of quantum mechanics are characterized by being linear and Hermitian. These characteristics ensure favorable behavior which is useful (as well as necessary) in the formulations of quantum mechanics. The benefits of linearity are obvious, e.g. it facilitates mathematical manipulation and superposition (see for instance § 5.8). Regarding Hermiticity, it ensures that the eigenvalues of the operator and the expectation value of the observable are real (see for instance § 4.5). Hermiticity also ensures mutual-orthogonality of eigenvectors (for distinct eigenvalues). However, it should be noted that non-Hermitian operators (e.g. creation and annihilation operators) are also used in quantum mechanics. To be more accurate we should have said: the quantum operators *associated with observable quantities* are Hermitian. Anyway, we generally mean this type of operators (and hence we implicitly assume Hermiticity) when we talk about operators in quantum mechanics (or about "quantum operators").

5. The operators of quantum mechanics are obtained systematically from the classical expressions of their corresponding observables. For example, from the classical expression of energy (i.e. $H = T + V$) we get the Hamiltonian operator (i.e. $\hat{H} = \hat{T} + \hat{V}$) which is the energy operator of quantum mechanics. In fact, from a more technical perspective, obtaining quantum operators from classical expressions follows certain substitution rules outlined in § 4.2 (and in points 8 and 9 of that section in particular). However, some quantum observables (e.g. spin) have no classical counterpart and hence their operators are not obtained by the substitution rules. We should also note that the substitution rules are generally restricted by the type of employed coordinate system.

6. As indicated earlier, two commuting quantum operators have the same set of eigenfunctions[190] and simultaneously-measurable (or compatible) eigenvalues.

7. The observables of non-commuting quantum operators are subject to the uncertainty principle while the observables of commuting operators are not subject to the uncertainty principle (see § 4.7.1). An example of non-commuting operators is the position and momentum operators while an example of commuting operators is the energy and momentum operators (where the details about these examples should be sought in the literature).

8. As mathematical objects, quantum operators are of different types such as differential operators and matrix operators (although the terminology of labeling the latter as "operators" may not be universal). In fact, even some non-operator-like mathematical objects (like the position vector $\mathbf{r}$ which is a multiplier) can be used as quantum operators (and hence $\mathbf{r}$ is used as a position operator that represents

---

[190] The reverse is also true, i.e. operators that have the same set of eigenfunctions commute.

the position observable in addition to its original use as a position vector although the two are usually distinguished from each other by putting a hat on the operator, i.e. $\hat{\mathbf{r}}$).
9. The set of all the eigenvalues of a quantum operator $\hat{O}$ is commonly known as the spectrum of $\hat{O}$. It is important to note that the spectrum of quantum operators can be continuous or discrete or a mix of both (where energy is a typical example for these cases since quantum systems can have continuous or/and discrete eigenenergies).[191]
10. Regarding the issue of measurement (see for instance § 4.2 and § 5.7), the outcome of a measurement conducted on a quantum system is an eigenvalue of the operator representing the measured quantity (i.e. observable) and the state (or wavefunction) of the system following the measurement is an eigenfunction of that operator (corresponding to that eigenvalue). In fact, this requires the eigenfunctions of the operator to be a complete set of basis functions (so that every possible outcome can be predicted).

4.5 Quantum Observables

We may define quantum observables as the physical attributes that can be obtained and determined by measurement (noting that the expected outcome of an impending quantum measurement may have certain ambiguities which will be investigated next).[192] In the following points we investigate observables in quantum mechanics and outline the general rules and principles that govern them and the ambiguities that surround them:
1. As explained earlier, an observable (or measurable) quantity O is represented quantum mechanically by an operator $\hat{O}$. The most common quantum operators corresponding to the most common observables are given in the following table:

Observable	Mathematical symbol	Operator
Position	$\mathbf{r}$	$\hat{\mathbf{r}}$
Momentum	$\mathbf{p}$	$-i\hbar\nabla$
Kinetic energy	$\frac{\mathbf{p}\cdot\mathbf{p}}{2m}$	$-\frac{\hbar^2}{2m}\nabla^2$
Potential energy	V	$\hat{V}$
Total energy	$H\left(=\frac{\mathbf{p}\cdot\mathbf{p}}{2m}+V\right)$	$\hat{H}\left(=-\frac{\hbar^2}{2m}\nabla^2+\hat{V}\right)$

2. The wavefunction (which contains all the information about observables) is not observable. Yes, its magnitude squared is observable in the sense of being a probability density for position.
3. Prior to measurement, there are two main types of (potential) ambiguity about observables in that measurement: probabilistic ambiguity (which is about the nature of the upcoming outcome) and quantitative ambiguity (which is about the exact value of the outcome and is governed by the uncertainty principle).[193]
4. Regarding the probabilistic ambiguity (when the wavefunction prior to measurement is a superposition of eigenstates), the observable outcome of an impending measurement (on a quantum system) is an eigenvalue of the operator representing that observable and the quantum state (of the system) after measurement is the eigenstate (or eigenfunction) corresponding to the obtained eigenvalue. Yes, if the system before measurement is in an eigenstate of that observable then there is no probabilistic ambiguity (to be lifted by measurement) and hence the outcome is certain which corresponds to that eigenstate.

---

[191] For example, the energy spectrum of free particle is continuous, the energy spectrum of a particle in an infinite square well potential is discrete, and the energy spectrum of a particle in a finite square well potential (or the energy level spectrum of atoms) is a mix of both.
[192] We use the non-technical term "ambiguity" to avoid confusion with other terms (noting that technical terms like "uncertainty" or "indeterminacy" are reserved for specific technical meanings). The purpose of this terminology should become clearer as we continue this investigation.
[193] Essentially, the probabilistic ambiguity is about the anticipated state while the quantitative ambiguity is about the anticipated value.

5. For a quantum system in a state represented by the (normalized) wavefunction $\psi(\mathbf{r},t)$, the expectation value $\langle O \rangle$ of an observable O of the system is given by:

$$\langle O \rangle = \int_{\text{all space}} \psi^* \hat{O} \psi \, dV \qquad (9)$$

Similarly, the expectation value of O^2 is given by:

$$\langle O^2 \rangle = \int_{\text{all space}} \psi^* \hat{O}^2 \psi \, dV \qquad (10)$$

This pattern (i.e. of sandwiching the operator between ψ^* and ψ in the integral to obtain the expectation value of the corresponding observable) generally holds, and for this reason this integral may be described as sandwich integral.[194]

6. Regarding the quantitative ambiguity, the uncertainty ΔO in the value of the outcome of a measurement of O is given by:

$$\Delta O = \sqrt{\langle O^2 \rangle - \langle O \rangle^2} \qquad (11)$$

where $\langle O \rangle$ and $\langle O^2 \rangle$ are given by Eqs. 9 and 10. However, if the quantum state prior to measurement is an eigenstate of $\hat{O}$ (i.e. $\hat{O}\psi_i = o_i \psi_i$ where o_i is an eigenvalue of $\hat{O}$ corresponding to the eigenstate ψ_i) then the value of the outcome is certain and it is equal to the eigenvalue o_i.[195]

7. If $\hat{A}$ and $\hat{B}$ are two operators then their commutator $\left[\hat{A},\hat{B}\right]$ is the difference between these operators in one order and in the opposite order, that is:

$$\left[\hat{A},\hat{B}\right] = \hat{A}\hat{B} - \hat{B}\hat{A} \qquad (12)$$

Two operators commute *iff* their commutator is zero.

8. The commutator of compatible observables is zero while the commutator of incompatible observables is not zero (in general). So, if A and B are two observables then $\left[\hat{A},\hat{B}\right] = 0$ if A and B are compatible and $\left[\hat{A},\hat{B}\right] \neq 0$ if A and B are incompatible. An example of compatible observables is the magnitude of angular momentum and its z component, while an example of incompatible observables is the position and momentum (of corresponding coordinates such as x and p_x). Also see § 4.7.1.[196]

4.6 Angular Momentum and Spin

Angular momentum and spin (which is seen as intrinsic or internal angular momentum) in quantum mechanics have some unusual (or non-classic) properties despite their similarity to angular momentum in orbital and rotational motions.[197] In this section, we summarize some of the quantum mechanical

---

[194] It should be noted that the expectation value represents the average value obtained by repeating the measurement on an ensemble of identically-prepared systems (and hence it is not the average value obtained by repeating the measurement on the same system because by conducting a measurement on a system the wavefunction collapses to one of its eigenstates and therefore the next measurement will not be on a superposition state). Some readers may feel that the term "expectation value" is rather misleading (and this feeling is justifiable). However, this term is part of the technical terminology of quantum mechanics and hence it should be respected (and treated as such). It is noteworthy that the expectation values of quantum measurements may be seen as representing classical observables (and hence the above formulation about expectation value provides a link between classical and quantum observables).

[195] This means that in this case the expected outcome of measurement has neither probabilistic ambiguity nor quantitative ambiguity (also see point 4).

[196] In fact, there are many details about the issues discussed in this section. Also, some of the given statements require more clarifications and restrictions. However, we present here (like elsewhere) what we need in this book for our main objectives and what falls within our scope.

[197] It is common in the literature to use "orbital angular momentum" to refer to angular momentum and distinguish it from spin (which is supposedly a type of angular momentum). However, we think "orbital" is not needed (because quantum spin is different from angular momentum despite their similarity) and it may even be misleading (because it may imply a sort of specific orbit which is rather restrictive and could suggest unduly classic-like orbital motion) and hence we generally avoid it in this book. Yes, in some contexts and situations (which are out of our scope) it may be useful or required.

4.6 Angular Momentum and Spin

results and principles related to angular momentum and spin of quantum objects (like atoms and electrons) which are generally needed in our investigations and discussions in this book (at least as useful background knowledge).

1. Angular momentum is quantized and it is measured in units of $\hbar$. Moreover, only one component of angular momentum (in addition to its magnitude) can be determined at a time because the components of angular momentum are incompatible observables (and hence no two components, such as the x and y components, can be determined simultaneously).[198]

2. The previous point also applies to spin in general, i.e. spin is quantized and it is measured in units of $\hbar$, moreover a quantum particle can have a determinate (or well-defined) spin along one orientation only. Also see point 4 of § 4.6.1.

3. Quantum numbers that quantify angular momentum and spin are integers and half integers (including zero). Also see points 8 and 9.

4. Angular momentum and spin may be combined (e.g. in atoms) separately or/and together according to certain rules.[199] Moreover, they are generally governed by similar rules and principles (some of which are indicated above). In fact, this should support the similarity of spin to angular momentum and its treatment as a type of angular momentum (i.e. intrinsic angular momentum).[200]

5. The angular momentum of a quantum particle may and may not be zero. Regarding the electrons in atoms, their wavefunction is spherically symmetric in the former case, and it has angular dependency in the latter case.

6. Spin is an intrinsic property and hence it cannot be considered (or written) as a function of space (or rather space-time).[201] Accordingly, for a complete description of the behavior and properties of quantum objects a non-spatial (or rather non-tempo-spatial) dependency is required (and this corresponds to spin). This means that the Schrodinger equation (which has only tempo-spatial dependency) is not sufficient for full description (since it has no spin dependency).

7. Spin is a quantum mechanical property that has no correspondence in classical physics (although its label suggests a resemblance to the classical spin like the rotation of a top around its axis).[202] The non-classical nature of quantum spin is exemplified by the following properties:
   - The magnitude of quantum spin is an intrinsic and non-changeable property of the particular type of quantum particle[203] (and accordingly quantum particles are classified according to their spin; see point 8).
   - Unlike classical spin, quantum spin is quantized.
   - Also see § 4.6.1 which includes some characteristic non-classical behavior of quantum spin (e.g. only one component of quantum spin can be determined at a time).

8. As indicated already, quantum spin is such an important intrinsic property that quantum particles are classified and distinguished (in their properties and behavior) by their spin. Accordingly, quantum particles are classified broadly as fermions and bosons where:
   - Fermions are particles of half-integer spin (such as spin-1/2 electrons) and consequently they are subject to the exclusion principle.[204]

---

[198] We should exempt the case when the angular momentum is zero.

[199] "Separately" means combining angular momentum with angular momentum or spin with spin (as in LS coupling), and "together" means combining angular momentum with spin (as in jj coupling). As indicated, they can also be combined separately and together (as in LS coupling).

[200] The association of spin with magnetic dipole moment may be seen as another support to this similarity (although the independence of charge in some cases may indicate otherwise or require other considerations).

[201] What we mean is that the space and time are not the actual factors that determine the spin and hence the dependency of spin on them is not the intrinsic or essential dependency that spin actually depends on (although spin can obviously be a function of space and time but in a rather casual way). This is unlike (for instance) the type of the dependency of potential energy on space and time.

[202] In fact, even (quantum) angular momentum has distinctive quantum properties and hence it is non-classical. However, we have no much interest in angular momentum in this book (unlike spin which is commonly referred to and used and hence it requires further investigation).

[203] The main focus here is on the simple subatomic or "fundamental" particles (like electrons and protons) rather than composite and "non-fundamental" quantum particles (like atoms and molecules and even composite nuclei).

[204] Note that "half-integer spin" (and its alike) is more general than "spin-1/2" since "half-integer" should include "1/2, 3/2,

- Bosons are particles of integer spin (such as spin-1 photons) and consequently they are not subject to the exclusion principle.

9. Along any spatial orientation, the spin of spin-1/2 fermions (like electrons) is quantized and it accepts only two (half-integer) values: $+\hbar/2$ and $-\hbar/2$. The conventional orientation to which the spin is referred by default is the (vertical) z axis and hence these values are commonly labeled as spin-up and spin-down (respectively). This labeling is usually used even when the orientation is not along the z axis (and we follow this misuse of words in this book).

10. Although spin is commonly seen as a Lorentzian effect (due to its link to the Dirac equation[205] rather than the Schrodinger equation), it may also be obtained from a non-Lorentzian treatment (and hence it is not necessarily Lorentzian).

4.6.1 SG Type Experiments

The experiment of Stern-Gerlach (which SG in the title stands for) is a well known historical experiment about quantum spin. The special importance of this experiment (and indeed the SG type experiments in general) which dictated the dedication of this subsection is that it demonstrates some characteristic quantum behaviors, and hence it plays a role even in the formation of the axiomatic framework of quantum theory (and the issue of measurement in particular). Moreover, it plays an important role in the interpretation of quantum mechanics. In fact, many of the experiments related to the interpretation of quantum mechanics are of SG type (including some which do not use spin in their underlying principles). The special importance of the SG type experiments in quantum theory is similar to the special importance of the double-slit type experiments (see § 1.6), and hence both experiments are referred to very frequently in the literature of quantum mechanics (and hence both have dedicated parts in this book).

In fact, we have no intention in this subsection to go through this experiment in its historical context. Instead, in the following points we will go through a discussion that outlines the main characteristics (related mainly to the setting, procedure and results) of this type of experiments. This should reveal and demonstrate key aspects of the behavior and attributes of quantum spin (as well as quantum aspects not related to spin specifically). But before that we draw the attention that in the following we use the symbols α and β to represent spin-up and spin-down states respectively (see point 9 of § 4.6). We also subscript these symbols with coordinates (i.e. of a rectangular Cartesian system) to indicate the orientation of the axis to which the spin is referred (e.g. α_x means spin-up state along the orientation of x axis).

1. It is an established experimental fact that if a beam of spin-1/2 particles (like electrons) moving in the y direction is passed through (a Stern-Gerlach apparatus generating) a non-uniform magnetic field with a gradient in the x orientation then the beam will split into two sub-beams: one diverted to the $+x$ direction and the other diverted to the $-x$ direction.[206]

    To explain this behavior quantum mechanically we say (see for example § 4.2): the (spin) wavefunction[207] of the particles in the original beam is a superposition of α_x state and β_x state and the measurement (by passing the particles through the apparatus) resulted in the collapse of this wavefunction to α_x eigenstate (for the $+x$ sub-beam particles) and β_x eigenstate (for the $-x$ sub-beam particles).

5/2, $\cdots$" while "spin-1/2" is restricted to 1/2. The essence of the exclusion principle is that two fermions of the same type cannot occupy the same quantum state.

[205] As indicated earlier (see point 8 of § 4.3), the Dirac equation is the Lorentzian equivalent (or version) of the Schrodinger equation.

[206] For the sake of simplicity and clarity, this description is rather loose and non-technical (noting as well that most technical details are irrelevant to our objectives). For example, we should impose the conditions that these particles are not interacting and they are not prepared in a particular spin eigenstate (i.e. α_x state or β_x state in our case). We should also note that "gradient in the x orientation" means that the inhomogeneous magnetic field of the Stern-Gerlach apparatus is in such a configuration that it imposes a force on the particles along the x orientation forcing them to deflect ("up" or "down") in that orientation (due to their magnetic dipole moment). We should also remark that this type of experiments may not be achievable directly on electrons (which we mentioned above) due to the existence of Lorentz force (and hence certain neutral atoms or monochromatic neutrons may be more appropriate to use in such experiments than electrons). Anyway, we have no interest in these issues and hence the readers interested in such details should refer to the literature.

[207] The use of "state function" here may be more appropriate (although we continue this use because it is more common).

2. Now, if one of the sub-beams obtained in point 1 (say the one diverted to the $+x$) is passed into an identical apparatus generating an identical magnetic field (i.e. of gradient in the x orientation) then all the particles in the sub-beam will divert to the $+x$ direction.
   To explain this we say: all the particles in the $+x$ sub-beam (which we obtained in point 1) are in an α_x eigenstate and hence the measurement should have a definite outcome (which is α_x).
3. However, if (instead) the $+x$ sub-beam obtained in point 1 is passed into an identical apparatus but this time the gradient of the magnetic field is along the z orientation then the $+x$ sub-beam will split into two sub-beams: one diverted to the $+z$ direction and the other diverted to the $-z$ direction.
   To explain this we say: the (spin) wavefunction of the particles in the $+x$ sub-beam is a superposition of α_z state and β_z state and the measurement resulted in the collapse of this wavefunction to α_z eigenstate (for the $+z$ sub-beam particles) and β_z eigenstate (for the $-z$ sub-beam particles).
4. Lastly, if the $+z$ sub-beam obtained in point 3 is passed into an identical apparatus but this time the gradient of the magnetic field is along the x orientation then the $+z$ sub-beam will split into two sub-beams: one diverted to the $+x$ direction and the other diverted to the $-x$ direction. So, although the $+z$ sub-beam (of point 3) is obtained from the $+x$ sub-beam (of point 1) whose all particles are in an α_x eigenstate, the particles in the $+z$ sub-beam are not in an α_x eigenstate any more (and that is why in the present point the $+z$ sub-beam splits into two sub-beams of α_x particles and β_x particles). This means that the z measurement (by passing the $+x$ sub-beam of point 1 through a z oriented magnetic gradient in point 3) has destroyed the α_x state of the particles of the $+x$ sub-beam of point 1 (i.e. the input to the z oriented apparatus in point 3). In other words, a particle cannot be in an α_x eigenstate and in an α_z eigenstate simultaneously, so it is either in an α_x eigenstate or in an α_z eigenstate. Therefore, we can determine only one spin component and hence any subsequent determination of another spin component will destroy our knowledge of the previous component. Also see point 2 of § 4.6.

4.7 The Principles of Quantum Theory

There are many principles (including some alleged principles) in the quantum theory. Examples of these principles are investigated in the following subsections. However, before that we discuss in the following points some general issues about these principles:
1. The examples of the principles that we selected in this investigation are closely related to the epistemology and interpretation of quantum theory (which represent the scope of this book). Moreover, they are widely used and quoted in the literature of quantum theory and hence they represent common and characteristic aspects of the theory. These reasons should explain why they are selected in this investigation.
2. There are two main types of quantum principles: one that we call "fundamental" and one that we call "marginal". The difference between the two types is that the former is part of the formalism and hence it is indispensable, while the latter is not. In fact, the marginal principles just reflect an aspect or feature of the quantum theory and its formalism without adding a new substance to the formalism. In other words, no new formal quantum mechanical rule is imposed by these principles. As we will see, the marginal principles are primarily related to other aspects of the theory (such as its interpretation or validation) although they may be used (and sometimes misused) in the formalism.
3. We discuss the issue of the principles of quantum theory within the present chapter (which is dedicated to the formalism) despite the fact that these principles are not necessarily part of the formalism because even the marginal principles are commonly and frequently used in the discussion and investigation of the formalism. So, they are generally considered or treated in the literature as part of the formalism. Moreover, from a structural perspective it is more appropriate to gather them in one part of this book (and the present part is the more appropriate for this taking into account the previous reason). Also see point 5.
4. There are some principles in the quantum theory which are very specific technically and rather restricted in validity. Although these principles are part of the formalism in its extended sense (and they are fundamental), they have not been investigated in detail in this book because they have no

particular significance (in themselves) to the epistemology of quantum physics according to our plan and considerations (and hence they are not within the scope of this book). An example of this type of principles is the exclusion principle which is very technical with no much significance (in itself) for the epistemology and interpretation of the quantum theory (within the above-outlined restrictions). Moreover, it is rather restricted in application (since it applies only to a specific type of quantum objects, i.e. fermions) and hence it does not represent or reflect a general feature of the theory.

5. Regarding the conservation principles (such as the principles of energy and momentum conservation), they are generally part of the formalism of the quantum theory, and accordingly they should be generally classified as fundamental (noting that some of them may be derived or instantiated from the main formalism of the theory). However, we do not investigate these principles here (see point 3). One reason is that some of them are inherited from classical physics and hence they are not quantum mechanical specifically (and thus they are not of particular significance to the epistemology of quantum physics). Moreover, some of them are technically too specific and of restricted validity (as well as having little significance for the epistemology and interpretation) and hence some of the reasons in the previous points (i.e. for not investigating here) also apply to them. Anyway, these principles are investigated broadly and as a whole (rather than individually and in detail) later on within our technical assessment of the quantum theory (see § 6.12) where we think this investigation belongs more appropriately.

4.7.1 The Uncertainty Principle

The essence of this principle is that the precisions of the measurements of certain couples (or pairs) of observable physical quantities (e.g. position and momentum) are correlated. This means that the degree (or level) of precision of the measurement of one observable in the couple is restricted by the degree of precision of the measurement of the other observable in the couple (within certain limits). In other words, we are not free to determine the degree of precision of each one of these observables independent of the degree of precision of the other observable since a minimum amount of uncertainty should remain in the combined measurements of these observables (where the restriction on the combined accuracy is represented by the product of the individual uncertainties of the two observable quantities and where the minimum uncertainty is quantitatively given by $\hbar/2$). So, if the uncertainty in the measurement of one of the observables (say position) is reduced then the uncertainty in the measurement of the other observable (i.e. momentum) will automatically increase to keep the minimum uncertainty in the combined measurements.[208] As indicated earlier (see for instance § 1.2), the observable pairs which are subject to this principle are described as incompatible observables (or quantities).[209]

There are many forms (or instances) of the uncertainty principle where each form involves a certain pair of incompatible physical observables. The most prominent pairs of incompatible observables are position-momentum and time-energy and hence the uncertainty principle for these pairs is given by the following forms:

$$\Delta x \, \Delta p_x \geq \frac{\hbar}{2} \qquad \text{(position-momentum form)} \qquad (13)$$

$$\Delta t \, \Delta E \geq \frac{\hbar}{2} \qquad \text{(time-energy form)} \qquad (14)$$

where x, p_x, t, E represent position, momentum (in the x direction), time, energy and Δ represents the uncertainty in their measurements while $\hbar$ is the reduced Planck constant (which is equal to $h/2\pi$ with h being the Planck constant).

The first equation (i.e. Eq. 13) means it is impossible to simultaneously measure the position x and momentum p_x of a quantum system with arbitrary accuracy, while the second equation (i.e. Eq. 14)

---

[208] Noting that the uncertainty principle is represented by an inequality (rather than equality) the above-indicated "reciprocity" should not be understood in its strict mathematical sense but as an indication to the existence of this lower limit and the restriction imposed by it.

[209] Other labels (such as "complementary") my be used in place of "incompatible". See footnote [16] on page 11.

4.7.1 The Uncertainty Principle

means it is impossible to simultaneously determine the energy of a quantum system and how long it is being in that energy with arbitrary accuracy. This is unlike classical physics where in principle we can have $\Delta x \, \Delta p_x < \hbar/2$ and $\Delta t \, \Delta E < \hbar/2$ (including infinite accuracy where both uncertainties in the couple are zero).

In the following points we outline some aspects of the uncertainty principle and its significance (noting that further investigation and assessment of this principle will be given in § 6.6):

1. As indicated earlier (see for instance § 4.4), in the case of compatible observables we have a common complete set of eigenfunctions and hence simultaneously-measurable eigenvalues (which means that the observables can be measured simultaneously with arbitrary accuracy) while in the case of incompatible observables we do not have such a set (and hence the uncertainty principle applies).
2. Relations similar to Eq. 13 hold for components other than x (i.e. $\Delta y \, \Delta p_y \geq \hbar/2$ and $\Delta z \, \Delta p_z \geq \hbar/2$).
3. Uncertainty relations do not hold between non-corresponding components of position and momentum (e.g. between x and p_y) and hence such observables are compatible.
4. The position-momentum couple and the time-energy couple in the uncertainty principle should reflect the association of space with momentum and time with energy (which is thoroughly discussed in the literature of physics; see for instance our book "The Mechanics of Lorentz Transformations"). Moreover, the prominence of these pairs should reflect the exceptional importance of the quantities in these pairs and their strong bond. In fact, we may say (to summarize the essence of this point): momentum is energy in space and energy is momentum in time.
5. As well as the specific (or special) forms that we referred to above, the uncertainty principle is formulated in a general form (which may be labeled as "generalized uncertainty principle") that supposedly includes all of its specific forms and instances. This general form is also justified and (supposedly) proved independently of any particular form or argument (and this should supposedly establish this principle more firmly). Also see point 12.
6. The non-existence of an uncertainty principle in classical physics (also see point 17) may be seen as an approximation, i.e. because Planck's constant is virtually zero at classical level (due to its extreme tininess) and hence we can have $\Delta x \, \Delta p_x = 0$ and $\Delta t \, \Delta E = 0$ (or rather $\Delta x \, \Delta p_x \simeq 0$ and $\Delta t \, \Delta E \simeq 0$). However, the non-existence may also be seen as a consequence of the fundamental difference between the two physics (see for instance § 3.7) and hence Planck's constant (and in fact the concepts associated with and behind Planck's constant such as particle-wave duality; see point 19) does not exist in classical physics. Also see point 7.
7. The uncertainty principle of quantum theory means that the limits on accuracy and resolution in quantum physics have a more fundamental reason than that in classical physics because the uncertainty in the quantum measurement is an essential principle of the observation process (potentially originating from the quantum phenomenon itself) and is not a symptom of not having perfect tools and equipment or strict procedures and protocols (as it is usually the case in the classical measurement and even in the quantum measurement but from a different perspective).
8. Historically, there were many controversies and differences in opinion about most aspects of the uncertainty principle (e.g. its validity, its justification and its nature) although this seems to be settled in the more recent times as if these controversies and opinions (or at least some of them) disappeared. We should also note that although this principle is commonly attributed to Heisenberg and is associated with his name, it seems that its roots go back in time before him although he may be the originator of this principle in the modern quantum theory.
9. As indicated, the uncertainty principle applies to certain physical properties (or rather pairs of properties) but not to others. So, incompatible observables (like the corresponding components of position and momentum) are subject to the uncertainty principle while compatible observables (like the z component of angular momentum and its magnitude) are not subject to the uncertainty principle. In fact, some physical properties (such as charge) are fixed at all times (regardless of any association with other physical properties) and hence they are not involved in any uncertainty relation.[210] So in brief, the uncertainty principle is restricted to certain pairs of observables where certain observables are not

---

[210] For example, the charge of electron and proton is e (in magnitude) regardless of anything else.

4.7.1 The Uncertainty Principle

involved in any of these pairs.

10. In general, the uncertainty principle is restricted to continuous variables (like position and momentum) and hence discrete variables (like baryon number or lepton number) are not subject to this type of uncertainty (i.e. the uncertainty of the uncertainty principle) although they (like any other physical quantity) can be subject to other types of uncertainty.[211]

11. As indicated above, the uncertainty of this principle is an intrinsic aspect of the quantum measurement unlike the casual or accidental uncertainty in classical physics which, in principle, can be reduced arbitrarily and even eliminated entirely. However, on inspecting and analyzing the claimed justifications and derivations of the uncertainty principle in the literature of quantum mechanics we can find a general type of arguments based on the claim that any measurement at the quantum scale requires interaction with the quantum object and hence (the complementary attribute of) the object will be affected by the measurement process (or by the measurement probe). For example (according to this claim), if we want to obtain a high-precision measurement of the momentum of an electron then we need to hit it by an energetic photon and this (i.e. the impact of the photon) will disturb the position of the electron and reduce the certainty in the measurement of its position. So, in effect the uncertainty principle (according to these arguments) is due to a practical factor at the quantum scale although it is formulated and presented as a theoretical factor. Thus, in principle if we assume the existence of a measurement process in which the measurement probe does not interact with the observed object and hence it does not disturb the "actual" attributes of the measured object then we can discard the uncertainty principle and therefore we can have in principle an infinitely-precise measurement.[212] Accordingly, if we follow this type of arguments in justifying the uncertainty principle then the uncertainty of this principle is no more than another form of casual or accidental uncertainty (and hence in essence there is no real difference between classical and quantum physics about uncertainty and its causes and nature). Also see § 6.6.1.

12. The uncertainty principle can be derived quantum mechanically (i.e. technically and formally) by using, for instance, quantum operators.[213] However, the uncertainty principle may also be justified (or "derived") by simple and "intuitive" (but non-technical) physical arguments. An example of these simple arguments (which are very common in the literature) is the following (which is about the position-momentum form of the uncertainty principle): because a quantum object (like electron) is a (matter) wave there should be an uncertainty about its position that is comparable to its wavelength (i.e. $\Delta x \sim \lambda$), and hence (as a particle) there should be (according to the de Broglie relation $\lambda = h/p$; see Eq. 16) an uncertainty about its momentum that is comparable to the reciprocal of its wavelength (i.e. $\Delta p \sim h/\lambda$) and hence $\Delta x \Delta p \sim h$ (where $\sim$ means comparable in size and where the subscript of p is ignored for clarity). Similar arguments may also be produced for this form as well as for other forms and instances of the uncertainty principle (involving other physical complementary couples). In fact, this type of arguments (despite their laxity and hence they should not be taken seriously but as pedagogical and illuminative) indicates (correctly) that the origin of the uncertainty principle is the dual nature (see point 19) of quantum objects which leads to uncertainties in the simultaneous measurements of complementary quantum observables that are based directly and simultaneously on this duality (see for instance § 1.6.1, § 5.3.1 and § 6.13.1). Also see point 2 of § 6.6.1 as well as point 13 in the present subsection.

---

[211] "Continuous" and "discrete" here are about the nature of the variable and hence discreteness due to quantization for instance may not be included. In fact, there are some details about this point and hence it should be understood broadly (as indicated).

[212] In fact, all we need to violate and discard the uncertainty principle is to allow the effect of disturbance to be arbitrarily small (even if the disturbance is not eliminated entirely).

[213] The uncertainty principle may also be justified mathematically (and rather more rigorously and generally) by appealing to Fourier analysis on wave construction. In fact, the literature in this regard is full of arguments and "proofs" (of different methods and flavors) in support of the uncertainty principle although most of these arguments and proofs are questionable (from various perspectives) and lack rigor (and sometimes lack even evident rationality). Anyway, this principle (in its physical context and significance) is essentially justified by its empirical success (or at least alleged success) within the limitations of indeterminism regardless of any theoretical reasoning or analytical proof (and this should apply generally to the entire quantum mechanics and even to science in general since "proofs" in science are largely about rationalization and conceptualization noting that the real proof in science is experiment and observation).

4.7.1 The Uncertainty Principle

13. The uncertainty principle is commonly rationalized by the paradigm of "wavepacket" where the uncertainty is represented by the spatial spread (or extent) over which the amplitude of the packet is significant. However, this model may represent the uncertainty in position (and its link to uncertainty in momentum) but it does not seem to represent (at least directly) other uncertainties such as the uncertainty in time or energy. So, this model (apart from being approximate and essentially qualitative) is limited in validity. This may be seen as another example for the limited (or questionable) validity of concepts like "wave" or "particle" or "wavepacket" to provide an exact and reliable models for the description of quantum objects and phenomena. Also see point 12.

14. Technically speaking, the uncertainty Δ in a random variable O is defined as the standard deviation of O and is given formally by:
$$\Delta O = \sqrt{\langle O^2 \rangle - \langle O \rangle^2} \qquad (15)$$
where the triangular brackets symbol $\langle \ \rangle$ means average (or mean) value. In words, the uncertainty in O is the square root of the difference between the average of the square of O and the square of the average of O.

15. The uncertainty principle is used to explain and justify many quantum phenomena. For example, it is used to explain the stability of atoms since the fall of electrons toward the nucleus (as supposed to occur considering the electrostatic force that arises from the Coulomb potential) will violate this principle since at the nucleus the electron will have precisely-known position and momentum (noting that this reasoning may be challenged because it assumes that the position and momentum of the nucleus are known precisely).[214] Another example is the use of this principle to determine the lifetime of the excited states of atoms and molecules and the natural line width of their spectral emission. Also, the zero-point energy in certain quantum systems (such as infinite square well and harmonic oscillator) is regarded as a consequence of the uncertainty principle (and hence the zero-point energy does not exist classically because the uncertainty principle does not exist classically).[215] In fact, the uncertainty principle is used even to explain macroscopic phenomena such as the conductivity in metals or the size of neutron stars (also see points 16 and 17).

However, in our view many (if not most) of these explanations and justifications are questionable and contestable and hence they should be treated with caution and critique (or at least they should be seen as rough and lax and could be accepted pedagogically and educationally but not technically and rigorously). In fact, this applies to most explanations and justifications in quantum theory. The real "explanations and justifications" are experiment and observation (although the "theoretical umbrella" should be useful in general despite its shakiness as long as we keep awareness of its actual and potential limitations to avoid being misled and misguided).

16. It is important to note that although the uncertainty principle does not apply (or its effect is negligible) in large-scale systems (which we commonly label as "classical" because they are subject to classical physics) it could still have (according to the mainstream physics) a significant and macroscopically-observable impact on such systems (such as certain types of stars) where the microscopic influence (of the uncertainty principle) underlies the macroscopic (or "classical") phenomena. In fact, the uncertainty principle should apply (as an underlying agent or factor) to many (if not most or all) of classical physical aspects as it is at the heart of atomic and sub-atomic systems which are the basic components and building blocks of all classical systems. Actually, this is a general feature of quantum phenomena

---

[214] In fact, the uncertainty principle is used to explain the stability of atomic ground states in particular (and hence further explanation and justification should be required).

[215] "Zero-point energy" means: the lowest acceptable energy solution of a given quantum system above the bottom (i.e. the zero-level) of the potential. The use of the uncertainty principle for explaining the zero-point energy may be justified by the claim that if the system is at the bottom of the potential then its momentum and position will be both zero in violation of the uncertainty principle. In fact, this justification can be questioned and challenged for a number of reasons. For example, the zero-point energy is quantized like the zero energy and hence if the former is allowed to have an uncertainty then the latter should also be allowed (noting that quantization and uncertainty are consistent and noting as well that the quantization itself is not justified by this rationale). In fact, it is not obvious that being at the bottom of the potential necessarily leads to arbitrary precision (e.g. zero momentum and position as claimed). Anyway, we will not go through these lengthy (and potentially useless) details although we should be aware of (and open to) such potential challenges as part of our general assessment of quantum physics and its epistemology.

where their direct effect may be negligible on a large macroscopic scale but their indirect effect can be significant where they underlie certain macroscopic phenomena.
17. The uncertainty principle is used by some to explain even some classically-explainable phenomena such as the diffraction of light (which has classical explanation). This is unlike the examples in point 15 (as well as point 16) related to macroscopic phenomena (e.g. conductivity in metals) where the uncertainty principle underlies the phenomena (and hence the phenomena are explainable only quantum mechanically although the phenomena themselves are classical, i.e. macroscopic). In fact, the diffraction example and its alike may indicate (as pointed out earlier) that the uncertainty principle has also some presence in classical physics (e.g. through its mathematical justifications like Fourier analysis of waves) and hence it is not exclusively quantum mechanical (although it is more prominent and common in quantum mechanics, moreover it applies to "particle-like" or "localized" objects).
18. Referring to point 13, the actual uncertainty limit depends on the form of the wavepacket. In fact, the $\hbar/2$ limit (which is usually seen in the uncertainty relations like those of Eqs. 13 and 14) is the lowest and it corresponds to a wavepacket of Gaussian profile.
19. A simple analysis of the uncertainty principle (especially within the contexts of its "proofs" and applications) should reveal that this principle is ultimately based (at least in its most prominent instances) on the particle-wave duality (see for instance § 5.3.1). So, the ultimate (gross) justification of this principle is this duality (noting the potentially limited applicability of this rationalization to certain observables).[216]
20. As explained above, the uncertainty principle is about the combined (or concurrent) measurement of complementary observables. This means that each one of these observables can (in principle) be measured independently with arbitrary accuracy.
21. It may seem obvious that the uncertainty principle is a fundamental principle, i.e. not marginal (see point 2 of § 4.7). However, the status of the uncertainty principle as a fundamental principle (and hence part of the formalism) may be disputed and may be claimed to depend on the school of interpretation of quantum mechanics (noting that it is usually and primarily associated with the Copenhagen school and its variants; see § 8.1). Anyway, considering the use and applications of the uncertainty principle in the current literature of quantum mechanics we think it is more appropriate to be classified as fundamental (rather than marginal) since many (school-independent) quantum mechanical conceptualizations and formulations depend essentially (whether directly or indirectly and whether questionably or not) on it and its implications and consequences.
22. As we will see (refer to § 4.7.3 and § 4.7.4), the various forms of the uncertainty principle may be seen as instances of the principle of complementarity (in its extended sense).[217] However, this should not link the status of the uncertainty principle (i.e. as fundamental or marginal) to the status of the complementarity principle because this instantiation is a matter of conceptualization (as well as being opinion- and school-dependent) rather than a matter of formalization and technicalization (or being based on a common consensus).

Also see § 6.6.

4.7.2 The Correspondence Principle

This principle means (according to its common version) that the predictions of quantum theory should agree with the predictions of classical theory in the limit of large quantum numbers where the classical predictions are verified.[218] The validity of the correspondence principle seems obvious because no correct

---

[216] As we will see (refer to § 6.6.1), there are two main methods (and hence two types of justification) for establishing the uncertainty principle: one method is based on the disturbance caused by the measurement of one property (e.g. position) on the other property (i.e. momentum), and the other method is based on the intrinsic indeterminism (due ultimately to the particle-wave duality). In fact, the above justification should be the case if the uncertainty principle is based on and justified by the particle-wave duality (not by the disturbance caused by measurement; see point 11 above and § 6.6.1). This should be inline with the fact that the disturbance argument is less representative of the essence and spirit of the uncertainty principle (as well as being less acceptable among physicists).

[217] In fact, this instantiation may also be linked inherently and specifically to the Copenhagen school.

[218] As we will see, the correspondence principle extends (according to the literature) beyond this.

4.7.2 The Correspondence Principle

theory should contradict the experimentally verified predictions and results of classical theory. However, there are some ambiguities about this principle and its role and applications as well as its significance and implications especially with regard to the relation between classical and quantum physics and if the latter is a more general theory than the former. Although the principle is generally accepted some of its details are (or at least can be) subject to debate and deliberation as well as potential controversy.

More clarifications about the correspondence principle and its role and applications (as well as significance and implications) are given in the following points:

1. Despite the claim of general validity of quantum physics, its actual domain of validity (according to our view) is the microscopic (or quantum) systems, while the domain of validity of classical physics is basically macroscopic systems (see § 3.7). This should raise a question about where the two systems overlap (so that the correspondence principle is applicable). In fact, some classical (or classically-based) results and formulations for some physical systems can be obtained quantum mechanically at appropriate limits (e.g. by convergence of quantum mechanical formulations at large quantum numbers) or proper conceptualization (e.g. by hypothesizing or interpreting certain parameters in a classical way as in the Ehrenfest theorem)[219] and these systems are where the correspondence principle is applicable. So, these systems are where the domain of validity of quantum physics supposedly extends to the domain of validity of classical physics (also see point 3 of § 2.6.2 and refer to § 3.7).[220]

2. In the spirit of the previous point, it is claimed that some branches and theories of classical physics can be obtained naturally from the quantum theory (e.g. it is shown in the literature that thermodynamics can be derived from quantum mechanics). So, how is this related to the correspondence principle? In fact, although some of these claims may be accepted (and hence potentially conceptualized by the correspondence principle) not all claims can be accepted and potentially conceptualized in this way. Anyway, if we inspect thermodynamics (as an instance of these claims), we can see that the quantum theory justifies and explains the macroscopic behavior of thermodynamic systems but the emerging thermodynamics laws and rules are not actually rules of the underlying quantum systems (i.e. atoms and molecules). So, in this instance quantum mechanics does not converge to classical mechanics at the macroscopic scale but rather it causes the emergence of new physics at the macroscopic scale. In other words, thermodynamics is a consequence (or symptom) of the underlying quantum mechanical laws but it is not an instantiation of these laws on a large scale.

3. As a requirement for the correspondence principle (according to the literature), certain quantum features (such as particle-wave duality or position-momentum uncertainty or energy quantization or probability oscillation) should be non-observable in classical systems because these phenomena do not exist classically. For example:
   • Matter waves of macroscopic objects (such as cars) should be too small to have observable consequences since these waves do not exist in classical physics.
   • The uncertainty in the position of macroscopic objects (when measuring their momentum) should be too small to be observable at the macroscopic scale.
   • Energy quantization (e.g. in harmonic oscillator system) is not observable because the discrete energy levels are too closely-spaced and hence they are not detectable (as discrete) at the macroscopic scale.
   • Oscillation of probability waves (e.g. in potential well) is not observable because these waves are rapidly oscillating (in space) at the macroscopic scale and hence the separation between consecutive peaks (or troughs) of waves is too small to be detectable.
   So, according to the literature these examples (which demonstrate and explain the absence of charac-

---

[219] Referring to § 3.7, the correspondence principle may be implicated or justified (in some of its instances) by the Ehrenfest theorem and its applications. We also note that examples for the aforementioned convergence at large quantum numbers (e.g. in atomic transitions) and the emergence of classical formulations (e.g. Newton's second law) from quantum formulations by the Ehrenfest theorem can be found in the literature.

[220] It should be noted that the common statement of the correspondence principle which sounds like the following: "The predictions of quantum mechanics have to converge against classical results for the limit of large quantum numbers" (see Demtröder in the references) does not give an impression or suggestion of the claim or potential claim (which is related to the claim of general validity of quantum physics) of its general applicability and extension to all physical systems of all sizes and scales.

4.7.2 The Correspondence Principle

teristic features of quantum physics in the macroscopic world and hence in classical physics) can be seen as instances for the correspondence principle in action and its extended applicability (and may be claimed to imply the general validity of quantum physics).

4. As indicated above, an important issue related to the correspondence principle is the relation between classical and quantum physics and if the latter is a more general theory, i.e. it includes (in some sense or form) the former and extends to its domain of validity. In the present point we investigate if the correspondence principle implies general validity (or applicability) of quantum physics. From our earlier investigation (as outlined in the previous points; also see § 3.7) which is based on an extensive inspection of the literature, we may classify the applications and instantiations (or potential applications and instantiations) of the correspondence principle into three main categories:

• Applications based on the convergence of quantum mechanical formulations to classical formulations (e.g. at large quantum numbers or by the Ehrenfest theorem). This category should represent the most obvious and less questionable instances of applicability of the correspondence principle (as well as being the most significant with regard to the possibility of general validity of quantum physics). However, this category is limited in application, e.g. not all quantum mechanical formulations converge to classical formulations at large quantum numbers and not all classical formulations can be obtained by the Ehrenfest theorem (noting that even those supposedly-valid instances, or at least some of them, are questionable from some aspects such as being based on questionable assumptions or conceptualizations or approximations). Accordingly, this category does not imply the general validity (or applicability) of quantum physics.[221]

• Applications based on the convergence of quantum mechanical predictions to classical predictions. Examples of this category are given in point 3. This category can be seen as an extension to the meaning and applicability of the correspondence principle in its original form. However, it should be obvious that this category does not imply the general validity (or applicability) of quantum physics because this category is primarily about the actual physics that govern the quantum and classical worlds and not about our formulation of these physics (i.e. it is about the phenomena and behaviors and not about the theories and formulations). To put it in a different way we can say: this category is about the consistency between the predictions of quantum mechanics and the predictions of classical mechanics and is not about the convergence of the formalism of quantum mechanics to the formalism of classical mechanics. Accordingly, this category implies the correctness of quantum mechanics (due to the consistency of its predictions with the predictions of classical mechanics) but it does not imply the general validity (or applicability) of quantum physics as a formal theory that extends in applicability to the domain of classical physics.

• Applications related to the motivation of classical physics by quantum physics (as in the above example of thermodynamics where its main laws are supposedly derivable from quantum mechanics). However, this is restricted to certain areas and branches of physics and hence it does not imply general validity of quantum physics. Moreover, this is not about the convergence of formulations of quantum physics to formulations of classical physics but it is about the emergence of new physics at the macroscopic scale from physics at the microscopic scale, and hence again it does not imply general validity (or applicability) of the formalism of quantum physics.

5. The correspondence principle is marginal (i.e. not fundamental; see point 2 of § 4.7). Its main role (or purpose) is to validate (or invalidate otherwise) the formalism of quantum theory in certain zones and areas, so in effect it is a "test principle". More specifically, the correspondence principle is in essence an epistemological principle (or rather an epistemologically-based physical principle). In fact, it is no more than an expression of the principles of reality and truth (i.e. the uniqueness of truth; see § 2.4.1) with the presumption of the extended (or allegedly general) validity of quantum mechanics to certain

---

[221] In fact, we will see later that this category does not imply the general validity (or applicability) of quantum physics for other reasons (e.g. this category shows the validity of classical physics in its domain due to this convergence rather than the applicability of quantum physics as it is as a formal theory and in its technical and conceptual capacity unlike, for instance, Lorentz mechanics in its convergence to classical mechanics where Lorentz mechanics is applicable as a formal theory and in its full capacity at the classical domain).

classical zones.[222]
6. Despite the fact that the correspondence principle is marginal (and despite the limitations on its applicability and significance as outlined or implicated in the previous points), it is important and useful from different aspects. For example:
   • It can be used (when applicable) for the test and check of quantum results in certain limiting cases (which are generally more obvious than other cases) and hence it can be used to scrutinize, analyze and validate (at least partly and tentatively) newly obtained quantum results which may be under suspicion or require verification.
   • It can be used (in reverse direction to the direction in the previous example) to anticipate quantum formulations from classical formulations.
   • It can improve the understanding of quantum physics through linking it to the (potentially more understandable) classical physics and this should be particularly useful for providing intuitive interpretation to quantum physics.
   • It can also improve the understanding of classical physics through linking it to the (potentially more fundamental) quantum physics and this should be particularly useful for providing justification and rationalization to classical physics.
   In brief, the correspondence principle is a bridge between classical and quantum physics which should provide useful correlations and enlightening insights in both physics and the worlds they represent.

Also see § 6.7.

4.7.3 The Complementarity Principle

Broadly speaking, this principle means that the wave and particle aspects of quantum objects and phenomena cannot be observed simultaneously, i.e. when we conduct a measurement on a quantum system then we either detect a wave behavior or detect a particle behavior. An obvious example of this is the double-slit experiment (see § 1.6) where the observation of wave (represented by the interference pattern) and particle (represented by the identification of trajectories) are incompatible and cannot occur simultaneously. As we will see, this principle is surrounded with ambiguities and controversies (noting that it was originally proposed by Bohr who seems to be obsessed with proposing and formulating raw and imprudent principles) and hence even its status as a principle should be questioned.

In fact, there is nothing in quantum mechanics that supports the complementarity principle, i.e. the formalism of quantum physics in itself does not prohibit the detection of simultaneous particle-wave behavior. Yes, if we extend and generalize the complementarity principle (see point 2 in the following) then the uncertainty principle (which is part of the formalism and is supposed to instantiate the complementarity principle) should impose limits on the precision of certain pairs of detected properties (see § 4.7.1) and hence the simultaneous detection of particle-wave may be prohibited (rather arbitrarily and abusively) by this pretext although the genuine instances of the uncertainty principle should still be justified rightly by the formalism and included in it (also see point 2 of § 4.7.4).

In the following points we investigate the complementarity principle from different aspects and perspectives:
1. Although we conceptualize and regulate complementarity here (see for instance the first sentence of this subsection) by observation and detection, some may conceptualize it more fundamentally by existence (i.e. in reality) and hence complementarity is primarily about (non-concurrent) existence of certain attributes of reality although it is revealed (as a symptom) by (non-concurrent) observation and detection of these attributes. In fact, this should depend on the school of interpretation that we follow (see chapter 8). Also see point 6.
2. The complementarity principle may be generalized and extended to mean that the observation of one physical property of a system makes another property unobservable.[223] For example, according to

---
[222] It should be noted that the correspondence principle may be used (marginally) in certain formal arguments. However, in our view this type of use does not justify its inclusion in the formalism of quantum theory and hence it is marginal.
[223] We note that this extension may be justified by the proposition that the uncertainty principle (in its various forms) originates from the particle-wave duality of quantum phenomena, and this duality (in a rather stretched sense that

4.7.3 The Complementarity Principle

the uncertainty principle the observation of position (precisely) makes the observation of momentum (precisely) impossible and hence this is a form of complementarity.[224] Accordingly, the uncertainty relations of the uncertainty principle (see for instance Eq. 13) are instances of the complementarity principle. In fact, we can find in quantum physics many other examples of complementarity in its extended sense. Also see point 2 of § 4.7.4.

3. "Complementary" (which "complementarity" is derived from or the other way around) suggests that the complementary variables (or observables) are needed both (or together or simultaneously) for complete description of the observed phenomenon.[225] This seems contradictory to the suggestion of the complementarity principle (and even the use of "complementary" in the uncertainty principle) where the two variables are "incompatible" in observation (i.e. they cannot be observed together or simultaneously). This seems to indicate a sort of confusion or lack of clarity about the essence and content of the principle of complementarity when it was initially labeled as such. In fact, this confusion and lack of clarity (and lack of proper labeling) is the source of contradictions and blunders in the literature about the complementarity principle and its meaning and significance. Also see points 3, 4 and 5 of § 6.8.

4. The complementarity principle has been challenged theoretically and experimentally (e.g. by the slit-grid experiment; see § 6.8.1). Anyway, regardless of the validity or invalidity of these challenges, this "principle" should be seen (at best) as a rule of thumb based on common observations rather than a strict principle or law. In fact, the complementarity in its extended sense (as explained in point 2) is certainly invalid as a general rule since there are many couples of physical properties that can be observed simultaneously. Yes, if we impose certain restrictions (to justify selective types of complementarity) then any selective complementary behavior will be justified essentially by the causes that underlie these restrictions (e.g. the uncertainty principle) and hence the complementarity will become a loosely-characterizing feature that represents a type of behavior rather than a principle (see § 5.9).

5. The complementarity principle indicates (indirectly) an important fact that is, the observed phenomenon depends on the type or method of observation, and this should reflect an important feature of quantum physics where the measurement (or the process of observation) determines the nature of the outcome in contrast to classical physics where the measurement is generally passive and inert (see § 1.3.3). In fact, this may even be seen as a type of indeterminacy since prior to observation the observed object has no definite identity (i.e. whether it is particle or wave) because its identity depends on the method of observation.[226]

6. The complementarity principle (and indeed the particle-wave duality itself which is at the root of this principle) seems to suggest (at least in our view) that the duality of quantum objects is not something that is about the nature of the observed quantum object by itself but it is about how we observe it and hence it is rather an attribute of the observation process rather than the observed object. Therefore, the proposition "quantum object is neither wave nor particle" should be as much true as the proposition "quantum object has dual particle-wave nature" (depending on whether we look from an intrinsic or extrinsic perspective). In fact, this should be inline with our previous proposal (see for instance § 1.3.2) that classical concepts and models like "wave" and "particle" should be considered as approximate prototypes when used to describe objects and phenomena that belong to the quantum world. Also see point 1.

7. Macroscopic objects behave either as waves or as particles (at least within our observational capabilities) and hence they are not subject to the complementarity principle (which is associated with

---

includes being disjoint or incompatible) is the essence of the complementarity principle.

[224] The use of "complementary" for describing the incompatible observables in the uncertainty principle may suggest this link between the complementarity principle and the uncertainty principle.

[225] In fact, this suggestion follows the most common usage of this word noting that "complementary" may be used in opposite meanings (see point 3 of § 6.8).

[226] In fact, this should be linked to our earlier suggestion that human knowledge is not pure discovery but it is a mix of discovery and invention since the (method of) observation determines the (type of) knowledge. We should also note that although the tone in this point suggests that the talk is about the original (rather than the extended) form of this principle, the essence of this point may be stretched to include the extended form.

duality). So, the practical domain of this principle is microscopic or quantum objects and hence this principle is strictly and distinctively quantum mechanical. This also applies to the extended version of this principle since the uncertainty principle does not exist classically (at least in its quantum content, spirit and applications). Also see § 5.9 as well as point 3 of § 4.7.4.

8. The complementarity principle is marginal (i.e. not fundamental; see point 2 of § 4.7). In fact, this principle is interpretative and explanatory in nature (or rather summarizing certain observations and representing a quantum feature) and may also serve the purpose of validation and test (like the correspondence principle; see point 6 of § 4.7.2).[227] It is worth noting that this principle is usually and primarily associated with the Copenhagen school, and this should confirm its status as interpretative and explanatory in nature (and being a feature). Also see § 5.9.

9. Some seem to reduce the essence of the complementarity principle to the claim that both complementary aspects (whether wave-particle or any incompatible pair of observables if the principle is generalized and extended) are needed for a complete description of the physical world. In fact, this should trivialize this principle and make it more of a (rather insignificant and naive) philosophical thought than a scientific principle (and this should be inline with most of our early observations, as well as our upcoming observations, about this principle).

To sum up and conclude, there are many ambiguities, uncertainties and controversies (some historical and some contemporary) about this principle and its meaning and application. In fact, these ambiguities, uncertainties and controversies are not restricted to the "complementarity principle" but extend even to the meaning of "complementarity". Also, some interpretations and opinions (found in the literature) make this principle trivial in content and significance. Hence, this (alleged) principle should be degraded to become (more adequately) a rule of thumb, or even a feature, rather than a strict principle. Anyway, in our view it is marginal (assuming its acceptance as a principle and discarding its potential association with the uncertainty principle in its technical sense and essence) and hence its role and significance in quantum mechanics is limited (despite its magnification by some due to unhealthy obsession or exaggeration or confusion with other principles like the uncertainty principle). Also see § 6.8.

4.7.4 Relationships between the Principles of Quantum Theory

We briefly discuss in the following points certain links and relationships between the principles of quantum theory which were investigated in the previous three subsections:

1. Regarding the relationship between the uncertainty principle and the correspondence principle, apart from the difference in their nature, practically they do not apply simultaneously (i.e. they can be seen as disjoint) because the uncertainty principle does not have (direct) tangible consequences on classical systems which the correspondence principle supposedly applies to.[228] So in brief, apart from this "disjoint" situation there is no inherent relationship between the contents of these principles since they are about two different things and with different objectives and roles (as well as different status as fundamental and marginal).

2. Regarding the relationship between the uncertainty principle and the complementarity principle (i.e. in its unextended sense), the two are different in content, role and objective although they have an obvious similarity since both are based on the idea of incompatibility (in observation) between a pair of physical observables. Yes, the uncertainty principle can be seen as an instance of the complementarity principle in its extended sense (see § 4.7.3) since according to the uncertainty principle the observation of one incompatible (or *complementary*) property (e.g. position) will render the other property (i.e.

---

[227] It should be noted that the complementarity principle may be used (marginally) in certain formal arguments. However, in our view this type of use does not justify its inclusion in the formalism of quantum theory and hence it is marginal. Yes, in its extended sense it may acquire its essence and authority from the uncertainty principle (and hence it is fundamental in this sense, i.e. as a representation of the uncertainty principle). It should also be noted that the complementarity principle is considered by some physicists to be at the core of quantum mechanics (apparently due to its relation to the uncertainty principle; see for instance point 2 of § 4.7.4). However, in our view complementarity (as such) is essentially marginal (and it is more appropriate to be regarded as a feature rather than a principle).

[228] This should apply more appropriately and meaningfully to the second category of the (potential) applications and instantiations of the correspondence principle (see point 4 of § 4.7.2).

momentum) unobservable. However, we should note that some physicists justify the complementarity principle by the uncertainty principle[229] and hence the former is founded on the latter, i.e. the former is a sort of featuring (or conceptualization or labeling) of the latter (if the former is extended) and an instance of the latter (if the former is not extended). Accordingly, the uncertainty principle is the primary and the complementarity principle is the secondary. In fact, this makes the complementarity principle (in itself) trivial and dispensable since it is no more than a sort of featuring or instantiation of the uncertainty principle (which should be sufficient in its content and role in the quantum theory). Anyway, if we identify the essence and content of each one of these principles (which is a matter of debate and opinion) then the relationship between them will be identified accordingly (and rather straightforwardly).

3. Regarding the relationship between the correspondence principle and the complementarity principle (i.e. in its unextended sense), we may claim (with awareness of clash with point 7 of § 4.7.3) that classical systems (which are supposedly the subject of the correspondence principle)[230] are the ideal instances for the validity and realization of the complementarity principle since the duality behavior is restricted to quantum systems. In other words, since duality does not exist classically then complementarity (in the sense of incompatibility) is realized automatically and inevitably. However, we should note the difference in the "realization" of the complementarity principle in quantum systems (where duality occurs and hence complementarity is realized alternately) and in classical systems (where duality does not occur and hence complementarity is realized uniquely). So overall, if we accept the complementarity principle then we can say: no system (whether classical or quantum) demonstrates dual behavior simultaneously (i.e. either because it does not demonstrate duality at all as in the classical case or because it does not demonstrate duality simultaneously as in the quantum case).

It should be noted that the complications (which are largely based on potential extensions and controversies) of the relationship between the uncertainty and complementarity principles (which we investigated in point 2) should be projected on (and considered in) the relationship between the correspondence and complementarity principles. The details (which are almost useless) could be worked out rather straightforwardly (and hence the reader should take care of this).

Also see § 6.6.4.

4.8 The Dynamics of Quantum Mechanics

The dynamics of a theory represents how it works and applies, or in simple terms: it is the theory in action. To summarize the dynamics of quantum mechanics and how it works during the lifetime of a quantum system we can say (following the mainstream conceptualization and interpretation): there are two main stages in the life of a quantum system: before measurement and at measurement.[231] **Before measurement** the system is governed by the time-dependent Schrodinger equation and hence it evolves deterministically in a predictable way according to this equation (noting that the system is generally in a superposition state during this stage).[232] **At measurement** the system is governed by the measurement principle (or postulate) where the system collapses non-deterministically to a particular state and hence the outcome of the measurement is subject to the rules of probability.

Accordingly, our program for describing the state of the system and predicting its behavior during the first stage (i.e. unitary evolution) is to find (with the help of Schrodinger's equation) a wavefunction that represents the (superposition) state and hence we can describe the system and predict its tempo-spatial evolution during this stage. For the measurement stage, we simply use this wavefunction to predict

---

[229] In fact, this requires a generalization or extension of the uncertainty principle.
[230] Again, the second category of the (potential) applications and instantiations of the correspondence principle is the more appropriate and meaningful in this context (see point 4 of § 4.7.2).
[231] We may also say: between measurements and at measurements. The former may be labeled as the unitary evolution stage and the latter as the collapse or measurement stage.
[232] As indicated, determinacy here implies predictability, i.e. if we know the wavefunction at a given time then we can (in principle) determine the wavefunction at any (past or future) time as long as no measurement occurs between the two times.

(probabilistically) the outcome of measurement where the amplitudes of this wavefunction are used to determine the probability of each particular outcome.

4.9 The Basic Elements of Formalism

In the end of this chapter, it is useful and important to identify (in a rather general and broad fashion) the basic elements that make the formalism of quantum theory. From the rather limited (but sufficient) investigation conducted in this chapter we may conclude that all (or almost all) the main elements of the formalism of the (current) quantum theory are encapsulated in its axiomatic framework (see § 4.2). However, we should take notice of the following remarks:

1. The Schrodinger equation (which is considered the main element or the heart in the formalism of quantum theory) is referred to in the axiomatic framework (and hence we did not ignore it in the above conclusion).
2. The common versions of the axiomatic framework (or at least some of them) may include some elements of interpretation (and hence this framework may not represent in its entirety the formalism).
3. We should also add to the axiomatic framework some principles (such as the principles of uncertainty, conservation, and exclusion). However, some of these principles may be inherited from classical physics (and hence they are not specifically quantum although they are still part of the formalism). Moreover, some of these principles are derivatives or corollaries from other (more fundamental) elements of the formalism (and hence they are not essential or basic elements).
4. We may also need to add to the formalism some extensions (related for instance to the quantum spin) which are not incorporated specifically in the axiomatic framework or considered in the Schrodinger equation (noting that we are mainly talking about the basic or *classic* quantum theory which is the scope of this book).

To sum up, if we tentatively-exclude or modify some (interpretative-like) elements in the axiomatic framework and include some elements (outlined above) to it then the axiomatic framework represents the essence of the entire formalism of quantum theory and contains (almost) all its basic elements. Accordingly, all the rest of quantum theory (excluding overt interpretation) are no more than mathematical and physical technicalities and details, preparatory materials, methodologies, instantiations, applications, ... etc. and hence they are not (essential or basic) elements of the formalism. This means that the overwhelming majority of the *formal* topics and investigations (e.g. applications related to typical potential fields or physics of atoms or approximation methods or ... etc.) in the standard textbooks of quantum physics do not belong to the formalism in this (rather severe and rigorous) sense (since they are not essential or basic elements) although they may be commonly seen as part of the formalism in an extended and relaxed sense.

Chapter 5
Characteristic Features of Quantum Theory

In this chapter we investigate the characteristic features of the quantum theory. These features are generally inferred by analyzing the formalism of the theory (which was investigated rather briefly in the previous chapter). However, the investigation in the present chapter generally follows a qualitative non-formal approach (inline with the nature of our investigation of the formalism in the previous chapter and according to our needs and objectives in this book). We should also note that although most of these features are specific to quantum theory (i.e. they are not found in classical theories or at least not as found in classical theories), they are not necessarily so.

5.1 Quantization

In simple terms, "quantization of something" means dividing it into discrete distinct units. Quantization of classically-continuous physical attributes (such as energy and angular momentum) of quantum systems may be the most distinctive feature of the quantum theory. In fact, quantization is the cause of the "quantum" label in the name of this theory (remembering that this theory started with the Planck quantization hypothesis).[233] Unlike classical systems which take a continuous spectrum of certain physical attributes such as energy and angular momentum, quantum systems usually take discrete values of these attributes. The significance of quantization is that at the quantum level and microscopic scale the world seems to be grainy in character as in matter since on this scale we have discrete atomic and sub-atomic particles (as packages of matter) and we have discrete packages of attributes like energy (which may be represented by energy levels of atoms or by photons for instance). More clarifications about quantization in the quantum theory are given in the following points (also see § 3.1):

1. Although quantization is a common feature of the quantum world and quantum physics, it is not universal, and hence some quantum physical attributes like position and time are, like their classical counterparts, strictly continuous.[234] In fact, even quantized physical attributes like energy can be continuous in quantum physics. So in brief, the difference between classical and quantum theories is that some (but not all) classically-continuous attributes are quantized (i.e. sometimes and under certain conditions) in quantum physics.

2. Quantization in quantum physics may not be as strict as "mathematical quantization" (where the accuracy is absolute). This is due to the existence of uncertainties in some physical parameters and attributes, and hence some types of quantum mechanical quantization are not strictly so. An example that can clarify the situation is the natural energy width of "quantized" excited atomic states which is reflected in the "quantized" energy of the emitted spectral lines in the form of natural line widths. In fact, this is what makes the absorption of radiation emitted by the same type of atoms possible.
Accordingly, we have in quantum mechanics non-quantized quantities (like position), strictly-quantized quantities (like spin or angular momentum quantum numbers) and loosely-quantized quantities (like excited energy levels and energy of spectral lines). In fact, all these three types may exist in a single quantum system and for a single attribute. For example, in atoms the ground state energy level is

---

[233] Historically, the quantum theory evolved and grew around the quantization hypothesis of Planck. Prominent examples of historical milestones in this regard include: photoelectric effect, Bohr's atomic model, Franck–Hertz experiment, Compton effect, and Stern-Gerlach experiment.
[234] It should be noted that there are theoretical attempts to discretize even space and time. Also, considering time as a physical attribute may be contested by some.

quantized sharply (i.e. with no uncertainty due to its infinite lifetime), the excited states are quantized loosely (i.e. with uncertainty due to their finite lifetime), and the unbound (or ionized) states are continuous.
3. Quantization (i.e. of inherently-continuous attributes) may also occur (occasionally and accidentally) in classical physics. For example, the frequency and wavelength of standing waves are quantized. However, this does not affect the status of quantization as a distinctive feature of quantum physics (noting that the types of quantization that occur in quantum physics are specific to this theory and do not occur in classical physics).

5.2 Indeterminism

Indeterminism (or indeterminacy) is one of the main characteristic features of quantum theory that distinguish it from other scientific theories. The essence of determinism (which is ingrained in almost all classical scientific theories) is that the future state of a given physical system is completely and definitely determined by its initial conditions and the factors that influence the system (such as mechanical forces) and hence the outcome of any experiment or measurement on the system should be completely determined with certainty once these conditions and factors are identified and accounted for. Accordingly, indeterminism means that this outcome is not definitely determined and hence there is an element of probability or uncertainty in the outcome.[235]

For example, in classical physics if we know the initial positions and velocities of the particles that make a given mechanical system (as well as all the factors and forces that influence the system) then from Newton's laws of motion we can (in principle at least) predict with definite certainty (i.e. deterministically) the configuration and evolution of the system at any time in the future. This is unlike the situation in quantum physics where knowing all the initial conditions and the factors that influence the quantum system (assuming this knowledge is possible at all) will usually lead (by employing the formalism of the quantum theory) to a probabilistic and uncertain type of prediction about the state of the system in the future and the outcome of a measurement conducted on the system to "reveal" its state.

As indicated, indeterminism is demonstrated (at least partly and in its most explicit form) by the probabilistic nature of quantum physics and its formalism (see § 5.2.2) where the outcome of quantum measurements is predicted probabilistically (rather than deterministically) through the use of the intensity (i.e. modulus squared $|\psi|^2$) of the wavefunction. It is also demonstrated in the uncertainty associated with the measured values of certain pairs of physical observables such as position and momentum (see § 5.2.3). Indeterminism in quantum physics is also demonstrated and instantiated in other shapes and forms such as superposition of states (see § 5.2.1) and indistinguishability of quantum objects (see § 3.1).

More clarifications about indeterminism in the quantum theory are given in the following points:
1. No scientific theory can be totally indeterministic; otherwise it will be useless for the description and prediction of physical reality. Accordingly, when we talk about indeterminism in quantum theory it should be understood to be with respect to certain aspects that are deterministic classically. For example, the evolution of wavefunction according to the Schrodinger equation is deterministic. In fact, even the collapse of wavefunction (through which the most characteristic form of quantum indeterminism is realized and demonstrated) is deterministic within the possible outcomes and available probabilities and according to their determinate values. This similarly applies to superposition and to uncertainty (as the two other main forms of indeterminism) since both are restricted and deterministic within the available superposed states and the central values to which the uncertainty applies (as well as in the range of uncertainty). So in brief, quantum physics like any other scientific theory, is broadly deterministic but with some distinguishing indeterminacy aspects when compared to classical physics (whose corresponding aspects are deterministic).
2. As indicated above, indeterminism in quantum mechanics has different shapes and forms. As we will see in the upcoming subsections, we have three main types of quantum mechanical indeterminacy:

---

[235] As we will see, indeterminism in the quantum theory has different forms and instances some of which may not fit exactly with the above characterization. Hence, this characterization should be considered broadly and as initial (but fundamental) identification to this concept.

indeterminacy prior to measurement (represented by superposition; see § 5.2.1), indeterminacy in the outcome of the measurement (represented by the probabilistic nature; see § 5.2.2), and indeterminacy in the value of the outcome of the measurement (represented by the uncertainty of the uncertainty principle; see § 5.2.3).[236] Also see points 2 and 8 of § 6.4.

3. It should be noted that indeterminism also exists (exceptionally) in some branches and applications of classical physics, e.g. in turbulent, non-linear and chaotic systems (although some of these can be questioned from various aspects such as being classical or not or being actually or apparently indeterministic).[237] Similarly, determinism also exists in quantum physics (noting that there are many deterministic aspects in the quantum theory as explained in point 1). So in brief, classical physics is primarily and characteristically deterministic (with some exceptions) while quantum physics is characteristically indeterministic in the sense of having non-classical types of indeterminism related to certain aspects of the theory (although the theory is also deterministic, i.e. broadly and from various aspects and perspectives). Anyway, the issue of determinism and indeterminism in classical and quantum physics is a big subject of investigation and debate considering for instance their forms and types and their instantiations and applications and hence its details should be pursued elsewhere.

4. With reference to the previous point, it is important to note that quantum indeterminism in quantum systems can propagate to classical systems that are affected by the quantum systems, as seen for instance in the thought experiment of Schrodinger's cat (see § 9.1.2) where the indeterminate nature of the quantum trigger makes the classical outcome (which is the death of the cat) indeterminate. In fact, this sort of propagation should be a common feature to quantum measurement (where quantum systems represented by the observed quantum phenomena are interfaced with classical systems represented by the macroscopic measuring equipment) when the quantum observable is indeterminate (e.g. by being in a superposition state or by having an intrinsic uncertainty in its value). However, as the origin and culprit of this indeterminism is the quantum system this type of indeterminism cannot be attributed to the classical systems and phenomena (which are actually or supposedly deterministic in reflecting the underlying quantum systems and phenomena and their indeterminism). Also see the last paragraph of § 2.15.3.

5.2.1 Superposition

According to the conventional quantum theory, when a quantum system is in a state of superposition (prior to measurement) it is actually "composed" of a number of sub-states, and hence the system does not have a unique and well-defined state. So, prior to measurement we generally do not know the exact state of the system other than being in a superposition (or mix) of a number of sub-states and hence the state of the system in this sense is vague and undetermined. It is obvious that this is a form of indeterminism that seemingly has no parallel in classical physics which implicitly presumes (following the mainstream literature) that physical systems have independent and determined reality and hence their states (in themselves and independent of any observation or measurement) are well-defined and completely-determined before and after measurement; the role of measurement therefore is just to reveal the already determinate and well-defined state.

An important consequence of quantum superposition is the probabilistic nature of quantum physics (see § 5.2.2) where the indeterminism of superposition (prior to measurement) is reflected in the indeterminism of the expected outcome of measurement and its probabilistic character. This correspondence of indeterminism before and after quantum measurement should be logically contrasted by the correspondence of

---
[236] It should be noted that uncertainty in quantum physics is not restricted to that of the uncertainty principle. However, the uncertainty of the uncertainty principle is the distinctive (or specifically-quantum) type of uncertainty.
[237] It is common in the literature to deny the existence of indeterminism in classical physics by distinguishing "classical indeterminism" from "quantum indeterminism" and demonstrating the difference in their natures. However, this does not negate the existence of indeterminism in classical physics even if it is of different nature compared to its quantum counterpart. So, this denial should be understood as negation of the existence of "quantum indeterminism" in classical physics although other types of indeterminism can exist in classical physics. Anyway, we do not need any distraction by this peripheral investigation which is out of scope (noting that our primary interest in this book is quantum physics and not classical physics).

determinism before and after classical measurement.

It is worth noting that quantum superposition (from a perspective different to the indeterminism perspective) may suggest multiple realities since the superposed states should have some sort of reality (in the classical sense of reality); otherwise there is no sensibility or rationality in the collapse of wavefunction (or "collapse of reality") to one particular state of the superposed states unless these states have a prior reality (or existence) that allows this specification to happen. The coexistence of these multiple realities may be conceptualized (or modeled) sensibly in the form of superposition of waves where the component waves have realities within the superposed wave. This conceptualization (which is essentially classical in nature since the superposition of waves is classical) should make quantum superposition similar (if not identical) to classical superposition. So, what makes quantum superposition distinct and characteristic to quantum theory is the wave-conceptualization of objects that are supposedly particles from a classical perspective. In other words, what makes this sort of superposition quantum is not the superposition itself but the underlying wave nature of particles (or rather the particle-wave duality). This should be inline with our observation (earlier and later) that the particle-wave duality is the origin of many (if not most or all) of the characteristic features of quantum physics.

However, it should be noted that the wave-conceptualization of superposition (which is based on the concurrent coexistence of realities in a mixed form as explained in the previous paragraph) may not be the only possible conceptualization of quantum superposition. For example, quantum superposition may be conceptualized (or modeled) in the form of a continuously-changing state among the superposed component states (like an alternately blinking light bulb or a light emitting diode that continuously changes its color or/and intensity). Accordingly, quantum superposition is not a combination of states that exist simultaneously but rather a continuously-evolving state in which only a single state of the "combination" is active at any instant of time.

It is worth noting that both these conceptualizations of quantum superposition (i.e. as coexistent waves or as continuously-evolving state) are based on the compatibility of reality before and after measurement, i.e. the emerging reality by measurement is embedded in the reality before measurement rather than being an incarnation from a different reality. This compatibility should optimally rationalize the transition of quantum state from its pre-measurement form to its post-measurement form. Other conceptualizations (of the type that maintains this compatibility as well as other types) to quantum superposition may also be possible. Also see § 5.8 and § 6.4 for investigations related to superposition.

5.2.2 Probabilisticity

The probabilistic nature of the quantum theory is one of the main features that distinguish this theory from other scientific theories. Non-quantum scientific theories are generally deterministic and hence the outcome of an observed physical process is specifically predictable if we know all the initial conditions and the physical factors involved in that process. Technically, the probabilistic nature of the quantum theory means that this theory essentially provides a statistical prediction of the outcome of measurements (in experiments and observations) conducted on a large number of non-interacting and identically-prepared systems (i.e. identical systems in the same quantum state). More clarifications about the probabilistic nature of the quantum theory are given in the following points:

1. The probabilistic nature of certain physical processes is not restricted to quantum phenomena. In fact, many classical phenomena (like throwing dice or tossing coins) have probabilistic outcome. However, what distinguishes classical probabilisticity from quantum probabilisticity is that the former originates from casual (or extrinsic) factors which are primarily related to us (such as our ignorance of the initial conditions or our negligence to do accurate measurements and calculations) rather than to the phenomena, while the latter originates from essential (or intrinsic) factors which are primarily related to the phenomena rather than to us. Hence, the classical process in itself and on its own is deterministic (although it is probabilistic relative to us) while the quantum process in itself and on its own is probabilistic. So, in principle we can predict the outcome of the classical process definitely and deterministically, while we can predict the outcome of the quantum process only probabilistically.[238]

---

[238] It should be noted that we are following in this the general consensus (or mainstream opinion) about this issue.

2. The probabilistic nature of quantum physics is reflected and formalized in various aspects of the quantum theory. For example:
   - According to the probabilistic interpretation[239] of quantum mechanics, the probability of detecting a particle represented by a wavefunction ψ in a small volume of space centered on a given point A in space is proportional to the value of $\psi\psi^* \left(= |\psi|^2\right)$ at A. This means that the wavefunction (which represents the physical state) is essentially a probability amplitude.[240]
   - When the wavefunction of a quantum system prior to measurement is a superposition of eigenstates, the observable outcome of a measurement on that observable is an eigenvalue of the operator representing that observable and the quantum state of the system after measurement is the eigenstate corresponding to the obtained eigenvalue where the probability of obtaining that particular outcome is given by the modulus squared of the amplitude (or expansion coefficient) of that eigenstate.
3. It is important (when talking about probability) to identify and understand the meaning of probability and appreciate its significance within the given context and circumstances. In fact, we may identify (or suggest) three main types of "probability" to be considered in this regard:
   - Subjective probability which is a psychological state (or feeling) experienced by a given thinker.
   - Objective (or physical) probability which is the presumed likelihood of the occurrence of a given event according to the laws and rules of physics (whether these laws and rules are independent of or dependent on the observer).
   - Mathematical probability which is the mathematical formulation and quantification of the subjective or/and objective probability.

   In our view, there may not be probability in reality (and hence no objective probability as such). So, the actual probability belongs to our perception and knowledge of reality and how we quantify the likelihood of occurrence (as seen from our viewpoint and perspective) of certain physical outcomes through the use of mathematical probability. Accordingly, the probabilisticity of quantum physics is more likely to be a characteristic feature of the theory (as classified here) rather than a characteristic feature of the physical phenomena. This should be inline with realism (which we incline to considering that the current quantum theory should not be the only possible quantum theory according to the non-uniqueness of science although this in itself does not guarantee the possibility of existence of a non-probabilistic quantum theory). Also see § 7.14.
4. Although the probabilistic nature is "probabilistic" (individually) it is deterministic (collectively). In other words, although the outcome of a measurement on individual (identically-prepared) systems is not known deterministically, the outcome of a large number of measurements on such systems (i.e. systems prepared identically and hence they are in the same quantum state) is known according to the (deterministic) rules of probability (as formalized in quantum mechanics).

Also see § 6.4.

5.2.3 Uncertainty

Uncertainty is another form of quantum indeterminism and is typically represented by the uncertainty principle which was investigated in § 4.7.1. As indicated earlier, the uncertainty that distinguishes quantum physics (and should be considered as one of its characteristic features as a form of quantum indeterminism) is that of the uncertainty principle, i.e. intrinsic uncertainty (noting that there are other types of uncertainty that are common to both classical and quantum physics). Also see § 6.4 and § 6.6.

---

[239] When we talk about the "probabilistic interpretation" or "probabilistic nature" in contexts like this we mean the commonly accepted interpretation of quantum mechanics which is based directly on its formalism and hence this type of "interpretation" is not specific to a particular school of interpretation (noting that these schools differ in rationalizing this probabilistic interpretation). As we will see in chapter 8, we may call this interpretation the "standard interpretation" or "basic interpretation".

[240] In fact, even the sandwich integrals in quantum mechanics (and their link to the corresponding expectation values) should suggest a probabilistic nature for the wavefunction.

5.3 Duality

This is another characteristic feature of quantum theory. There are many aspects of duality (and even multiplicity) in quantum physics. In the following subsections we investigate three types of duality: particle-wave duality, formalism duality and classic-quantum duality.[241]

5.3.1 Particle-Wave Duality

Quantum objects like photons and electrons behave as localized particles as well as waves and hence they have particle-wave dual nature. For example, diffraction and interference of light indicate the wave nature of photon while the photoelectric and Compton effects reflect its particle nature.[242] Similarly, the electron behaves like a localized particle but also shows wave properties like diffraction (e.g. in Davisson-Germer experiment). In fact, the paradigm of particle-wave duality is extended to all material objects (whether quantum or classical)[243] by the de Broglie hypothesis and wave equation which assign a wavelength λ to any object of momentum (of magnitude) p, that is:

$$\lambda = \frac{h}{p} = \frac{h}{mv} \tag{16}$$

where h is the Planck constant and m and v are the mass and speed of the object. This relation is commonly known as the de Broglie equation.

More clarifications about the particle-wave duality are given in the following points:

1. The de Broglie equation (i.e. Eq. 16) supposedly reflects particle-wave duality in its most formal and explicit way. In fact, Eq. 16 also provides the required link between the wave nature (represented by λ) and the particle nature (represented by $p = mv$) of physical objects where Planck's constant h plays the role of mediator in this link. In this regard, it is worth noting that it is claimed that even the energy-frequency relation $E = h\nu$ reflects particle-wave duality (possibly by considering it in a proper context).

   Anyway, there are questionable aspects about the justification of the de Broglie hypothesis and the derivation of his matter wave equation (Eq. 16). The least that can be said is that the arguments leading to these results are not sufficiently rigorous and this should cast a shadow on the "wave" nature of matter waves and if they are really and exactly waves (like classical waves) to justify the alleged duality. This should be inline with what we indicated elsewhere in this book (see for instance § 3.2) that classical and matter waves are different types of entities and hence the concept of matter "wave" (on which the entire quantum mechanics rests since it is all about *wave*function) should be treated with caution and it requires close inspection and assessment since it could be behind many of the epistemological troubles of quantum physics and the difficulty of finding consistent and rational interpretation to its formalism (refer for instance to § 7.15). Also see point 2 of § 3.2 as well as point 2 in the present subsection.

2. From Eq. 16 we can see that the wavelength of a particle is a function of its momentum which is dependent on the observer and his frame of reference. So, according to Eq. 16 a particle can have any wavelength (finite or infinite) depending on its observer.[244] In fact, according to this equation the

---

[241] Another example of duality in quantum physics is the duality of incompatible observables in the uncertainty relations. This also applies to complementarity in its various interpretations and identifications (see for instance § 4.7.3, § 4.7.4 and § 6.8). In fact, we may even consider the formalism-interpretation mix as a duality aspect of the quantum theory (see for instance § 7.10). Superposition of states (see § 5.2.1) may also be seen as another form of "duality" (or rather multiplicity) in quantum physics.

[242] In fact, the particle nature of photon is observed more directly and tangibly in double-slit type of experiments, for instance, where the photon is detected as particle (i.e. localized at the detection screen).

[243] In fact, this represents the mainstream view. However, the universality of the particle-wave duality (which is supposed to originate from the de Broglie hypothesis) and its extension beyond quantum systems can be challenged at least for practical reasons and considerations. For example, if the (alleged) wave of a macroscopic object has a wavelength of 10^{-50} m then it is meaningless to talk about its wave at least because it is beyond our detection capabilities (as well as having no tangible consequences). In fact, such waves are like metaphysical objects (if they are not really so).

[244] This obviously includes having different wavelengths (with respect to different observers) at the same time.

5.3.1 Particle-Wave Duality

wave nature of particles (or rather the material ones like electron) is not an intrinsic property since it depends on the observer and frame of reference. This should cast more shadows on the reality and authenticity of this alleged type of wave and hence on the alleged duality (at least in its commonly recognized conceptualization). Also see point 5 of § 3.2.

3. As a requirement for the correspondence principle (see § 4.7.2), particle-wave duality is not observable in classical (or macroscopic) systems since matter waves are not detectable classically. In fact, this can be seen from Eq. 16 by noting that Planck's constant is classically zero (because it does not exist at all since it is not conceptualized, or because it is practically zero since it is very tiny) and hence on classical scales λ is zero (which means that classical material objects have essentially a single nature, i.e. "particle"). Also see points 3 and 4 of § 4.7.2.

4. It is claimed that the oddity of the particle-wave duality is exaggerated in the literature, and hence simple explanations can be proposed to justify this duality rather easily. In fact, the particle-wave duality may be explained by a number of proposals using different approaches. However, we may classify these proposals into two main types:

   • **Duality in detection**: i.e. the duality is actually an attribute (or symptom) of observation and measurement rather than an attribute of the observed quantum object and its nature. Although this type of proposals may offer a simple explanation and could remove some of the perplexity of duality, it does not provide a fundamental explanation to the source of this duality in detection, i.e. what distinguishes the nature of the quantum objects that makes them demonstrate this duality in detection?

   • **Duality in nature**: i.e. the duality is actually in the observed quantum object itself where its nature can be seen as partly-particle and partly-wave. An example of this type of proposals is the wavepacket model where the quantum object is a particle-like in its relative localization and is a wave-like in its diffusion. Another (potential) example is the pilot wave model where the particle (representing the quantum object) is guided by a wave and hence the particle in its trajectory and its spatial distribution (as a member of an ensemble of similar particles) is controlled by the wave that associates it.[245] The proposals of this type usually face more difficult challenges not only in the explanation and visualization of their models but also in the generalization of these models and their ability to provide convincing explanations of all aspects and in all instances and circumstances of this duality (see for instance point 8). In brief, the existing models of this type are not as simple and convincing as might be thought and hence the oddity seems to be genuine and real.

   Anyway, whether we can or cannot explain the particle-wave duality in a qualitative way (by envisaging it in a known pattern or visual form), the observed duality may be explained rather easily (but generically and evasively) by the scale consideration (see § 1.3.2) because our patterns and models are based on the macroscopic (or classical) scale and hence they are not necessarily compatible with the physical phenomena on the microscopic (or quantum) scale. In other words, there may not be a classical pattern (such as "wave" or "particle") that can represent and describe a given quantum physical phenomenon. Accordingly, we can say: quantum objects are neither waves nor particles and hence they cannot be visualized classically (which means that this duality has no reality in the quantum world or at least the observed duality may not reflect the reality at the quantum level). In fact, this should be inline with the first type of proposals (i.e. duality in detection) because the observed duality is detected classically by using macroscopic (or classical) equipment and hence the duality belongs to the classical world rather than the quantum world (noting that the classical observation may not necessarily reflect the nature of the reality of the quantum world).

5. As explained earlier, the complementarity principle (see § 4.7.3) restricts (or regulates) the detection of the particle and wave natures of quantum objects. So, if we believe the complementarity principle then particle-wave duality does not mean coexistence of particle and wave (at least in detection and

---

[245] The classification of the pilot wave model into the second type may be challenged by the claim that the object is actually a particle but it has a wave guide or associate. If this challenge is accepted then we may need to make a third type to which this model should be assigned (noting that we may even assign this model to the first type with some modification or tolerance). We should also draw the attention to the conceptual challenge to "duality" according to this model (as will be explained in footnote [249]).

5.3.1 Particle-Wave Duality

may even cast a shadow on the coexistence in reality and in their very nature assuming that the duality in nature is real).

6. As indicated earlier, "wave" in "particle-wave duality" is supposed to be a "matter wave" or "probability wave" and hence it should not be understood in its classical (or physical) sense like electromagnetic wave (see § 3.2).[246] However, we may classify quantum particles into two main types: those that can be at rest (or have variable speed or observer-dependent speed) such as electrons and those that cannot be at rest such as photons.[247] We may claim that the wave nature of the latter is intrinsic and physical (since it cannot be at rest) while the wave nature of the former is not intrinsic and not physical since it is observer-dependent (and hence its wave nature, as well as its wavelength, depend on the frame of the observer; see point 2). In fact, this could lead to having two types of particle-wave duality (where these two types differ in the type of "wave" and whether it is physical wave or matter wave; also see point 4 of § 7.11). Refer also to § 1.5.

7. Noting the difference between classical and quantum particles (see § 3.3), it seems that "particle" in "particle-wave duality" should also be non-classical (i.e. with respect to some attributes).

8. There are various (formal and "visual") models that were developed within the quantum theory to represent quantum objects and explain their particle-wave duality. Apparently, the most dominant and prominent model of this kind is the "wavepacket" which is "localized" like particle and dispersed in space like wave and hence it represents both characters simultaneously. However, this model should be seen as a mathematical device to facilitate the formulation (as well as certain aspects of interpretation) of the theory and hence we should not read too much into it. For example, elementary quantum particles are not divisible (unlike wavepackets which are used to describe them since wavepackets can be split). Also, the width of wavepackets is susceptible to change with time (unlike genuine particles). There are other deficiencies and limitations in this model (some of which are discussed elsewhere in this book). So, wavepackets do not represent quantum objects and their duality really and exactly (although they are useful mathematical, and sometimes even visual, devices for modeling certain aspects of quantum objects and their duality).

Another prominent (but seemingly less dominant) model for representing quantum objects and visualizing and rationalizing their particle-wave duality is the "pilot wave" proposal which depicts quantum objects as particles guided by waves. In fact, this model may underlie and explain the observed trend that quantum objects are particles in detection and waves in propagation and distribution[248] (and hence it seems rational from this perspective). It may also be seen as the most classic-like duality model (and hence it seems more intuitive and in harmony with common sense). However, it also has its own deficiencies and limitations (conceptually as well as practically). For example, according to this model we have a classical-type particle guided by a quantum-type wave (i.e. non-physical matter wave) which does not seem very sensible or have an obvious physical mechanism (also see point 6).[249] Also, the seemingly-classical nature of particle in this model may not be consistent with some quantum aspects and attributes (e.g. indistinguishability or intrinsic uncertainty; see for instance § 3.3) which are endorsed by experimental evidence. There are other criticisms and challenges to this model (some of which are discussed elsewhere in this book).

Anyway, these models are generally incompatible (conceptually or/and practically) and hence they cannot be used alternately and as and where we wish and need (i.e. by using one model here and another model there). For example, the wavepacket model is obviously incompatible with the pilot

---

[246] As we will see in point 7, this should apply even to "particle" although this may be less obvious (see § 3.3). It is worth noting that having a "physical" particle with a "non-physical" wave should cast a shadow on the very concept of particle-wave duality and its sensibility (at least in its common perception).

[247] We may also classify them more simply as massive and massless.

[248] "Distribution" is a reference to the fact that the particle is more likely to be where the amplitude (or intensity) of the wave is stronger.

[249] In fact, this may put a question mark (conceptually) on the "duality nature" of quantum objects because the particle character belongs to the objects individually while the wave character (at least in one of its conceptualizations and interpretations) belongs to ensembles of them (noting that matter wave is a probability or statistical wave). Also, the particle character is physical and observable directly while the wave character is not physical and not observable directly (especially in some situations like in single-particle experiments where the objects are emitted individually and one at a time) and hence these characters cannot belong to a single object of "dual nature".

5.3.1 Particle-Wave Duality

wave model if these models are supposed to reflect and represent the underlying reality and provide sensible interpretations and rationalizations. However, some physicists seem relaxed about using these models as they wish and need with no regard to compatibility and consistency (although this may be understandable and justifiable as long as these models are used and treated as formal tools and devices without giving them realistic and interpretative value and significance). In fact, there are many examples of conflicting models used in quantum theory to describe and rationalize different aspects of quantum phenomena (and particle-wave duality is just an instance for such models). So, such models should be treated with caution and should not be accepted and embraced literally and as representing the "reality" of these objects although they can be useful (i.e. conditionally and as mathematical devices or educational prototypes).

So to sum up and conclude we say: there are two dominant models for the particle-wave duality and how it should be envisaged: wavepacket model (which imagines the quantum object as a wavepacket that is "localized" but with uncertainty) and pilot wave model (which imagines the quantum object as a particle guided by a wave). Both models are untenable (as well as being incompatible) and can be challenged (at least as representatives of reality and in their generality) since they cannot explain this duality in all instances and circumstances of quantum phenomena and they lead to inconsistencies in certain cases and situations when treated as realistic models.

9. It is alleged that there is a logical inconsistency in the particle-wave duality where it is argued (for instance) that in the double-slit type of experiments[250] the wave nature requires the object (e.g. photon) to pass through both slits while the particle nature requires the object to pass through only one slit, and this is logically inconsistent (noting that if the object passed from slit A only then it did not pass through slit B while if it passed through both slits then it passed through slit B and this is obvious contradiction). However, in our view there is no contradiction as long as we understand "wave" and "particle" as approximate classical models used to describe the behavior of the quantum object. So, the alleged contradiction (e.g. in the double-slit experiments) is fundamentally based on assuming actual and exact "particle" and "wave" in their classical sense (where particle cannot be wave and vice versa).[251] In other words, passing through a single slit is based on having an exact classical particle (which cannot be wave) and passing through both slits is based on having an exact wave (which cannot be particle) and hence we have a contradiction. Otherwise, if "wave" is understood as "wave-like" behavior and "particle" is understood as "particle-like" behavior (especially if this is restricted to detection; see point 4) then there is no contradiction because we do not have actual and exact wave and particle in their classical sense to have a contradiction. So, the culprit of this alleged contradiction is the literal conceptualization (or classical understanding) of "wave" and "particle"; otherwise the quantum object is neither wave nor particle in the literal and classical sense of these words.

It is worth noting that the complementarity principle (see § 4.7.3 and § 6.8) seems to address (at least partly) this supposed contradiction since it negates the simultaneous realization (at least in detection) of "particle" and "wave". In this context, we may also draw the attention to what we noted earlier that a general trend that seems to characterize the particle-wave duality behavior of quantum objects (although perhaps not strictly) and could help in addressing this alleged contradiction is that they are detected like localized particles and propagate (and hence distribute spatially) like dispersed waves (as seen for instance in the double-slit experiment; see § 1.6)[252] and hence "particle" and "wave" in general are not realized simultaneously and from the same perspective to contradict each other. This should also be inline with the fact that their "wave" is a "matter" or "probability" wave and hence it is not physical or detectable (unlike ordinary or classical waves like sound and light; see § 3.2). So, if "particle" and "wave" are of different nature (i.e. physical and statistical) and they belong to different

---

[250] In fact, this type of arguments may be more pertinent to certain types of double-slit experiments, particularly those of single-shot type where particles are released individually one at a time (see for instance § 1.6.3) although this should depend on the conceptualization and interpretation of these experiments.

[251] In fact, if "particle" and "wave" are understood in their exact classical sense then the contradiction (or inconsistency) should be in the duality itself regardless of any experiment or argument like the above. This is because in this case "particle"-"wave" will be something like "localized"-"non-localized" or "massive"-"non-massive".

[252] As indicated earlier, this trend may be rationalized by the familiar proposal (for explaining the particle-wave duality) that quantum objects are particles guided by (probability) waves.

entities (i.e. individual objects and ensemble of objects) as well as being not concurrent (as this trend seems to suggest) then there should be no contradiction.

To conclude this lengthy and messy investigation we say: quantum objects are apparently neither particles nor waves (i.e. in a strict classical sense). However, both the "particle" and "wave" models (which are essentially classical) are useful and needed in the description of their behavior (possibly complementarily and alternately, rather than concurrently and simultaneously, and possibly in detection rather than in nature), where the general trend is that they are detected (individually) as particles and propagate and distribute (collectively) as waves. It seems that there is no duality model (i.e. for explaining and envisaging the particle-wave duality in a classic and tangible way) that is sufficiently rational and general to be adopted and accepted although the existing models are generally useful (within certain restrictions and limitations) but they are generally incompatible. Also see § 6.13.1.

5.3.2 Formalism Duality

The duality of formalism refers to the existence of two main types of formulation to the quantum theory: wave mechanics and matrix mechanics. The first is represented by the Schrodinger equation (and hence its mathematical formulation is largely based on the differential and integral calculus) while the second is represented by the Heisenberg matrix approach (and hence its mathematical formulation is largely based on the mathematics of matrices and linear algebra) noting that both these mechanics are conceptualized on the style of linear algebra (i.e. in its more general formulation than what is given in elementary linear algebra). It is shown in the early days of quantum mechanics that these two formulations are equivalent and can be merged into a single more general version that includes both. In fact, this can be seen as a rather simple example of the non-uniqueness of science (see § 2.4.3).

It is noteworthy that wave mechanics is more common in use than matrix mechanics. This is because wave mechanics is easier to digest and less ambiguous and abstract. Moreover, it has other favorable properties like being more compatible with classical physics (due for instance to its similarity to the mechanics of waves and reliance on the differential and integral calculus) and hence it can make use of classical formulations and methods, as well as lending itself more easily to these formulations and methods, which are usually more familiar to physicists and more user friendly as well as being more intuitive and digestible. It is worth noting that matrix mechanics is so abstract and mathematical that it is not really interpretable compared to wave mechanics (noting that both are difficult or impossible to interpret for other reasons in addition to mathematical and theoretical abstraction; see for instance § 7.15).

5.3.3 Classic-Quantum Duality

Despite its revolutionary and novel nature, quantum theory is based on layers upon layers of classical physics (such as concepts, formulations and principles) and hence quantum theory is not purely quantum. Accordingly, we may describe quantum mechanics as semi-classical theory. This hybrid nature (or inhomogeneity) which is due to historical reasons (see for instance § 7.1) as well as scale considerations (see § 1.3.2) has an impact on many aspects of the theory and its interpretation in particular.

Anyway, this hybrid nature has positive, as well as negative, effects and consequences. Perhaps the most positive impact of this inhomogeneity is the (partial) intuitivity brought in by the imported elements of classical physics, while the most negative impact of this inhomogeneity is on the interpretability of the theory since it makes the theory difficult, if not impossible, to interpret consistently and homogeneously (or rationally). More details and assessment about this feature will be given later on (see for instance § 7.2).

5.4 Quantum Interference

The wave nature of quantum phenomena is reflected in (and demonstrated by) the phenomenon of quantum interference which is a distinctive quantum feature. The essence of this phenomenon is that the probability of a given quantum event A is not in general equal to the sum of the probabilities of the individual

5.4 Quantum Interference

mutually-exclusive events (or routes) by which A can happen, and hence in quantum interference the sum (or addition) rule of probability is not followed. As a result, an interference pattern is generated where the "probabilities" are combined constructively in some regions and destructively in other regions. For example, in the double-slit experiment (say of electrons; see § 1.6) we get a diffraction pattern when only one of the two slits is open but we get an interference pattern when both slits are open. This means that the statistical distribution of the electrons (i.e. on the display screen) in the latter case is not the sum of the statistical distributions of the electrons in the former cases (i.e. when only the left or the right slit is open).

More technically and precisely, in quantum interference we do not add the probabilities of the individual waves but we add the waves first (and hence they combine and interfere) and then get the probability from this combined (or interfered) waves. Now, in quantum mechanics the probabilities are given by the intensity of quantum waves (where this intensity is quantified by the probability density which is the modulus squared of the wave, i.e. $|\psi|^2$). This means that in quantum interference we do not add the probability densities $|\psi|^2$'s (which are real and measurable) but we add the amplitudes ψ's of the interfering quantum waves (which are complex and not measurable) then we get the probability from this combined (or interfered) amplitude ψ_c by taking its modulus squared (or intensity) $|\psi_c|^2$. For example, if we have two interacting (or interfering or coherent) quantum waves ψ_a and ψ_b then the probability of their combination is not given by $|\psi_a|^2+|\psi_b|^2$ but by $|\psi_a + \psi_b|^2$. The justification of all this is simple that is: the cancellation (or destruction) required by interference can only occur through the interaction between the amplitudes (which are complex) when they are added (or combined) noting that the probability densities are real non-negative and hence if they are added no cancellation can occur (and hence no interference pattern will emerge).

Quantum interference is characterized (or supposedly so) by a number of features that have no classical parallels. For example, in the double-slit experiment (as a typical example for quantum interference; see § 1.6), quantum interference (supposedly) occurs only if we do not detect which slit the electron passes from.[253] So, even when both slits are open we do not observe quantum interference if we identify (by passive measurements that do not disturb the electrons) from which slit the individual electrons pass. This means that quantum interference is conditioned by our ignorance of the actual path (and hence it is destroyed by our knowledge). As we see, this cannot be understood or explained classically and intuitively where passive observations should not affect the outcome of measurement (noting that physical reality in classical physics is supposedly independent of observation). In fact, this feature is exceptionally weird and non-classical since purely-subjective factors (like the knowledge or awareness of the observer) are not supposed (classically) to influence the observed physical phenomena.[254] We also note that self-interference (supposedly) occurs in quantum interference which is another weird and non-classical feature (at least according to some of its conceptualizations and interpretations). So, quantum interference is distinguished from classical interference by these features (as well as other features), and hence it is different in nature and character from classical interference. Also see § 1.6.2, § 1.6.3 and § 6.13.2 as well as § 6.5.

---

[253] "Electron" here is just an instance of quantum objects used in such experiment.

[254] It should be noted that the focus here (when we use "non-classical" and "classically") is on classical physics which is based (at least implicitly) on the assumption that the observation (as a subjective process) is totally independent of the observed phenomena and hence observation (as such) is an entirely passive process and cannot influence the observed phenomena. Therefore, we exclude scientific (or allegedly scientific) branches like parapsychology from this discussion since the very subject of investigation in such branches is the possible effect or influence of factors like observation or consciousness on external physical phenomena. In fact, these branches of investigation are not classical in any sense and hence their exclusion should be clear with no need to this note (so this note is made for the sake of more clarity). Moreover, they are not conventional "scientific" branches in any sense and hence they should not be compared or associated even with quantum mechanics which also includes similar odd features such as potential effect of consciousness on wavefunction reduction. We should also note that the (alleged) effect of knowledge and observation in quantum phenomena may extend beyond quantum interference and hence it may not be specific to quantum interference (i.e. within quantum physics). Also see § 1.6.2, § 1.11 and § 6.13.2.

5.5 Quantum Tunneling and Scattering

In simple terms, quantum tunneling (which is also known by other names like barrier penetration and tunnel effect) is the ability of a quantum object to "jump over" (or "penetrate through") an energy barrier where the energy of the object is less than the height of the energy barrier and hence this process is classically forbidden because it implies negative energy (see point 1).[255] So, if a quantum particle (e.g. electron) with energy E is confined in a well of finite potential V where $E < V$ then it is possible (quantum mechanically but not classically) for the electron to jump over this potential (or penetrate through the barrier) and (possibly) escape from the well (or more generally be found outside the well regardless of escaping or not).

Although this phenomenon is not observed in daily life (which is the domain of applicability of classical physics) due to its highly improbable occurrence on the classical scale (or due to the invalidity of quantum mechanics in the classical domain), it is (supposedly) observed extensively in quantum systems and is used to explain many quantum phenomena some of which can have direct impact on classical (or macroscopic) phenomena. For example, alpha particle decay of radioactive elements is explained by quantum tunneling. It is also believed that quantum tunneling underlies some large scale physical processes such as the generation of energy in the stars by thermonuclear fusion. Quantum tunneling is also at the root of some very important technological applications (such as the transistor and similar electronic components).

A quantum phenomenon that is similar (and related) to quantum tunneling in its characteristically quantum mechanical behavior is quantum scattering (i.e. off potential barriers and steps). Classically, if the energy of a particle incident on a potential barrier is greater than the height of the barrier then the particle will pass through (or over) with no scattering. However, quantum mechanically such a particle can scatter off the barrier (as well as can pass the barrier), i.e. the probability of reflection at the barrier is not zero (unlike the behavior of classical particles).[256] As indicated, this characteristic quantum behavior of scattering at potential barriers (or rising steps) also applies at dropping potential steps, i.e. when the potential height falls abruptly (see point 9).

As we will see (refer for instance to the following points as well as to § 6.9), this distinct feature of quantum tunneling and scattering of particles should suggest a "wave" characteristic of quantum "particles" (or rather a particle-wave dual nature). This should be inline with the proposition (which we indicated earlier) that many of the characteristic quantum features originate from (or correlate to) this duality. In the following points we outline some of the properties of quantum tunneling and scattering:

1. From a classical viewpoint (or at least so), we should distinguish between acceptable negative energy (like the potential energy of a potential well or the total energy of a bound system) and unacceptable negative energy. In fact, the unacceptable negative energy is the kinetic energy not the potential or total energy. The reason is that the kinetic energy $\frac{1}{2}mv^2$ is essentially defined by the product of two non-negative quantities (i.e. m and v^2) and hence it cannot be negative.[257] On the other hand, the potential energy is conventional (in sign and magnitude) and hence it can be negative. This also applies to the total energy since it includes a conventional part which is the potential energy.
   Unacceptable negative energy should also include the rest energy mc^2 since mass cannot be negative (noting that rest energy may not be classical in a certain sense and according to certain justification of the mass-energy relation; see our book "The Mechanics of Lorentz Transformations"). So in brief, (classically-) acceptable negative energy is potential energy or what includes potential energy.
2. Quantum tunneling can occur only if the quantum barrier is not infinitely high, i.e. only when the confining potential is finite.[258]

---

[255] This may also be conceptualized (or expressed) in terms of conservation of energy instead of prohibition of negative energy (noting that the two are equivalent and essentially the same). It should be noted that quantum tunneling is essentially about massive quantum objects (also see point 3).

[256] It should be noted that classically if the energy of a particle incident on a potential barrier is lower than the height of the barrier then the particle will reflect off the barrier with no possibility of penetration, while quantum mechanically such a particle can penetrate the barrier (as well as can reflect off the barrier). However, here we do not consider this with the scattering phenomenon because it can be classified within the tunneling phenomenon.

[257] This should apply even to the Lorentzian kinetic energy (noting that the Lorentz factor is positive).

[258] There are some details in this regard about delta function barrier which should be sought in the literature (see for

3. The probability of tunneling decreases by increasing the width of the barrier, the mass of the (tunneled) object and the deficit energy (i.e. $V - E$).
4. As suggested by the terminology used in the literature for describing quantum tunneling, the envisaged mechanism for tunneling is either by surmounting (or jumping over) the quantum barrier or by penetrating (or going through) the barrier. It should be noted that although the actual quantum barriers are not tangible barriers like walls we use this distinction for conceptualization and presentation purposes where we exploit the tangible barrier analogy to express certain abstract ideas (as if the "jumping" and "penetrating" represent different abstract mechanisms corresponding to the tangible mechanisms). We should also note that the tunneled quantum object can be envisaged (during the tunneling process) either as a particle or as a wave (and possibly both). So, we actually have several possible "pictures" and scenarios for envisaging and depicting quantum tunneling (some of these "pictures" and scenarios will be investigated further subsequently). In fact, these considerations (or at least some of them) should also apply to quantum scattering with pertinent amendments and adaptations.
5. We may distinguish between two types of tunneling: in one type the quantum object jumps over (or penetrates through) the barrier and appear on the other side of the barrier (i.e. in a classically allowed region), while in the other type the quantum object just penetrates partially (or diffuse) inside the barrier (i.e. without reaching a classically allowed region). An example of the former type occurs in the case of a quantum potential well with finite height and finite width while an example of the latter occurs in the case of a quantum potential well with finite height and infinite width (or indeed even in the previous case where the quantum object just diffuses without escaping from the potential well and reaching a classically allowed region on the other side of the barrier).
6. Quantum tunneling may be explained and justified (at least in some of its instances) by the uncertainty principle where the barrier surplus energy ΔE is supposedly exceeded for a short time Δt (which allows the tunneled object to escape).[259] However, this scenario apparently violates the energy conservation principle (which is forbidden classically) although it avoids the requirement of negative energy (which is also forbidden classically). Other propositions and scenarios about the justifications and mechanisms of quantum tunneling can also be found in the literature.
7. No energy is consumed in the penetration of quantum barriers, and hence the penetrating particle emerges (if it tunneled through and escaped the barrier) with the same energy. This may suggest that this process is not really a penetration process (imitating a similar classical process), but it is about the probability of existence somewhere (i.e. on one side of the barrier or on the other side). This (arguably) may also endorse the suggestion that quantum particles behave during tunneling like waves.
8. It should be noted that tunneling (or penetration) occurs classically in the mechanics of waves (where classical waves can tunnel through or penetrate barriers) and hence tunneling in itself is not a characteristic quantum phenomenon. Yes, what is characteristic to quantum mechanics is the tunneling of particles (or rather particle-like objects) which does not occur in classical mechanics (also see § 6.9). To envisage and justify this classically we can say: classical barriers are continuous (i.e. non-porous) objects and hence by the localization property of particles (see the reference to occupancy in point 1 of § 1.4) classical particles cannot exist inside the barrier (and hence they cannot penetrate it), while waves are not localized and hence they can "penetrate through".[260] We may also envisage this in terms of "jumping" where classical particles cannot jump over a barrier when their total energy is less than the height of the barrier and hence it is not sufficient to allow this jump, while this does not apply to waves and hence they can "jump over". We may also express this in the language of optics by saying: classical barriers are opaque to particles but (partially) transparent to waves (and hence waves can "transmit" or "pass" through barriers while particles cannot). All these forms of visualization and justification are useful in different contexts and circumstances for making analogies and correspondences which are very useful especially in the rationalization and interpretation of these quantum phenomena.

---

instance Griffiths in the References).

[259] This may also be expressed in terms of the position-momentum form of the uncertainty principle.

[260] This should be understood as analogy; otherwise potential barriers are not necessarily material barriers (as will be discussed later).

5.6 Quantum Entanglement

9. Like tunneling, scattering in itself is not a characteristic quantum phenomenon (as it occurs classically for both waves and particles) although quantum objects demonstrate some specifically quantum mechanical characteristics in their scattering (such as the reflection of "particles" incident on a potential barrier lower than their energy, or their reflection at the edge of a potential ditch). In fact, this behavior demonstrates their wave nature (or rather dual nature) since the reflection is a result of an abrupt change (i.e. by stepping up or stepping down) in the potential regardless of the energy of the "particles" and if it is higher or lower than the potential at the boundary of step and regardless of the shape of the change in the potential and if it is in the form of a rising-up step or a dropping-down step. This behavior should remind us of the (partial) reflection of light waves (according to classical wave optics) when they meet an abrupt change in the refractive index of the medium which they propagate through.[261]

Also see § 6.9.

5.6 Quantum Entanglement

Roughly speaking, entanglement refers to the situation when a composite physical system is made of (two or more) correlated subsystems such that the state of one subsystem can be used to infer the state of the other subsystem(s). For example, let have a physical system made of two particles whose total angular momentum is zero and let separate these particles without affecting their angular momentum. Now, if we measure the angular momentum of one of these particles and find that it is equal (considering its signed magnitude with respect to a given direction) to $+\hbar$ then we should know spontaneously (i.e. without measuring the angular momentum of the other particle) that the angular momentum of the other particle is $-\hbar$. So in brief, entanglement refers to a type of correlation between two (or more) subsystems that makes their states "interlaced" or "interdependent". In fact, deep understanding and appreciation of entanglement in its quantum sense requires more technical conceptualization and formulation as well as specific examples of entangled quantum systems and how their states are described quantum mechanically and how they behave. Some of these technicalities and examples will be given and inspected next (as well as in the upcoming parts of the book; see for instance § 6.10 and 9.3.1).[262]

In the following points we outline some of the properties of quantum entanglement as well as its meaning and significance (somewhat technically):

1. In quantum entanglement, the entangled objects are bound together in a mutual state represented by a "superposition" of the states of the individual objects, and hence we cannot identify or distinguish the states of the objects individually and separately. In other words, it is impossible to separate (or "factorize") the wavefunctions of the individual objects since the entire (entangled) system is represented by a single wavefunction that reflects the state of the entire system as a whole and includes all the objects involved in the entanglement. A direct and obvious consequence of this is that no measurement can be performed on a single entangled object without affecting the other object(s) involved in the entanglement (since all the objects share a single wavefunction), and hence by performing a measurement on one entangled object the wavefunction of the entire system will collapse and thus all the entangled objects are affected spontaneously and simultaneously (or at least that is what is supposed to happen).

---

[261] In fact, the behavior of optical waves at refractive index boundaries (e.g. whether the waves reflect partially or not) depends on a number of factors (e.g. the angle of incidence) which are out of scope and hence they are not discussed or indicated here. What is of interest to us is the basic analogy with waves in their reflection behavior.

[262] As we will see, quantum entanglement is not a trivial or ordinary phenomenon and hence its examples are rare and exceptional in their occurrence and nature. In fact, the creation of entangled quantum systems requires specialized and sophisticated techniques (i.e. they cannot be picked off the shelf or found in daily life). We should also note that the above description of entanglement is simple and introductory as well as being generic and not specif to quantum entanglement . A more technical description of entanglement in its quantum sense may sound like this: quantum objects are entangled when their state cannot be factorized into (or expressed as) a product of single-object states, i.e. the (entangled) system as a whole is described by a single wavefunction rather than by separate wavefunctions corresponding to the individual objects. So, the bond between the components of the system is too strong to be represented by a simple "combination" of independent state functions. More clarifications about the meaning and significance of quantum entanglement will be provided as we progress in our investigation. Also see point 13.

For example, if the wavefunction of two electrons is given by:

$$\psi = \frac{1}{\sqrt{2}} \left(\alpha_1 \beta_2 - \beta_1 \alpha_2 \right) \tag{17}$$

where the subscripts $1, 2$ refer to the electrons while α, β refer to their spin state (i.e. spin-up and spin-down), then these electrons are entangled because we cannot factorize this wavefunction into a product of two wavefunctions each representing one of the two electrons. So, a spin-measurement on electron 1 (for instance) whose outcome is spin-up means that the wavefunction collapsed to the state $\alpha_1 \beta_2$ and hence the electron 2 should be in the state of spin-down. On the other hand, if the outcome of a spin-measurement on electron 1 is spin-down then the wavefunction should have collapsed to the state $\beta_1 \alpha_2$ and hence the electron 2 should be in the state of spin-up.

2. It is important to note that entanglement always occurs (or materializes) through certain physical properties of the system such as spin (of entangled electrons) or polarization (of entangled photons). So, if two electrons, for instance, are entangled through their spin then their spins are correlated or interlaced in the above-described way and sense. Hence, it is important (for a proper description and full identification of quantum entanglement) to identify the physical property (and perhaps properties) by which the entanglement is materialized. Therefore, when we talk in this book about quantum entanglement or entangled objects it should be understood that the entanglement is with respect to a given physical property of the entangled objects (such as their spin or their polarization) even if no explicit reference is made to such a property.

3. Quantum entanglement is not affected by the spatial separation of the components of the entangled system. So, if B and C are two entangled quantum objects and they were moving apart, they remain entangled as long as their wavefunction is not reduced by measurement. As we will see, this (in association with the assumption of global, abrupt and instant collapse of wavefunction) should lead to the violation of the principle of locality (see § 1.8).

4. While the global state of the entangled system is generally determined and can be identified,[263] the states of its individual parts are not determined and cannot be identified (as long as the system is entangled). For instance, the global spin-state of an entangled quantum system consisting of two spin-$1/2$ particles created by the spontaneous split of a single particle with zero spin is well known (i.e. the particles should have opposite spin since their total spin should be zero by the conservation of angular momentum), but the spin-states of the two particles are not determined and cannot be identified while the system is still entangled (i.e. without conducting a measurement).[264] Similarly, if we generate a pair of entangled photons (e.g. by using the technique of parametric down conversion), then although the global polarization state of the entangled system as a whole is determined and known (e.g. it has perpendicular or identical polarizations for its two components), the absolute polarization of the individual photons is undetermined and unknown (as long as they remain entangled).

5. Having a single wavefunction for the entire entangled system means that the collapse of the wavefunction when occurs (and if occurs, i.e. by a local measurement on one of the entangled subsystems) is a global process that occurs instantaneously (and applies to the entire wavefunction over its entire spatial extension) due to the single and simple nature of the wavefunction (see point 18 of § 4.1). In other words, there is no meaning of the collapse of "one part" of the wavefunction (i.e. where the measurement took place) but not of the other part (i.e. where the measurement did not take place) because the wavefunction has no parts (to allow this partial collapse to happen) but it is a single entity. To put it negligently we can say: when "one part" of the wavefunction collapses, all the "other parts" collapse without delay.[265] An obvious consequence of the global and instantaneous collapse is that we can influence a remote part of an entangled system instantaneously by conducting a measurement

---

[263] In fact, being determined and can be identified may be conditioned by measurement (i.e. if we conduct a measurement) although realism should dictate otherwise (i.e. it is determined and can be identified even before measurement).

[264] In fact, "spontaneous" events may not be really spontaneous as they may be induced by background fields for instance. However, this should not affect their status as spontaneous in some sense (e.g. not induced by human intervention); moreover this is irrelevant to our purpose.

[265] In fact, "parts" belong to the entangled system but not to its wavefunction.

5.6 Quantum Entanglement

on the local part (i.e. where we are located) of the entangled system.[266]

6. As a result of quantum entanglement (and the global collapse of wavefunction), the outcome of measurement on an entangled object (in a given entangled system) is totally dependent on (and correlated to) the outcome of measurement on the other entangled object (in that system). This is because a measurement on one part of an entangled system (e.g. a system of two spin-1/2 particles created from the spontaneous split of a single particle of zero spin) should determine the state of every part in that system. For instance, if B and C are two entangled objects then a measurement conducted on B will not only determine (i.e. by its outcome) the state of B but it will also determine the state of C, and hence the outcome of any subsequent measurement on C will be known in advance since the state of C (following the measurement on B) is entirely determined by the measurement on B. More specifically, if B and C are two entangled electrons whose total spin is zero then a spin-measurement (say along the z axis) on B whose outcome is spin-up will determine the outcome of any spin-measurement (along the z axis) in the future on C as spin-down even though the outcome of a measurement on C before the measurement on B could be spin-up or spin-down with equal probabilities.

7. As indicated earlier, the instantaneous collapse of the wavefunction of an entangled system by conducting a measurement on one entangled object in the system should imply violation of the principle of locality (see § 1.8) if the entangled objects are separated spatially. In fact, this principle should be violated (in one of its interpretations) even if the collapse of wavefunction is propagated superluminally (but not instantaneously). As we will see (e.g. in § 6.10), the violation of the principle of locality could lead to serious scientific and epistemological problems to modern physics (especially for the relativity theories) and can have serious philosophical consequences (e.g. with regard to our understanding of space and time).[267]

8. From a practical perspective, quantum entanglement is very important (at least tentatively) to many novel and evolving practical applications related for instance to quantum communication, quantum encryption and quantum computation.[268] In fact, most of these applications are built from the beginning on quantum entanglement and how it can be exploited, conditioned and adapted to achieve certain objectives and purposes. Quantum entanglement is similarly important from a theoretical and conceptual perspective especially for its epistemological and philosophical significance and consequences (such as for the interpretation of quantum physics). As indicated above (see point 7), its impact extends beyond the quantum theory and its attachments and accessories to reach other disciplines of physics such as the relativity theories.

9. Referring to points 7 and 8, the instantaneous interaction between the entangled objects (e.g. by conducting a measurement on one of them) requires the existence of a global time (so that the states of the two objects at that specific global time is determined and being well-defined). This should be an obvious violation of special relativity which denies the existence of global time. So, entanglement (in its quantum sense and implications) violates special relativity (at least) from two aspects: by allowing action at a distance (or infinite or superluminal speed of interaction) and by implicitly-accepting the existence of a global time. In fact, the concept of "global time" should be implicit to (and ingrained into) the very concept of wavefunction as a single and simple entity of global nature in space-time (see point 5 above as well as point 18 of § 4.1).

10. As indicated earlier, entanglement of quantum objects is achieved by specific and highly-specialized techniques (such as optical parametric down conversion for polarization-entangled photons). So, it is not a trivial and commonplace phenomenon that can be met in daily life or in primitive laboratories. In fact, this should be inline with our view that quantum mechanics is restricted in validity, i.e. its

---

[266] Although "we can influence", we generally cannot determine the type of this influence and its specific nature because the outcome of the measurement (which determines the type of influence) is probabilistic. However, in our view future research should look for possibilities beyond this limitation (which are generally banned by the current quantum theory).

[267] See for instance our books "The Mechanics of Lorentz Transformations" and "General Relativity Simplified & Assessed".

[268] In fact, the involvement of quantum phenomena (like quantum entanglement) with their weird non-classical nature in these novel and evolving practical applications of quantum mechanics should suggest that these applications are not only about achieving what is achievable classically in a more efficient way (e.g. by increasing the speed of communication or making it more secure or increasing the capacity of storage of information), but they are (primarily or at least partly) about achieving what is not achievable classically.

validity does not extend to the macroscopic world (noting that no quantum-mechanically entangled macroscopic objects seem to exist). Also see § 3.7.

11. As indicated earlier, quantum entanglement may involve more than two quantum objects, and hence it is not necessarily a binary relation or condition (unlike duality or complementarity or uncertainty relations for instance).

12. The wavefunctions of a number of quantum objects may overlap and interact (e.g. electrons in atoms). However, this is not the same as (although it may seem similar to) quantum entanglement where the objects have a single wavefunction that represents the entirety of all the objects involved in the entanglement. So, quantum entanglement is fundamentally different from interaction (of objects) and overlap (of their wavefunctions) and may be considered as a more stringent condition for the bond between the involved components or objects of a composite quantum system. In this regard we may make a simple analogy where entanglement is like a single solid piece of material (say wood or tissue) while interaction (with no entanglement) is like a number of pieces of wood or tissue glued or stitched together.

13. Entanglement (in its basic and generic sense as correlation between physical objects) is not specifically quantum, and hence entanglement (in this sense) can occur even in classical systems. However, the two types of entanglement are fundamentally different. "Classical entanglement" may be conceptualized as an entanglement in our knowledge, i.e. as correlation of the knowledge about the state of one of the entangled objects to the knowledge about the state of the other entangled object rather than being a correlation between the two states. This means that the states of classically-entangled objects are independent of each other in reality, and this is unlike the states of quantum-mechanically entangled objects which are not independent of each other in reality. Yes, what justifies classical entanglement (and leads to entanglement in our knowledge) is a historical correlation (or "entanglement") between the two objects early in their life. For example, if a classical object A is split spontaneously (under the influence of internal forces) into two classical objects, B and C, then the classical spins of B and C are correlated due to their dependence on the classical spin of A, i.e. the sum of the spins of B and C equals the spin of A due to the conservation of angular momentum. However, this correlation is essentially a historical event happened at (and restricted to) the moment of creation of B and C due to their relation to A, and hence B and C will evolve in the future independently as two stand-alone systems. In other words, their spins are not correlated any more apart from this historical correlation which leads to a subsequent correlation in our knowledge about their spins whose total should remain equal to the spin of A (as long as they are not influenced individually by external torques) and hence if we know the spin of B then we know the spin of C and vice versa. In fact, the difference between quantum and classical entanglement is at the root of Bell's argument and the failure of his inequality. The characteristic and fundamental difference between quantum and classical entanglement will be understood and appreciated more during our future investigations (see for instance § 9.3.2 and the appended section § 10.3). In fact, this fundamental difference is what makes quantum entanglement specifically quantum mechanical and a distinctive feature to the quantum theory.

Also see § 6.10.

5.7 Quantum Measurement and Collapse of Wavefunction

Measurement, in its quantum sense and significance, is one of the most prominent characteristic features of quantum theory. Its role and physical significance, as well as its various interpretations, are different to those of classical theory. Also, measurement is the most troubling and controversial issue in the theoretical framework of quantum mechanics (and its interpretation in particular). In fact, there are many ambiguities and controversies about its nature and role in the quantum processes (i.e. as subject of observation) as well as its position within the formalism.[269]

---

[269] It is worth noting that we investigate quantum measurement and wavefunction collapse in this chapter (which is about quantum features) rather than in chapter 4 which may seem more appropriate since the issue of measurement is dealt with in the axiomatic framework of quantum mechanics and hence it is part of it (see § 4.2) because the issue of measurement has no generally-accepted "formulation" (if there is any formulation at all) and hence it more appropriately belongs to

5.7 Quantum Measurement and Collapse of Wavefunction

As explained earlier (following the mainstream literature), quantum measurement (or observation) leads to the collapse of wavefunction (i.e. according to the dominant interpretations and conceptualizations of quantum mechanics) from being a linear combination of sub-states each of which is a possible outcome of the measurement to a specific state that represents the actual outcome of the measurement.[270] This "magic" effect of measurement (i.e. collapse of wavefunction) requires rationalization and justification as well as further examination and assessment (see for instance § 6.11). However, before we go through any detailed assessment we need to clarify some aspects about quantum measurement and wavefunction collapse in the following points:

1. Quantum measurement is essentially an interaction between a quantum system (i.e. the measured) and a macroscopic system (i.e. the instrument of measurement).[271] In fact, it is the most direct and intimate interaction and contact between our macroscopic world and the observed microscopic world. This conceptualization is based on the premise that any quantum measurement should be conducted through the use of a macroscopic device that can be interrogated and read directly by the observer; otherwise the measuring device will be part of the observed quantum phenomenon and hence its "outcome" requires a measurement through the use of a macroscopic device. So, quantum measurement, by its definition and nature, is based on the concept of two worlds (i.e. macroscopic and microscopic) interacting to produce certain effect (which is the observation or the outcome). This is very unlike classical measurement where it is entirely contained and defined within a single world (i.e. macroscopic). This sort of classical "homogeneity" and quantum "inhomogeneity" is what distinguishes the two types of measurement fundamentally and mostly. Also see § 3.4 as well as § 3.6.

2. Although the issue of quantum measurement (in its generic sense) is part of the formalism (see for instance § 4.2), its essence and role is subject to opinion and personal choice and hence most of the detailed aspects discussed in the literature about measurement belong to interpretation although some of which (at least) are incorporated in the formalism and regarded (or alleged) to be part of it (as seen for example in the mainstream conceptualization and formalization of quantum theory which largely follow the views of Copenhagen school). In fact, some of these detailed aspects (like wavefunction collapse) have no known physical mechanisms or explanation and may not even have tangible meaning or sense (to qualify them to be part of the formalism) and hence they are no more than verbal (rather than substantial or physical) explanation or description of these aspects (or at best they reflect the conceptualization of some individuals and their expressions of their personal impressions). So in reality, the formalism is ambiguous about measurement and hence any measurement-related issue in the theory should be handled and assessed from a practical perspective and thus if accepted it should be on the basis of its empirical merit and success (within the limitations of indeterminism which should diminish this success). Also see § 7.10 as well as § 7.6.

3. The "collapse of wavefunction" (which may also be labeled as "reduction" or "projection") may be characterized by the following (noting that we generally follow the dominant views in the literature):
   - It is a totally quantum mechanical paradigm that has no classical parallel (and indeed even "wavefunction" is so).
   - It is an irreversible process (see point 7 of § 4.1).

---

the interpretations and controversies of quantum theory (rather than to its formalism) and represents a characteristic feature of this theory in its odd and non-classical nature.

[270] More technically and explicitly, quantum measurement reduces the wavefunction of the quantum system (which is subject to measurement) from being a superposition of eigenstates (in general) to a particular eigenstate (which is one of the superposed eigenstates) of the measurement operator (where the observable outcome of measurement is the eigenvalue corresponding to this particular eigenstate), and this reduction is what is called the collapse of wavefunction. The probability of collapse to a certain eigenstate is equal (assuming normalization) to the modulus squared of the expansion coefficient of that eigenstate in the superposition wavefunction.

[271] This is how measurement is conceptualized by some physicists, and we think this is one of the appropriate ways (but not the only way) for conceptualizing measurement. In fact, this should be inline with our view that quantum theory is not of general validity (refer for instance to § 3.7 and § 4.7.2) and hence classical and quantum worlds follow different laws in general. It is worth noting that "instrument of measurement" is not necessarily artificial and man-made. For instance, a naturally-occurring deposit of minerals that makes the emission of alpha particles or gamma rays visible (through scintillation) is an "instrument of measurement". Accordingly, pure observation (with no man-made arrangement or intervention) is a type of measurement in this sense.

- It is subject to (or governed by) the measurement postulate rather than the Schrodinger equation (see § 4.2 and § 4.8).[272]
- It is probabilistic (rather than deterministic) in nature (see point 4).
- It occurs globally, abruptly and instantly (see point 5).
- It is specific to certain schools of interpretation and hence it is denied by some schools. However, it is recognized within the most dominant schools; moreover it is used more widely in the language and terminology of quantum mechanics (even among those who do not recognize it in its conventional sense). In fact, in its loose sense it is a characteristic feature of quantum theory that extends across all (or almost all) schools of interpretation (rather than being a characteristic feature of certain schools) although it is controversial from certain perspectives such as the mechanism behind it or the formalism (if any) that governs it.
- The specific form (or nature) of the collapse of wavefunction depends on the type of measurement (i.e. what is the measured property), and hence if we measure momentum then the wavefunction will collapse to a state of a given value of momentum while if we measure spin (of spin-1/2 particles) then the wavefunction (or rather state function) will collapse to an "up" state or "down" state.[273] In brief, the collapse corresponds to (and represents) the nature of the measurement and its objective and hence it is measurement-dependent.[274]
- The collapse of wavefunction is a "collapse of state" (and hence it is an objective physical process) and not a "collapse of knowledge", i.e. a subjective observational process (as if the state of the system before measurement is the same as its state after measurement but we were ignorant before measurement and hence the role of measurement is to reveal this unknown state to us and remove this ignorance).

4. As indicated earlier (see for instance § 5.2.2 as well as point 3 and the preamble of this section), the collapse of wavefunction is generally a random occurrence in the sense that the actual outcome of the measurement (which associates or leads to the collapse) can be predicted probabilistically and statistically but not deterministically and definitely. So, prior to measurement on an individual quantum system we can (in general) only predict that the outcome will be one of a number of possibilities each with a certain probability. Yes, prior to measurement on an ensemble of large number of identically-prepared systems we can predict the outcome definitely and deterministically (in a classical statistical sense). Also see points 4 and 6 of § 4.5.
5. According to the common understanding (and seemingly the dominant view in quantum physics), the collapse of wavefunction is a global and instantaneous[275] "occurrence" and hence it is non-local, i.e. it leads to violation of the locality principle (see for instance § 1.8, § 4.1 and § 6.11).
6. As indicated above, there are many opinions about quantum measurement and its nature and role in quantum physics. In fact, most of these represent different conceptualizations of measurement although the basic physical processes and elements of measurement are generally not subject to controversy. For example, some consider it (or conceptualize it) as an interaction between the microscopic (quantum) system and the macroscopic (classical) measuring apparatus and hence they build their view and analysis on this aspect, while others consider the quantum event (e.g. emission of an alpha particle) as a trigger for the macroscopic response of the measuring device (e.g. click of Geiger counter) and this response is the measurement. Yet another group of physicists and philosophers focus their attention and analysis on the consciousness of the observer and its (alleged) role in the collapse of the wavefunction, i.e. the quantum system assumes a definite state by the conscious attention of the observer and this

---

[272] Actually, quantum measurement is the subject to the measurement postulate (and so subsequently the collapse which is a conceptualization or aspect of quantum measurement according to the mainstream opinion).

[273] In such contexts, "up" and "down" refer to two opposite directions rather than the conventional (vertically-oriented) up and down. This use of "up" and "down" is justified by the convention of using the z orientation (which is conventionally vertical) as a standard reference for spin orientation. See point 9 of § 4.6.

[274] In fact, this phrasing may suggest a single wavefunction that represents all the properties and observables of the system. An alternative view is to have an observable-dependent wavefunction and hence any quantum system can have multiple wavefunctions each corresponding to a certain property and observable. We should also note that the above presentation is rather simple and non-technical (to avoid unwanted complications); a more technical presentation can be given using the concept and terminology of operators (as outlined earlier).

[275] It may also be described as "abrupt" or "discontinuous" or something like these (where the meaning and physical significance of all these descriptions should be obvious).

5.7 Quantum Measurement and Collapse of Wavefunction

123

attention is the essence of measurement.[276]

Although most of these opinions shed light on different aspects of the measurement process (and hence they contribute positively to the analysis and understanding of measurement), no one of these opinions can provide full and satisfactory explanation to this process and its status and role (i.e. within the framework of quantum mechanics). Moreover, there are some extreme non-realistic (and even irrational) opinions based, for instance, on subjective methodologies and idealistic philosophies which despise rationality, sensibility and intuitivity. In fact, quantum measurement is where the most ugly sides of the quantum theory reside.

7. It is important to distinguish between classical and quantum measurement from the perspective of inertness (see § 1.3.3). Measurement in classical physics is essentially a passive process although accidental or casual disturbances can occur where the measuring equipment disturbs the observed phenomenon (e.g. the immersed thermometer changes the water temperature that to be measured or the connected voltmeter changes the potential difference that to be measured), but these disturbances can (in principle) be eliminated (by caution and proper preparations) or at least minimized to become practically non-existing. On contrary, quantum measurement is in essence not passive because even if we assume (arguably) that we can eliminate accidental or casual disturbances (such as the disturbance caused by the "measuring photon" on the "measured electron") we cannot eliminate the intrinsic "disturbance" of measurement represented by the reduction (or collapse) of wavefunction which supposedly is the essence of measurement (or at least an inherent effect and unavoidable consequence of measurement).[277] Also see § 3.4.

8. An issue that seems unclear-cut among physicists is whether the collapse of wavefunction (or whatever "collapse" may be conceptualized in as a representative of the essence of quantum measurement) can occur without measurement (e.g. by a physical disturbance not caused by human observation or intervention).[278] In fact, any opinion or judgment about this issue (and its alike) should depend on a number of factors. For example, it should depend on the view towards realism and whether or not the quantum world and its phenomena exist in a well-determined form regardless of observer and observation (or measurement). It should also depend on what we mean by (and convene about) "collapse" and "measurement" (e.g. whether "measurement" is equivalent or similar to "observation" or not). So, this issue is not determined or can be judged unconditionally and in itself. However, it may seem obvious that quantum measurement (as understood and identified in the mainstream literature) is not the only way to bring about a change in quantum systems and cause "collapse". For example, we have many spontaneous and observer-independent quantum phenomena and events such as spontaneous atomic transitions or spontaneous split of particles or quantum events caused by collisions and scatterings of thermodynamic origin rather than by external interventions (although these, or some of them, may not be classified as "collapse" conventionally and technically).[279]

9. Whether the collapse of wavefunction (as an effect of measurement representing its outcome) can be erased (and hence the outcome of a measurement that has already been performed can be altered subsequently) seems to be a possibility according to some investigations and opinions.[280] In fact, there

---

[276] In fact, there are serious attempts to correlate consciousness to quantum mechanics, e.g. by explaining consciousness quantum mechanically or by justifying the perplexities and dilemmas of quantum mechanics by consciousness. See for example § 1.11 and point 7 of § 6.11.

[277] We note that although the paradigm of "collapse" is associated with certain interpretations, its essence exists in most (if not all) schools of interpretation. Anyway, it is part of the current and widely-accepted axiomatic framework of quantum mechanics and its applications.

[278] This issue may also be posed in other ways, e.g. whether the existence of a conscious observer is a necessary condition for the realization of measurement (although this should depend on the conceptualization of measurement and collapse and the role of factors like consciousness). Refer for instance to § 1.11 and point 7 of § 6.11.

[279] Phenomena like these could be seen as support to "quantum realism" in general (and even to the realistic conceptualization and interpretation of quantum measurement) since quantum events can take place in a definite form independent of any observation or intervention by an intelligent or conscious being. In fact, the denial of this type of events (or rather interpreting them non-realistically) should lead to absolute idealism where even the existence of the physical world outside the limits of observation (e.g. with regard to space, time, perspective, etc.) could be negated and denied. Also see footnote [264] on page 118.

[280] The reader should note that this is not the same as reversing the collapse (which we denied in point 3) although this

are certain experiments (usually variants of double-slit experiment) that are designed and conducted specifically to test this possibility where the results seem to endorse this possibility (according to the claims). If so, then this could (allegedly) be an evidence in support of some weird quantum mechanical features like the effect of knowledge and delayed cause (refer for instance to the subsections of § 1.6).
10. We should finally note that there are other characteristic quantum features that may be considered by some to be part of the measurement problem in quantum theory and hence they may be classified as measurement issues. An example of this is the intrinsic uncertainty (of the uncertainty principle) which is associated with quantum measurements. However, in our view such issues should be investigated independently and outside the investigation of quantum measurement (as we did for example by investigating the uncertainty principle elsewhere) because despite their association with measurement (as being demonstrated by measurement or being symptoms of it) they belong to different (and possibly more fundamental) aspects of quantum physics and hence they should be assigned to their proper place and investigated where they belong more appropriately.

Also see § 6.11.

5.8 Linearity

As seen before, quantum theory is (generally) linear (as demonstrated for instance by the linearity of the Schrodinger equation and the linearity of the quantum operators), and this has some favorable consequences such as:
• The use of linear algebra (with its relative ease and elegance) as a proper mathematical tool for quantum mechanical formulations (see for instance § 1.2).
• The superposition of states and solutions, e.g. new solutions of Schrodinger's equation can be obtained as linear combinations from other solutions (see for instance § 4.2).
• The possibility of normalization by a constant when relevant (see point 16 of § 4.1).

Anyway, the advantages and virtues of linear theories over non-linear theories should be obvious in general and in their practical aspects in particular (e.g. ease of application). However, linearity on the other side may represent a "simple" and "primitive" (and possibly "approximate") property or feature and hence linearity may be blamed for the failure of the (linear) theory to capture certain physical characteristics and aspects. Accordingly, non-linearity may be proposed or introduced to a linear theory as a solution to certain limitations and handicaps of the theory or as a means for extending or elaborating or improving the theory, and this (as we will see) happened to the quantum theory.

In the following points we investigate further the linearity of quantum theory:
1. Whether linearity is an intrinsic and absolute feature of the theory or rather an approximation seems to be a debatable issue. Yes, non-linearity is proposed as a fix to certain problems in the theory or as a support to certain interpretations, and hence from this perspective the latter choice may seem more sensible (and thus non-linearity is adopted unconditionally by some). Anyway, the non-linearity of quantum theory is considered (in some quantum investigations and theorems especially those related to some schools of interpretation) as a possibility.[281] Accordingly, the linearity of quantum theory can potentially be an approximation (rather than being an exact feature and absolute property). However, it seems that the mainstream opinion among quantum physicists is that linearity is an original and exact feature of quantum theory, or at least this should be adopted as a working assumption until it is proved wrong (and hence we keep assuming linearity as long as there is no evidence against it). For example, if some phenomena cannot be explained linearly then this could be an evidence in support of non-linearity (and hence we adopt non-linearity), but if a school of interpretation proposes non-linearity as a fix to some of its troubles or a justification for its claims and allegations then this is not an evidence (and hence we do not adopt non-linearity although it should remain a possibility).

---

may depend on the meaning of reversibility.

[281] In fact, introducing non-linearity onto the basic formulation of quantum theory (as represented by the Schrodinger equation) was considered by some as a possible solution to some of the most difficult and chronic problems in quantum mechanics and its interpretation (e.g. the problem of measurement).

2. Linearity is not specific to quantum theory since linearity can be found in some branches of classical physics as well (e.g. classical electrodynamics or classical wave theories). Hence, linearity is a characteristic feature of quantum theory in itself not as compared to other theories (unlike quantum entanglement for instance which is specific to quantum theory and hence it is a distinctive and unique quantum feature that has no classical parallel).
3. In our view, the linearity of quantum theory (which is talked about frequently in the literature) should be restricted to the "before measurement" (or "between measurements") stage where the quantum evolution is governed by the (linear) Schrodinger equation (see § 4.8). As for the "at measurement" stage, the situation seems unclear (as the issue of measurement itself is not clear) and hence it could be a subject of debate and opinion (assuming that the linearity or non-linearity of quantum measurement theory is sensible at all in the absence of a specific measurement-formulation to be described as such although some measurement-formulation related to certain schools should be taken into account).[282] Also see point 4.
4. Inline with what we said already (see the preamble and point 1 as well as point 3), linearity was proposed as a cause for the problem of measurement in quantum theory, and hence non-linearity was proposed as a solution or fix to this problem.[283] The reason (according to one argument in support of this proposition) is that a linear "measurement operator" will reproduce the superposition state and hence the (problematic) paradigm of collapse is needed then to justify the non-superposition of the outcome of measurement. Accordingly, if the "measurement operator" selects a single and specific eigenstate (from among the superposed eigenstates) then the paradigm of collapse is not needed to justify this outcome since this outcome is produced from the formalism of the theory directly and explicitly. However, this action (i.e. the selection of a single and specific eigenstate) can only be achieved by non-linear "measurement operator" and hence non-linearity is proposed to solve the problem of quantum measurement and avoid the problematic paradigm of wavefunction collapse (and its alike). However, this argument in support of non-linearity (as a solution to the measurement problem) may be questioned and challenged.

5.9 Complementarity

This is another distinctive quantum feature whose essence is that certain pairs of observable physical attributes (e.g. wave-particle) and quantities (e.g. incompatible variables in the uncertainty principle) cannot be observed or detected (and perhaps even exist) concurrently. This feature was investigated (as a principle rather than as a feature) earlier and will be investigated further later on (see § 4.7.3 and § 6.8). As indicated earlier (see point 4 of § 4.7.3), it may be more appropriate to classify complementarity as a quantum feature (as we do here) rather than a quantum principle (and hence its more natural position is here rather than in chapter 4).

5.10 Other Features

There are other features of quantum theory that deserve consideration and investigation (although they are not given special attention here due to their specialized nature and limited significance to our scope and objectives noting that some may be included under some other more general features that have already been investigated). In fact, there are many such features (represented potentially by concepts, principles,

---

[282] In fact, any "collapse" (or "collapse-like") conceptualization of quantum measurement should make the quantum theory for "at measurement" stage essentially (although not necessarily formally) non-linear (and this may be supported by arguments like the one in point 4).

[283] The required non-linearity is usually achieved by adding a non-linear term to the Schrodinger equation. It should be obvious that if this amendment to the Schrodinger equation is adopted then the entire quantum theory (i.e. in both "before measurement" and "at measurement" stages) becomes non-linear (and this is a fundamental change to the quantum theory that incurs many direct and indirect consequences and implications of formal and interpretative nature). It is worth noting that according to certain schools of interpretation (see § 8.3) linearity is sufficient to explain and address even the measurement problem and hence the entire quantum theory (i.e. in both "before measurement" and "at measurement" stages) is linear (which should make the linearity fan happy and this could be one reason for the considerable acceptance of the bizarre many-worlds interpretation among respected physicists).

5.10 Other Features

etc.) that have no analogue or parallel in classical physics (at least in their specific quantum mechanical sense). Some examples are:

- The concept of zero-point energy, e.g. in quantum systems like harmonic oscillator and particle in infinite square well.
- The concept of exchange symmetry (or exchange force or exchange energy).
- The exclusion principle.
- The feature of indistinguishability (see § 3.1).
- The effect of knowledge or consciousness in the quantum process (which essentially belongs to the issue of measurement and collapse; see for instance § 1.6.2 and § 5.7) and self-interference (see for instance § 1.6.3).
- Certain conservation principles such as the conservation of probability or parity (see § 6.12).

However, it should be noted that some of these "specialized features" actually or potentially belong to the interpretation rather than the formalism (and hence they are prime subject to debate and opinion). In fact, some of them may even be questioned and challenged as valid quantum features (although they should still be considered as features of the theory in general since they are part of the debates and investigations of this theory and have supporters among physicists and philosophers of science). We also note that some of the above examples (and their alike) may not belong to the basic quantum theory which is the (primary) subject of investigation of this book.

Chapter 6
Technical Assessment of Quantum Theory

In this chapter we assess the quantum theory from technical perspectives related primarily to the formalism of the theory (noting that another type of assessment related to general aspects will be conducted in chapter 7). However, it should be noted that the assessments of the individual aspects are largely epistemological in nature and try to evaluate the significance of various parts and aspects of the formalism and its characteristic features and assess their implications and consequences.[284] So, in the upcoming sections and subsections of the present chapter we will conduct this technical assessment (rather thoroughly but considering the limitations on the intended scope and available space). However, before that we should draw the attention to the following remarks:

1. One of the most important aspects that should be considered in the assessment of quantum theory is its compliance with realism and the identification of any potential aspects in the theory that can be seen as direct or indirect threats to realism (as summarized primarily by the principles of reality and truth; see § 2.4.1 and § 2.5.2). This also applies to similar fundamental features and principles of science like causality (see § 1.7).

2. From a practical (or empirical) viewpoint, quantum mechanics is a very successful theory and it is vindicated by a huge amount of experimental and observational evidence. In fact, it is created and developed on the basis of experiment and observation and hence it represents an extraction (or formulation) of what is already been observed and tested (rather than being a subsequently-vindicated theorization). The huge technological advances and applications that are based on this theory are another type of evidence for the success of this theory and its undeniable validity. So, we can safely say that this theory is "practically correct" and hence all we need to consider in our assessment, beside its epistemological significance and interpretation (which we investigate in the upcoming parts and chapters), is its optimality as a theoretical structure and if it is a good "reflection of reality" (or potentially even the best possible reflection of reality according to the current state of science) or not. Accordingly, our reservations and question marks (which we express frequently here and elsewhere) about the quantum theory should be understood and appreciated within this context (i.e. not targeting the empirical validity of the theory or demeaning its great achievements).

3. The evidence in support of quantum mechanics is generally seen as evidence for the formalism and not for any particular school of interpretation such as the Copenhagen and many-worlds interpretations (see § 8.1 and § 8.3). However, there are evidence (or alleged evidence) against certain schools, e.g. Aspect's experiment which is based on Bell's theorem and is claimed to refute hidden-variable interpretations (or rather some of them); see § 9.3.2, § 9.3.3 and § 10.4. Also, there are recent attempts to find or create evidence in support of certain interpretations (as well as against other interpretations). In this regard, we should mention the recent slit-grid experiment (see § 6.8.1) which some claimed to refute the Copenhagen and many-worlds interpretations and support the transactional interpretation (see § 8.4). This issue will be investigated further later on (see for instance chapter 8).[285]

4. In principle, it is possible that the limitations imposed on the observations and measurements (in the form of probabilisticity and uncertainty principle for instance; see § 4.7.1 and § 5.2.2) are intrinsic to the (current) quantum theory but not to the physics of quantum systems. This means that in

---

[284] In fact, most of the sections of this chapter have titles identical to titles of other sections in the previous chapters (mostly chapters 4 and 5) where in the upcoming sections we assess the contents of the preceding sections.

[285] In fact, even some of the general experiments of quantum mechanics (i.e. those which are commonly recognized as having no significance for endorsing specific interpretations) have been seen by some as evidence for specific interpretations.

principle we may be able to find a totally-deterministic quantum theory without probabilisticity and without uncertainty principle. In other words, the probabilistic nature and the uncertainty principle (and their alike) are features of the existing quantum theory and not necessarily features of the physics of quantum systems (as represented for instance by a different theory or a different formulation of the quantum theory).[286] In our view, many of the confusions and controversies in this regard originate from the implicit assumption of the uniqueness of science, i.e. there is only one possible theory about quantum phenomena and hence if the existing quantum theory requires probabilisticity and uncertainty principle then theses features (and their alike) are intrinsic to the physics of quantum phenomena.

So in general, we should distinguish in any assessment between the aspects and features (whether technical or general) of the (current) quantum theory and the aspects and features of the (actual) physics of the quantum world. In fact, this distinction is generally not easy (especially if we believe in the uniqueness of science; see § 2.4.3).

5. We note that partial assessment (to various technical aspects of the quantum theory) has been done previously in this book where and when it was felt to be appropriate or necessary. So, we can consider this chapter to be about *further* technical assessment. This similarly applies to chapter 7.[287]

6.1 Wavefunction

We assess here the quantum mechanical concept of "wavefunction" and its role and significance in the quantum theory. This assessment (as well as some further investigation) is presented in the following points:

1. As discussed before, the wavefunction (or matter wave) is a sort of mathematical device rather than a physical entity. In fact, the complex nature of wavefunction (as well as other considerations and indications) should be a sign for the "mathematical device" nature of the wavefunction since physical entities are naturally represented by real quantities because they are observable (see § 3.2; also see § 7.4). However, being a mathematical device should not mean that it is entirely fictitious or it does not reflect an aspect of reality (as a probability wave). Nevertheless, being a mathematical device (even though reflecting an aspect of reality) may indicate that using the "wavefunction" model in the description of reality is not an ideal way for representing reality (also see point 2 of § 2.5.4).

2. Apart from being complex (and hence not observable, at least directly) which compromises its physical reality, the wavefunction faces other challenges to its reality. For example, the collapse of the wavefunction (in a rather mysterious way) by measurement put question marks (at least by some physicists) on its reality and if it really represents a physical entity. In fact, there are other signs and indications for the non-physicality of wavefunction (see for instance § 3.2 as well as the next points).

    However, it is important to note that some physicists have a different opinion such as considering the wavefunction a *physical* pilot wave. There are also more explicit and less compromising opinions about the physicality of wavefunction. In this context we may quote the following (from "Interpreting the Quantum World" by Asher Peres): "... the wave function is a genuine physical entity, not just an intellectual tool invented for the purpose of computing probabilities. Even if the wave function is not an ordinary physical object, it still has ontological meaning: it represents the factual physical situation, not only our subjective knowledge of nature".

3. It is claimed that the wavefunction should be associated with ensembles of (independent, identical and identically-prepared) quantum objects rather than with individual quantum objects unless the wavefunction represents an eigenstate of an observable corresponding to a given eigenvalue.[288] Although

---

[286] In fact, even the current quantum theory in its standard probabilistic interpretation (as represented by Born's rule) may be questioned, i.e. is the standard interpretation of the wavefunction as a probability amplitude the only possible interpretation?

[287] We should also note that there is some intentional repetition mainly for the purpose of making each section and part self-contained (as much as possible and with observation of other considerations) or for the purpose of developing the subject gradually and from different perspectives and in different contexts (which helps in understanding and memorizing).

[288] In fact, this is based on following the dominant interpretations (or more specifically by excluding the hidden-variable scenario). We should note that if the association with ensembles is accepted then it should (for the sake of generality and consistency) include even the case of eigenstate. We should also note that this claimed association is based on (or

6.1 Wavefunction

this interpretation of wavefunction seems sensible and useful it does not seem to be valid in general (e.g. for the wavefunction of a single quantum object that is not a member of an ensemble).[289] Moreover, it does not seem to be able to address all the problematic issues about wavefunction (noting that in some cases and circumstances an assignment of wavefunction to individual objects is required or at least seems sensible if not necessary).[290]

In fact, if this interpretation is accepted then this may be another indication to the non-physical nature of wavefunction since its association with ensembles (rather than individual objects) of rather generic and subtle nature makes it less physically-specific (in comparison to its association with individual objects).

4. The association of wavefunction with ensembles of quantum objects rather than with individual quantum objects (as explained in point 3) may also make the paradigm of "wavefunction collapse" redundant (also see point 13 of § 6.11).

5. The wavefunction of a given quantum system depends on the state characterized by that wavefunction, and hence a given quantum system can have different wavefunctions depending on the characterized state. For example, the wavefunction that characterizes the spin of a quantum system is not the same as the wavefunction that characterizes its energy or momentum.[291]

6. As well as the dependency of the wavefunction of a given quantum system on the state characterized by that wavefunction (see point 5), the wavefunction of a given quantum system and for a given characterized state can have different mathematical forms and representations (e.g. in configuration space and in momentum space). In fact, this (as well as the dependency) should be inline with the nature of wavefunction as a mathematical device.

7. In our view, non-locality is ingrained in the concept of wavefunction from different aspects and perspectives. For example, the instantaneous collapse of the wavefunction is one of these aspects. In fact, even the statistical interpretation of the wavefunction (as a probability amplitude of position over the entire space or universe) should imply non-locality because no time delay is assumed for the propagation of this probability throughout space (since no time delay is assumed for the "propagation" of the wavefunction and any localized influence on it).[292] Accordingly, we may say: wavefunction is non-local in its emergence as in its collapse (i.e. it emerges globally and instantaneously and collapses globally and instantaneously). So, the paradigm of "wavefunction" in its quantum mechanical sense (and according to its commonly-accepted standard interpretation as a probability amplitude) is intrinsically non-local.

8. It is worth noting that the Born interpretation of wavefunction provides a "discrete" probabilistic interpretation to the "continuous" wavefunction and hence it is also useful from this perspective (in addition to its usefulness in providing a basic physical interpretation to the wavefunction) since it avoids potential conflicts and inconsistencies between various models and formulations in quantum theory. For instance, Born's interpretation was used historically to avoid a potential clash or contradiction between the wave mechanics of Schrodinger and the matrix mechanics of Heisenberg.

9. If we have to apply the paradigm of "wavefunction" to classical macroscopic objects (assuming such

---

justified by) the probabilistic nature of the quantum mechanical predictions. Also, the association of wavefunction with ensembles (rather than individual objects) should be inline (and in spirit) with the Born statistical interpretation of wavefunction.

[289] In the case of a single quantum object the probabilistic interpretation, for instance, is inevitably related to an individual object.

[290] The above-mentioned case (i.e. "an eigenstate of an observable corresponding to a given eigenvalue") may indicate the necessity and sensibility of this in all cases.

[291] As indicated earlier, there is another conceptualization to the wavefunction in which it is considered as a unique entity (for its system) but it represents various aspects of the system depending on the actual consideration. In fact, the wavefunction in this conceptualization could be seen as more physical than in the above conceptualization.

[292] For example, if a quantum system (say an atom) exists at certain position and certain time then its components (e.g. the electrons of the atom) should be localized "in reality" (as can be identified by measurement) which is not entirely consistent (in terms of locality) with the probabilistic interpretation since the components may be allowed subsequently (on the basis of probability) to be at positions that cannot be reached without violation of locality within a certain time frame. Yes, this inconsistency seemingly depends on the implicit assumption of "localized in reality" which may be disputed.

application is valid in principle) then either we apply this to the macroscopic objects as they are (e.g. wavefunction of cat or ball) or as composite objects made of large number of quantum microscopic objects. However, we should note that in the first case there are theoretical and conceptual difficulties while in the second case there are practical difficulties. Also see point 3 of § 6.7.

Also see § 4.1.

6.2 The Axiomatic Framework of Quantum Mechanics

As explained earlier, the axiomatic framework of quantum mechanics (as embedded in its postulates) represents the backbone of the quantum theory and the essence of its formalism. In the following remarks we outline some assessment aspects of this framework and its significance to the quantum theory:

1. The axiomatic framework of quantum mechanics is not agreed upon and hence there are controversies and differences about the number and content of the postulates as well as on their status and if some of them could or should be proved or not. In fact, the axiomatic framework differs not only between individual physicists but even between different schools of interpretation (see chapter 8) where the interpretation of some schools requires the inclusion or exclusion or modification of certain postulates. So in brief, the axiomatic framework is subject to debate, controversy, opinion and interpretation.

2. The most controversial aspect of the axiomatic framework is the postulate of measurement. For example, according to some opinions and interpretations the entire issue of measurement should be excluded from the quantum theory and its formalism (including its postulates), while according to other opinions the issue of measurement is at the heart of the quantum theory and its axiomatic framework. There are also many controversies and differences in opinion about the conceptualization and formalization of measurement and this is reflected directly and indirectly on the axiomatic framework.

3. The axiomatic framework (including the Schrodinger equation which is referred to in the postulates) represents the essence of almost the entire formalism of the quantum theory. However, some extensions and appendices (such as some principles; see § 4.7 and § 6.12) should be added. We should also add some detailed issues related for instance to the formulation of spin (which is not incorporated in the Schrodinger equation).

4. As indicated earlier, the postulates of quantum theory may contain elements of interpretation (e.g. the collapse of wavefunction and the Born statistical interpretation) although this could be subject to debate and opinion. So, although these postulates are supposed to provide the technical and formal framework for the quantum theory they also embed (definitely or tentatively) certain elements of interpretation. This can be inferred, for instance, from the inclusion of paradigms like "wavefunction collapse" whose interpretative nature and significance cannot be denied. This may also be deduced from the fact that these postulates differ in number, nature and content between different individuals and schools of interpretation, e.g. the postulate of measurement is denied by some schools while "wavefunction collapse" may be denied or interpreted differently by different schools.

Also see § 4.2.

6.3 Schrodinger's Equation

Schrodinger's equation is commonly regarded as a postulate of quantum theory (and hence it is part of its axiomatic framework). However, it may also be "derived" from other principles of the theory following a different route of theorization and formulation. But it is worth noting that such derivations are questionable from the perspective of having real substance rather than being an exercise in ordering and structuring. In the following points we try to assess Schrodinger's equation and evaluate its role in the quantum theory:

1. One of the major and obvious limitations of the (time-dependent) Schrodinger equation is that it describes the tempo-spatial evolution of the quantum system prior to measurement (or rather between measurements), but it cannot describe the process of reduction (or collapse) of the wavefunction as a result of measurement or predict the outcome of measurement. This can be easily seen from the Schrodinger equation (refer to Eq. 6) which has no explicit or implicit reference to measurement or the

6.3 Schrodinger's Equation

collapse of wavefunction by measurement. In fact, "collapse" and its alike are interpretative aspects although they are ingrained in the current quantum theory and seen as part of its formalism. Hence, these aspects are not part (at least directly and intrinsically) in the formalism as represented by the Schrodinger equation although the concept of "collapse" is part of the mainstream axiomatic framework and hence it is seen as part of the formalism beyond the Schrodinger equation (but noting that this framework seemingly contains elements of interpretation).

2. Whether the linearity of quantum mechanics (as a result of the linearity of Schrodinger's equation; see § 4.3) is genuine or it is an approximation is not very clear although it is generally assumed (and even declared explicitly by some) to be genuine. However, it is worth noting that certain formulations (mainly related to the interpretations of quantum mechanics) are based on assuming non-linearity by adding non-linear term(s) to the Schrodinger equation.[293] Anyway, the linearity of Schrodinger's equation should be seen as a kind of limitation in this equation if the linearity of quantum mechanics is actually an approximation (rather than being an intrinsic and genuine feature of the theory). Also see § 5.8.

3. As well as the limitations indicated in the previous points (i.e. its restriction to "between measurements" and potential limitation by linearity), the Schrodinger equation has other limitations. In fact, the limitations of this equation (as a basis for quantum physics) are almost limitless. For example, it has no exact (analytical) solution except for the simplest of quantum systems and hence most of the solutions of the Schrodinger equation are obtained numerically or approximately (e.g. by using simplified quantum models or perturbation methods) or tentatively or empirically or ... etc. Also, this equation does not include "non-classical" effects and elements (such as those attributed to Lorentz mechanics) and hence even the "exact" solution of the simplest systems generally requires many corrections (e.g. accounting for spin or ambient fields) as seen for instance in the many corrections introduced on the solution of the hydrogen atom as discussed extensively in the standard textbooks of quantum mechanics and atomic physics. In fact, this equation is more of an ansatz (or starting point or an initial attempt) than a realistic formulation to realistic quantum systems.

4. The limitations of the Schrodinger equation (as well as its solutions, i.e. wavefunctions) are not limited to the rather formal aspects that we outlined in the previous points but they extend to its physical content and substance due to the presence of many sources of ambiguity and haziness. For example, there is no clue or indication in the equation itself to the statistical nature of its solution as a probability amplitude (and that is why we needed the standard probabilistic interpretation of Born) or the physical nature of the tempo-spatial evolution of the system and its wavefunction. In simple words, we may describe the Schrodinger equation as superficial (or shallow in its physical content) rather than fundamental (or substantial), and this is reflected (at least partly) in the many epistemological ambiguities and deficiencies that we face when we try to interpret quantum physics and make sense of its formalism. The fact that this equation is a postulate (rather than being genuinely-derivable from more basic and rational principles and facts) may also reflect the "empirical" (or "trial and error") nature of this equation and its theoretical deficiencies and limitations. These ambiguities and limitations make even the application and use of the formalism itself require a sort of "initial" or "basic" interpretation (in the form of the standard or probabilistic interpretation of the wavefunction represented by the Born interpretation). In fact, this feature distinguishes quantum physics from all other major scientific theories where the interpretation of the formalism in the other theories is generally seen as an additive or bonus rather than a necessity and requirement for the formalism and its application. In other words, the formalism in the other theories rests on a set of well-defined and well-understood (at least in principle and in general) paradigms and concepts rather than a vague "wavefunction" which no one knows (at least initially and before the proposal of Born's interpretation) what it means.[294]

---

[293] It is worth noting that linearity is essential to many aspects and features of the current quantum theory, and hence non-linearity (if adopted) should require a major change to the theory. For example, superposition (which is a very important feature of the current quantum theory) is directly linked to the linearity (and homogeneity) of the Schrodinger equation and this feature will be compromised by non-linearity.

[294] When Schrodinger proposed his equation (using a rather "empirical" approach based on the de Broglie matter wave

5. As indicated earlier, the Schrodinger equation has extended forms or improved versions (like the Dirac and Pauli equations) to include, for instance, Lorentzian and spin effects and address some of the limitations of this equation. However, these extensions and improvements are essentially about addressing some limitations related to its scientific content and hence other instances and types of limitation (such as those indicated earlier like its restriction to "between measurements") are not usually addressed even by these extended forms and improved versions.[295] So in brief, the extensions and improvements introduced by advanced quantum mechanical branches (like quantum field theory) on the basic quantum theory (as a result of extensions and improvements on the Schrodinger equation) address only some of these limitations.[296] Anyway, our interest in this book is restricted to the Schrodinger equation in its basic form (and hence we will not discuss issues beyond this such as the interpretation of quantum mechanical branches based on these extensions and improvements).
6. A feature that is worth noting in the Schrodinger equation (see Eq. 6) is the association of energy with time. Also see point 4 of § 4.7.1.

6.4 Indeterminism

The main issue in the assessment of indeterminism (in its quantum sense and interpretation) is whether it affects the definiteness of reality, and this could affect realism as well as causality (at least in its classical and familiar meaning which is represented by the principle of causality; see § 1.7). In fact, quantum indeterminism has three main forms: superposition, probabilisticity and uncertainty.

As for **superposition** (see § 5.2.1), it should affect (at least in some of its interpretations) the definiteness of reality and this should compromise a (potential) attribute of the "existence of reality" principle (if we accept this attribute).[297] However, as indicated earlier (see for instance § 2.5.2) the damage to this attribute may not have a serious and harming epistemological consequences if we modify our epistemology to accommodate a reality that may not be completely definite. Yes, there are possibly other damaging and harmful (epistemological) aspects to this choice such as violating causality. We should also note that if quantum superposition suggests multiple realities then this (in some of its interpretations and conceptualizations) could compromise the principle of uniqueness of reality (see § 2.4.1).

As for **probabilisticity** (see § 5.2.2), it should affect causality (in its classical and familiar interpretation; see § 1.7) and hence if we want to keep causality then we may need to reformulate the causality principle to accommodate "probabilistic causality" where a superposition state can cause a probabilistic outcome (i.e. there is a correlation between a cause and a "fuzzy" or probabilistic effect). However, this amendment to the causality principle should not fix the problem entirely because the exact outcome of the cause still requires a cause (unless we accept a form of a hidden-variable choice). In fact, there are possibly other damaging and harmful (epistemological) aspects to this choice of modifying the causality principle.

As for **uncertainty** (see § 5.2.3), it should affect (at least in some of its interpretations) the definiteness of reality (and hence we repeat what we have said about superposition in a previous paragraph). This issue will be investigated further in § 6.6.

In the following points we investigate and assess quantum indeterminism further:
1. As we will see later (refer to point 2 of § 7.6), some forms of indeterminism should tarnish the empirical success of quantum mechanics, i.e. the empirical success of quantum theory is limited by indeterminism

---

hypothesis) he interpreted the wavefunction as something related to charge density. This confusion should reflect the level of ambiguity and uncertainty associated with this equation. The "obviousness" of the physical meaning and significance of the wavefunction in modern times is the result of intensive education and indoctrination.

[295] In fact, some schools of interpretation have formulations that supposedly address such types of limitation.

[296] In fact, the epistemological and interpretative issues and limitations of quantum mechanics are the least addressed (if addressed at all) by these extensions and improvements. So, effectively many of the problematic issues about the epistemology and interpretation of quantum mechanics equally apply even to these branches because these issues are deep-rooted in the fundamental principles of quantum mechanics which propagate and permeate throughout these branches. So we can say: despite the restrictions on our scope in this book (since we are interested in the basic quantum theory) many of our criticisms and assessments related to the epistemology of quantum physics and its interpretation have more general validity and they apply even to these branches.

[297] As indicated earlier (see § 5.2.1), quantum superposition could also suggest multiple realities since the superposed states should have some sort of reality.

6.4 Indeterminism

since indeterminism imposes limits on the descriptive and predictive power of the theory. However, by the principle of non-uniqueness of science (see § 2.4.3) it is possible in principle to find a different "quantum theory" that is more successful empirically (by lifting at least some of the limitations imposed by indeterminism) as well as more successful epistemologically (by having a rational and consistent interpretation which the current quantum theory does not have). Also see point 4.

2. It is important to note that there are other forms and demonstrations of indeterminism in the quantum theory (i.e. other than the aforementioned three forms). For example, indistinguishability of quantum particles (see § 3.1) is a form of indeterminism. So, the above three forms of indeterminism represent the main and most prominent forms of quantum mechanical indeterminism but not the only forms. As we can see, the effect of these types of indeterminism are not the same (neither in the actual physics of the quantum world nor in our quantum mechanical formulations).

3. Whether quantum indeterminism belongs to reality (i.e. the reality itself is not definite) or to our knowledge (due to our ignorance despite, or regardless of, the determinism of reality) is not only a controversial issue but it is also dependent on the type and instance of indeterminism. For example, superposition may suggest indeterminism in reality while certain uncertainties may suggest indeterminism in knowledge. Anyway, the nature of indeterminism is essentially an interpretative aspect (although it usually creeps to the formalism through dominant interpretations).

4. It may be impossible for the (current) quantum theory to be deterministic (because indeterminism in its various forms is embedded and implanted in it) as this may have been shown for instance by the Bell argument which leads to the supposed refutation of local hidden-variable theories (see § 8.2 and § 9.3.2). However, this does not rule out the possibility that the quantum world and quantum phenomena can be described by a totally-deterministic physical theory (which is different from the current quantum theory) especially if we lift the condition of locality (noting that we do not believe in the principle of locality; see § 1.8). In fact, this is based on our firm belief in the non-uniqueness of science (see § 2.4.3). Accordingly, if this is accepted then many of the disputes (as well as the philosophical and epistemological debates and arguments) should disappear naturally. However, the problem is that the (tacit) view of "uniqueness of science" is dominant and seems to have an overwhelming acceptance among scientists and philosophers of science. From our perspective (and regardless of uniqueness or non-uniqueness of science), quantum mechanics cannot be the final and ultimate theory about the quantum world, and hence lifting (some or all types of) indeterminism in the present quantum theory by a future quantum theory should be seen (at least) as a possibility if not a necessity.

   Anyway, in our view quantum mechanics (despite its indeterminism) cannot be a threat to realism in general. Yes, if quantum mechanics is the only possible theory then it possibly (but not necessarily) can be a threat to realism (or rather to some aspects and instances of realism at the quantum level); see § 7.14. This should also apply to causality in general (see § 7.12).

5. Indeterminism in classical systems can be of classical type (or origin) and of quantum type (or origin).[298] The former is seen in purely-classical experiments, for example, where the outcome is probabilistic (in a classical sense) such as when we throw a coin expecting a head or a tail. The latter is seen in quantum measurement where indeterminism at the quantum level is reflected in indeterminism at the classical level for the classical or macroscopic outcome of the measurement, e.g. whether the light signal of the measuring device will be green or red according to the quantum outcome.

6. Quantum indeterminism is embedded in the axiomatic framework of the quantum theory (e.g. through the statistical interpretation of wavefunction or through the measurement postulate) and hence the paradigm of indeterminism may be seen as part of the formalism. However, it should be remembered that the quantum (unitary) evolution between measurements is deterministic (although some forms of indeterminism, e.g. superposition, may still exist during this stage).

7. It is important to note that quantum-type indeterminism can also be found in non-physical sciences like biological and social sciences (although with different extents, conditions, reasons, etc.). In this regard we refer the reader to the "intrinsic uncertainty" (as a form of characteristically-quantum type of indeterminism) found in certain branches of science and even in daily life (see for instance point 7

---

[298] We should take note of the conceptual distinction between classical indeterminism and quantum mechanical indeterminism (as explained earlier).

of § 6.6). So, "quantum indeterminism" is not entirely and exclusively quantum (although it may be so when compared to corresponding aspects in classical physics).

8. An issue related to the previous point is whether or not determinism is violated by freewill (and hence freewill is a form or an instance of indeterminism in the physical world). Moreover, if determinism is violated by freewill, does freewill lead to violation of causality (at least in its classical or familiar sense) or not? These issues have been investigated earlier in § 1.7 and § 2.15.

6.5 Quantum Interference

Quantum interference was analyzed and assessed (rather casually) from some aspects within our previous investigations. The important point to note here is that quantum interference (as defined earlier; see § 5.4) is the essence of the "wave" nature of quantum objects and phenomena, and this should be inline with the probabilistic interpretation of wavefunction. Hence, if we consider the probabilistic significance of quantum interference we can appreciate why quantum waves should not be seen as physical waves (like the electromagnetic waves for instance).

On the other hand, if we note that in quantum mechanics the probability density is represented by the square of the modulus of the wavefunction (i.e. $|\psi|^2$) then we can see (from the perspective of quantum interference) that what actually happens in the double-slit experiment (and its alike of quantum interference examples) is to add the amplitudes of the quantum waves (or probability waves) rather than their probability densities. This means that in quantum interference the quantum waves interact to produce interference (instead of adding the probability densities). This may suggest that wavefunctions have a kind of physical reality that allows them to interact and combine (although this reality is not similar to the reality of classical waves) and hence wavefunctions may be more than just what "probability amplitudes" (or "mathematical devices") might suggest. Also see § 5.4 as well as § 6.1.

6.6 The Uncertainty Principle

As we know, the uncertainty principle imposes a limit on the precision of simultaneous measurements of certain pairs of variables (i.e. incompatible or complementary variables), and hence it is an instance or form of indeterminism which we investigated earlier (see for instance § 6.4). In fact, the uncertainty principle was investigated rather extensively earlier in § 4.7.1, as well as in other positions and contexts. In the following subsections we investigate and assess this principle further from different perspectives. However, before that we would like to draw the attention to the following points:

1. The rationalization of the position-momentum form of the uncertainty principle (see Eq. 13) is rather obvious and intuitive (according to the wavepacket model and the associated technicalities of the Fourier analysis for instance). But the rationalization of the time-energy form (as well as other forms) of this principle is less obvious and intuitive and could be ambiguous (although they are usually justified by similar physical and mathematical arguments including Fourier analysis).[299] A simple and "intuitive" rationalization of the time-energy form may be given by the following argument (as well as similar arguments): if we have a rapid change in the observable (and hence the window of observation Δt for a significant change is small) then ΔE should be large (and vice versa).[300] Similar simple and intuitive rationalizations may also be suggested for the other forms of this principle. However, these rationalizations should be treated as imitations (or even "educational artifacts") rather than strict and rigorous scientific arguments of general validity.[301] The ultimate "rationalization" of all forms is the experimental evidence (and the formulation that is supposedly based on this evidence) which will be investigated in point 3.

---

[299] According to the Fourier analysis, there are restrictions on the simultaneous determination of well-defined time and well-defined frequency of a signal (since they have conflicting demands).

[300] For example, in the stationary states the window of observation Δt is "infinite" (since there is no change) and hence ΔE is zero.

[301] Noting that t is a parameter rather than a dynamical variable (unlike other variables in the uncertainty relations), this may be seen as another source of troubles and an indication to the problematic nature of the time-energy form of the uncertainty principle.

6.6 The Uncertainty Principle

2. The meaning of Δt in the time-energy form of the uncertainty principle (see Eq. 14) and what it stands for depends on the physical situation and the observed phenomenon that is subject to this principle (or supposed to be so). This opens the gate sometimes for rather arbitrary use and inappropriate exploitation of this form and hence compromises the rigor (and even soundness) of some quantum mechanical arguments. This may also apply (but less overtly and excessively) to other forms of this principle. Also see § 6.6.2 and § 6.12.

3. It seems that so far there is no sufficient precision in the quantum measurements to test and verify this principle experimentally and directly, e.g. by simultaneous measurements of two incompatible observables to see if the measurements can or cannot violate this principle (where in the latter case a necessity should be shown). Accordingly, the uncertainty principle is more of a "theoretical paradigm" (that should be used tentatively and cautiously) than an established principle, and hence it should be treated as such. Yes, the theoretical derivations of this principle (in its general form as well as in its various specific forms) may be endorsed "experimentally" but indirectly if they are based on experimentally-validated principles and premises (and this does not seem to be the case in general). Also, the principle (in its various specific forms) may also be vindicated experimentally (but indirectly) by the applications and instantiations that are based on it theoretically.

4. The experimental value of the uncertainty principle is limited because other types of uncertainties and errors (e.g. from the imperfections of the measuring equipment) are usually much larger than the uncertainty of this principle and hence they mask and dilute its effect. Therefore, the value of this principle is essentially theoretical (i.e. in calculations and arguments) as well as in instances and applications based on this principle (where this principle for instance underlies a large scale phenomenon; see point 16 of § 4.7.1).

5. As indicated earlier (see for instance § 3.5), casual forms of uncertainty (due for instance to limitations on the accuracy of the measuring devices or to experimental and calculational errors and approximations) are common to both classical and quantum physics, while the intrinsic form of uncertainty (due to the uncertainty principle) is restricted to quantum physics (or that is what is supposed to be; see point 7).

6. There are many attempts (mostly attributed to Einstein) to refute and invalidate the uncertainty principle by designing and employing thought experiments. Regardless of the validity and invalidity of the uncertainty principle in itself, most of these attempts are worthless and can be easily challenged and refuted. For example, some of these thought experiments essentially appeal to our classical intuition (e.g. about reality) and hence their value is based on the value of such intuition (which is questionable in the quantum domain). Moreover, thought experiments (in general) have no such magic power that enables and entitles them to determine how the physical reality should be or how it must behave.[302] However, it is fair to say that some of these attempts may be a response to similar worthless thought experiments and theoretical arguments to establish and validate the uncertainty principle (and hence their worthlessness is justified by the worthlessness of their opponents).

7. Although the uncertainty principle (as such) does not exist outside the quantum theory (and hence it may be regarded as a characteristic feature of the theory), the paradigm of "intrinsic uncertainty" which characterizes this principle is common outside the quantum theory (even in our daily life).[303] For example, the value of a house (of certain type and conditions) in London is (say) £300k with an intrinsic uncertainty of ±10k. This means that as long as it is sold between £290k and £310k there is no gain or loss (due to this intrinsic uncertainty) but there will be gain or loss if it is sold

---

[302] In fact, we have strong reservations about the use of thought experiments in science especially when they are used to establish or judge premises and propositions of experimental nature. Thought experiments are prone to illusion, confusion, contradiction, etc. since they (by their nature) are not subject to any rule or restriction of physical reality and hence they can easily drift and stray. Yes, rigorous theoretical thinking that is based on obvious facts and acceptable premises and hypotheses (whether validated experimentally or justified theoretically) is acceptable but even this should be subject to experimental tests and validation (noting that thought experiments generally do not fall into this category). Also see § 7.8.

[303] In fact, the rationalization (or "proving") of the uncertainty principle by the Fourier analysis (which is essentially a classical theory of waves) should indicate the existence of instances of the "intrinsic uncertainty" paradigm (if not the paradigm itself or the principle) in classical physics.

outside these limits (say £280k or £320k). This is unlike an ounce of gold (at given time, location, caratage, etc.) whose value is sharply defined (down to the smallest defined currency unit such as cent or penny) although there may be some extrinsic (or accidental) uncertainty due for instance to the transaction (whose fee may be included in the value and may vary by type and broker). So in brief, although the uncertainty principle is (seemingly) a characteristic feature of quantum mechanics, "intrinsic uncertainty" (which characterizes this principle) is not.

8. The uncertainty principle may be challenged by quantum entanglement where the complementary variables in this principle can (supposedly) be determined simultaneously by measurements on the entangled objects of the entangled system (i.e. by measuring one variable on each entangled object) as in some forms of the EPR argument (see § 9.3.1). However, the sensibility and validity of this challenge should depend on a number of controversial factors such as the nature of entanglement and the potential role of hidden variables as well as the justification of the uncertainty principle. This issue will be clarified further in the future (see for instance § 9.3.2 and § 10.3).

Also see § 4.7.1.

6.6.1 Justification of the Uncertainty Principle

We can find many formal and informal arguments and "proofs" for justifying and establishing the uncertainty principle. Also, there are many opinions and controversies about these justifications and their rationales. In the following points we try to investigate and assess these justifications:

1. In the literature of quantum physics there are two main types of general arguments for establishing and justifying the uncertainty principle:

   • The arguments that are based on the impact of the measurement probe on the observed object (where one incompatible quantity is disturbed by the measurement of the other incompatible quantity). This type of arguments can be easily challenged by claiming that this sort of uncertainty is a symptom of the measurement process and hence in principle it is possible to find a method of measurement in which the (non-measured) quantity is not disturbed (at least substantially and significantly) by the process of measurement.

   In more details, on inspecting the literature of quantum theory about the uncertainty principle we can find many claims (explicit and implicit) that this principle is based on the premise that any measurement at the quantum level requires interaction with the quantum object and hence the object will be affected by the measurement process (or by the measurement probe). For example, some authors justify the uncertainty principle by arguments like this: if we want to obtain a high-precision measurement of the momentum of an electron then we need to hit it by an energetic photon and this (i.e. the impact of the photon) will affect the position of the electron and reduce the certainty in the measurement of its position.

   This sort of arguments may be seen as pedagogical and educational, i.e. to make the principle more acceptable and digestible and inline with our physical intuition and common sense. However, this argument (and its alike) should lead (if taken seriously and literally) to the conclusion that in effect the uncertainty principle is due to a practical limitation related to the process of measurement at the quantum scale (although it is seemingly formulated and presented as a theoretical requirement), and so in principle if we assume the existence of a "perfect" measurement process in which the measurement probe does not interact with the observed object (or at least the effect of interaction is negligible) and hence it does not affect the "actual" attribute of the measured object then we can discard the uncertainty principle and thus we can have an infinitely-precise (or arbitrarily-precise) measurement.

   • The (more fundamental) arguments that are based, for instance, on the Fourier analysis where the uncertainty is seen as an intrinsic aspect of the observed phenomena and not because of a practical or procedural limitation on the measurement.[304] These arguments show that the uncertainty does not originate from the disturbance caused by measurement but from the very nature of the quantum

---

[304] In fact, the first type of arguments is about measurability while the second type is about definability (or even realizability). It may also be justified to use "intrusive" and "non-intrusive" to distinguish between the two types of arguments for establishing the uncertainty principle.

objects and their wave (or duality) character.

Now, the uncertainty principle in its strict quantum interpretation and justification (as inferred for instance from analyzing the formalism from which it comes) is supposed to originate from the wave nature of quantum objects and hence it is an essential (or fundamental) theoretical requirement rather than a marginal (or accidental) practical requirement created by the measurement process and its limitations. Accordingly, if we have to accept any argument for justifying the uncertainty principle then it should be of the second type. Yes, if we accept the claim that inertness of observation (see § 1.3.3) of quantum systems is impossible (at least because of the scale factor; see § 1.3.2) as well as the claim that the disturbance caused by measurement cannot be made practically and arbitrarily negligible, then even the arguments of the first type may be accepted. However, both these claims may be contested and challenged. Anyway, the arguments of the first type can be accepted (unconditionally) as pedagogical and educational rather than rigorous and authentic.[305] Also see points 11 and 19 of § 4.7.1.

2. A thorough inspection of the literature about the justifications and proofs of the uncertainty principle should reveal that most (if not all) the alleged justifications and proofs are questionable. This equally applies to the technical and non-technical (or simple) ones (some of which are indicated in point 12 of § 4.7.1 as well as in point 1 of § 6.6). Accordingly, if there is any real justification to this principle it should be its empirical success (with a broad and generic rationalization by the particle-wave duality noting the potentially limited applicability of this rationalization to certain observables).

6.6.2 The Uncertainty Principle and Creation

The uncertainty principle is interpreted in the literature of quantum theory as a principle that allows a "brief" violation of the conservation principles of momentum and energy (and possibly other principles) where the uncertainty in momentum Δp and the uncertainty in energy ΔE mean (according to this interpretation) the possibility of violation of momentum conservation by an amount Δp over a spatial uncertainty of Δx and the possibility of violation of energy conservation by an amount ΔE over a temporal uncertainty of Δt. In our view, the essence of the conservation principles (at least in some of their instances) is the rejection of the possibility of creation and annihilation and hence any theory (or principle) that violates the conservation principles (even though "briefly" or "by tiny amount") should be classified as a creation theory (or principle) and hence it should become non-physical or metaphysical (see § 2.5.4). So, as long as a violation of a conservation principle occurs there is no difference between being "briefly" or "by tiny amount" and not being so.

As a result of the possibility of a "brief" violation of the conservation principles, the uncertainty principle is used to justify the non-physical and illusory paradigm of virtual particles which may also be seen as a creation theory where these alleged particles emerge from nothing in time Δt thanks to the uncertainty ΔE in energy. In fact, some of these alleged particles can be converted to real particles by strange mechanisms (refer for instance to the literature of Hawking radiation which we discussed briefly in our book "General Relativity Simplified & Assessed"). In brief, the uncertainty principle is misused by theoretical physicists to create and justify unfounded illusions and fantasies with no regard to the requirement of the conservation principles which seem to be accepted in general by quantum physicists (see § 6.12).

Another result of the possibility of a "brief" violation of the conservation principles (and the emergence of virtual particles which was referred to in the previous paragraph) is the bizarre conceptualization of space (or rather space-time) by some theoretical physicists as something like a boiling quantum soup where bubbles of quantum objects (e.g. virtual particles) emerge from nothing and disappear to nothing everywhere in space and at any instant in time thanks to the uncertainty principle.[306] In fact, these (as well as similar proposals and ideas) are just a few examples of the illusory ideas and fantasies that invaded modern physics where theoretical physicists gave themselves the right to create whatever they

---

[305] We note that the disturbance arguments are generally less acceptable among physicists (as well as being less representative of the essence and spirit of the uncertainty principle).

[306] Illusory objects (like boiling quantum soup and virtual particles) may be replaced (realistically) by the existence of real physical objects (like background electromagnetic fields and cosmic particles).

like of baseless fantasies using things like the uncertainty principle even if these fantasies lead to violations to supposedly-accepted principles. Also see point 2 of § 6.6 as well as § 6.12.

6.6.3 Uncertainties about the Uncertainty Principle

Despite its apparent simplicity, there are many question marks and mysteries about the uncertainty principle and its characteristics and significance. For example:
• Why uncertainty occurs between certain pairs but not others and between certain variables but not others (see § 4.7.1) noting that from purely mathematical and physical perspectives (regardless of observations) such extensions are possible?
• What is at the root of the relationship between the uncertainty principle and the commutation of operators (and why)?
• Is there any significance in the units of $\hbar$ which is in the uncertainty principle (i.e. why the uncertainty in the common forms of the uncertainty principle is in terms of action)?
• Is there any significance in the quantification (in order) by $\hbar$?

Although we can find many (explicit and implicit) attempts to address (at least some of) these issues and answer these questions (and their alike) we do not find treatments and answers that are deep enough and convincing enough.[307] So, anyone interested in the uncertainty principle (and in quantum mechanics in general) should carefully inspect these aspects and analyze their significance and should not take this principle at its face value.

In fact, there are other types of uncertainties in the uncertainty principle (which we may call disguised forms of uncertainty and they belong to its essence and content). Some of these disguised forms of uncertainty are indicated earlier. For example, there is uncertainty about the meaning of Δt in the time-energy form of this principle (see point 2 of § 6.6 as well as § 6.6.2 and § 6.12). Another example is the issue of justification of the uncertainty principle and if it should be the disturbance caused by measurement or the very nature of the quantum objects and their wave (or dual) characteristic (see § 6.6.1). A third example is whether the uncertainty principle can allow violations to (at least some of) the conservation principles or not (see § 6.6.2 and § 6.12). These uncertainties (or at least some of them especially the fundamental ones like the latter) cast a shadow on this principle and make it one of the troublesome aspects of quantum mechanics and its interpretation.

In this context, we should remember that one of the proposed roots and origins of the uncertainty principle is the particle-wave duality nature of quantum objects (or the wave aspect of this duality; see § 6.6.1) and since this nature itself is questionable (at least as an exact model rather than an imitate and approximate model; see for instance § 3.2) then it is no surprise that the uncertainty principle itself should be a subject of uncertainties and suspicion. We should also remember that the physicists in the early days of quantum mechanics struggled to digest this principle and accommodate it formally and epistemologically (although it was digested rather easily by the next generations of physicists thanks to education and indoctrination) and this should be another indication to its problematic nature. So in brief, these many uncertainties about the uncertainty principle and its nature, justification, significance, limitations, ... etc. should be a cause for concern and must call for careful inspection and revision to this principle.

6.6.4 Relationship between the Uncertainty Principle and Other Principles

The relationship between the uncertainty principle and other principles is discussed in different parts of the book (and hence the reader should refer to these parts). For example, the relationship between the uncertainty principle and the correspondence and complementarity principles was discussed mostly in § 4.7.4. Similarly, the relationship between the uncertainty principle and the conservation principles (like

---

[307] For example, in an attempt to address some of the issues raised in our questions above we find an attempt like "the act of measuring one observable drives the system into different states from those into which the system is driven by measuring the other observable". But this attempt does not address the main issue because we can still ask why the system is driven in this specific way? In brief, most of these attempts are of formal and technical nature rather than substantial and physical nature.

energy and momentum conservation) was discussed (from some aspects) earlier in § 6.6.2 and § 6.6.3 and will be discussed further in § 6.12.

6.7 The Correspondence Principle

The correspondence principle, despite being marginal practically and has almost no position in the formalism, is pivotal conceptually and has an important position in the epistemology of quantum mechanics. This is because this principle represents and reflects tacitly the idea of quantum physicists to quantum mechanics and its domain of validity as well as its relation to classical mechanics (which, in our view, is a dubious and problematic issue). In the following points we assess the correspondence principle and investigate its potentially-misleading role in defining the validity domain of quantum mechanics:

1. A common misunderstanding (or misleading aspect) about the correspondence principle is that (by the principles of uniqueness of reality and truth as well as the assumption of the validity of quantum mechanics) quantum physics should converge to classical physics in classical systems. However, in reality this could be the case if we have another assumption that is the validity domain of quantum mechanics extends to classical systems. So, if we accept this assumption (even though implicitly) then we may accept the aforementioned premise while if we do not accept this assumption then the premise becomes questionable or invalid. In fact, we should even need another assumption that is the uniqueness of science (which is questionable; see § 2.4.3) because if science is not unique then we can have two different theories that are equally valid in the same domain (i.e. the classical domain in our case).

2. The correspondence principle is seemingly based on the implicit presumption of the validity of quantum mechanics beyond the domain of quantum systems. However, this presumption can be easily challenged in its generality (and is challenged earlier and later in the book). Accordingly, classical physics is valid primarily only in its domain (which is macroscopic systems) while quantum physics is valid primarily only in its domain (which is quantum systems) where the two domains are either disjoint (with no common zone) or they have only a marginal common zone.[308] In fact, the latter (i.e. having a marginal common zone) seems to be the case if we note that the claimed instances of common applicability of both physics are actually restricted to certain large scale quantum systems which are on the border between macroscopic and quantum (or microscopic) scales.[309] For example, we cannot find a single sensible (direct) application of quantum mechanics to a macroscopic system, e.g. determining the trajectories of projectiles (such as balls) or the orbits of celestial objects[310] or determining the wavefunction of a classical system (e.g. car) or its collapse. Yes, there are claims and attempts to extend the validity of quantum mechanics to macroscopic systems but none of these claims and attempts are credible or productive. In fact, apart from the prohibitive conceptual and theoretical

---

[308] Quantum physics is like classical physics in the possibility of being restricted in validity to a restricted domain, i.e. as classical mechanics is restricted to macroscopic systems quantum mechanics can be restricted to quantum systems. In fact, if classical systems are seen as a limiting case to quantum systems on the large scale limit, then it is equally legitimate (in principle) to consider quantum systems as a limiting case to classical systems on the small scale limit (and hence if we have a special type of physics to one type of systems then it is reasonable to have another special type of physics to the other type of systems). The correspondence principle actually discriminates (or is based on a discriminatory view) against classical mechanics and is in favor of quantum mechanics when it assumes that quantum mechanics is valid beyond its prime domain (which is quantum systems) where this discrimination may be justified by the pretext that quantum physics supersedes classical physics and is proposed as rectification to the failures of classical physics.

[309] We should note that similar to the correspondence principle where quantum physics extends to areas in the domain of classical physics, there are areas in the domain of quantum physics where correct results are obtained by applying classical physics (especially in the early stages of development of quantum physics as seen for instance in the Bohr atomic model and the old quantum theory; in fact this similarly applies to classically-based modern "quantum theories"). This should endorse the idea that there are common marginal areas and isles where both classical and quantum physics are applicable and valid (see § 3.7 and § 4.7.2). This should also be linked to the premise that quantum mechanics is in essence a semi-classical theory since it contains many classical elements, and hence the relation between the correspondence principle and the issue of classic-quantum duality should be considered in this context (see § 5.3.3 and § 7.2).

[310] In this regard, we note the conceptual, as well as the formal and practical, difficulties for such determinations (see for instance footnote [144] on page 72).

6.7 The Correspondence Principle

difficulties, it is completely unrealistic to apply quantum mechanics on macroscopic systems (even if we can discard or manage those difficulties) where even a small macroscopic system may contain 10^{10} quantum sub-systems (e.g. atoms and molecules) each of which could have many possible quantum states (with other possible complications such as entanglement or interaction with external agents or observational errors and uncertainties). Also see § 3.7 as well as point 9 of § 6.1.

3. In our view (with reference to point 2), applying quantum mechanics directly on a macroscopic system as it is (i.e. not as a composite of quantum sub-systems) is not sensible conceptually or theoretically. So, any potentially-legitimate application of quantum mechanics on a macroscopic system should be through its constituents of quantum sub-systems, and this is obviously not viable practically regardless of being theoretically legitimate and sensible or not. In fact, we should note (with regard to the theoretical legitimacy and sensibility) that the current quantum theory is largely formulated and tested on small quantum systems with small number of quantum components, and hence its theoretical legitimacy and sensibility on large systems (whether as they are or as composites with large number of quantum components) can be questioned and challenged. Yes, some quantum models are formulated and tested on large systems with large number of quantum components but they generally represent simple and regular (or periodic) patterns (such as large regular crystals) and hence this cannot be extended and generalized (at least easily and naturally) to large systems with large number of quantum components when they do not represent simple and regular (or periodic) patterns (like cars or trees). Also see point 9 of § 6.1.

4. Along the lines of the previous points, the following question may be raised: how the physics of macroscopic systems (as expressed in the form of classical physics) should then be correlated to the quantum physics of its constituents (e.g. atoms and molecules)? Should the physics of classical systems be seen as representing a sort of average behavior and properties of the underlying quantum systems? If so (or any other possibility) then how the correspondence principle (assuming it has such an extended validity at least in the marginal zone) emerges from this mess?
In fact, questions like these have no simple, unique and definite answer and they do not attract sufficient attention and interest in the literature. However, we can say: the answer depends on several factors such as the type of quantum system and the theory that is supposed to represent it as well as the perspective of investigation and analysis. Some of these issues have been discussed or approached earlier and will be investigated further later on. Also see § 3.6, § 3.7 and § 4.7.2.

5. As indicated earlier, the correspondence principle (in one of its implications) represents the spirit of the principle of uniqueness of truth (see § 2.4.1 and § 4.7.2).[311] However, if the expectation of the correspondence principle is to produce the classical formulation (and not just the agreement with the classical prediction) then we need more than the principle of uniqueness of truth to justify this (see point 4 of § 4.7.2). In fact, we may need the assumption of the uniqueness of science (rather than the non-uniqueness of science; see § 2.4.3) to justify this expectation.

6. There are many question marks about the significance of the correspondence principle and its epistemological implications. For example, does the correspondence principle represent the emergence of realism (or "classical" physical reality) out of the fuzziness and indeterminism of the quantum world? Can this provide a link or relation, for instance, between quantum superposition or the uncertainty principle and the classical conceptions and formulations and hence make the correspondence principle more sensible? These questions (and many other questions) have no definite answer and they depend on the adopted philosophical views and epistemological interpretations and opinions.

7. It is very important to note the (subtle) difference between two issues: the issue of convergence of one theory to another theory at a certain domain or sub-domain, and the issue of validity and direct applicability of a theory at a certain domain or sub-domain. For example, the formalism of quantum physics may converge to the formalism of classical physics in the classical domain or in part of the classical domain (and hence quantum physics produces the classical formulations at that classical limit). However, this does not necessarily mean that the formalism of quantum physics can be applied directly

---

[311] In fact, the correspondence principle in its formal application rests on the general (or extended) validity of quantum mechanics to include non-quantum systems, and in its epistemological significance rests on the principle of uniqueness of truth.

to the classical domain or sub-domain (where theoretical or conceptual or practical considerations can prevent such an application).

To distinguish between the two cases (i.e. where direct application is possible and where it is not) we give two examples. One example is the convergence of Lorentz mechanics to classical mechanics where (despite and regardless of this convergence) Lorentz mechanics can still be applied directly (and hence it is valid) in the classical domain. The other example is the (potentially-alleged) convergence of quantum mechanics to classical mechanics where quantum mechanics (in our view) is not applicable directly (and hence it is not valid, at least for practical reasons and considerations if not for theoretical and conceptual ones) in the classical domain or sub-domain.[312]

However, it seems that there is a confusion among quantum physicists about this issue and hence they may consider the (potentially-alleged) convergence of quantum mechanics to classical mechanics (e.g. by the Ehrenfest theorem) as an instance or evidence for the general validity and applicability of quantum mechanics in the classical domain, whereas in reality what is applicable and valid in this case is not quantum mechanics but classical mechanics which quantum mechanics (supposedly) converges to. Also see points 1 and 2 of § 7.3.

Also see § 4.7.2.

6.8 The Complementarity Principle

This principle was proposed originally by Bohr and advocated by his followers (mostly the adherents of the Copenhagen school of interpretation). Therefore, it may be more appropriate to call it "Bohr's complementarity principle" or "Copenhagen's complementarity principle". As indicated earlier (see § 4.7.3), this principle in our view is marginal.[313] Moreover, it is more of a rule of thumb (or even a feature) than a strict principle. In fact, this principle has been challenged theoretically by numerous arguments and thought experiments. It was also challenged recently by some experiments such as the slit-grid experiment (see § 6.8.1).

We think this (alleged) principle has been given in the literature much more weight (e.g. in investigations, debates, arguments, etc.) than it deserves. Moreover, its significance (whether true or alleged and definite or tentative) should be more interpretative than technical, and hence it should belong to the interpretation of quantum mechanics more than to its formalism (noting that it is used and misused in both). However, if this principle is treated as a strict principle and it is regarded as part (or at least a companion) of the formalism then it could have far reaching consequences (e.g. in the analysis and conclusions of the double-slit experiment and assessing certain quantum aspects and features like self-interference).

In the following points we outline some of our observations about this principle:

1. There are many ambiguities and uncertainties about this alleged principle regarding its meaning, applications, limitations, etc. and all these contribute to the many controversies and debates about this principle. In fact, these ambiguities and uncertainties (and the associated controversies, etc.) diminish the significance and value of this alleged principle and make it almost useless (despite the opposite claims).

2. The mess and confusion about this alleged principle should be partly attributed to the status of "particle" and "wave" (as well as their "duality") models which are no more than approximations and imitations and hence they may provide as much confusion as (supposed) clarification and rationalization. If we take notice of the question marks on the models of "particle", "wave" and "duality" (as reliable models in representing the quantum reality) then the mess and confusion about the meaning and application of the complementarity principle should become more understandable and justifiable. Also see § 4.7.3.

---

[312] In fact, this could be understood and explained by the conceptual homogeneity between classical and Lorentz mechanics (where they employ almost identical familiar macroscopic concepts and models) and the conceptual inhomogeneity between classical and quantum mechanics (noting their conceptual difference and incompatibility). Other (related or unrelated) practical considerations may also participate in this explanation and supplement it.

[313] If the uncertainty relations are instances of the complementarity principle then this principle is not essentially marginal (although it should then be marginal, at least in these instances, because it is redundant since it is replaceable by these relations).

6.8 The Complementarity Principle

3. As indicated before, there are serious ambiguities about this principle and this affects its value and how it is used in different (and sometime contradictory) meanings and contexts. In fact, a simple inspection of the meanings and synonyms of the word "complementary" (which "complementarity" is derived from) should reveal that it has opposite meanings (or suggestions)[314] and hence it contains intrinsic haziness and ambiguity. Accordingly, this word (as well as its derivatives like "complementarity") is not suitable for use in the terminology of science since this terminology requires high standards of clarity, specificity and technicality (especially when the essence and content of what the terminology is supposed to attach to are intricate and confusing). Also see points 4 and 5.

4. To give a better idea about the miserable state of the complementarity principle and how it was born and nourished we give the following quote from Bell (see "Speakable and unspeakable in quantum mechanics" in the References):

    He (i.e. Bohr) thought that 'complementarity' was important not only for physics, but for the whole of human knowledge. The justly immense prestige of Bohr has led to the mention of complementarity in most text books of quantum theory. But usually only in a few lines. One is tempted to suspect that the authors do not understand the Bohr philosophy sufficiently to find it helpful. Einstein himself had great difficulty in reaching a sharp formulation of Bohr's meaning. What hope then for the rest of us? There is very little I can say about 'complementarity'. But I wish to say one thing. It seems to me that Bohr used this word with the reverse of its usual meaning. (End of quote)

    So, it seems that even the originator (i.e. Bohr) did not understand his principle properly to suggest a proper name for it and hence he used "complementarity" with the reverse of its usual meaning (excluding, of course, the possibility of bad intention to mystify this principle and make it inaccessible to others)![315] Also see the next point as well as point 3 of § 4.7.3.

    We should also note that expressions like "not only for physics, but for the whole of human knowledge" or "the Bohr philosophy" in the above quote should endorse our suggestion of the "feature" nature (rather than the "principle" nature) of complementarity.

5. For the sake of clarity (as well as for the sake of saving the uncertainty principle and keeping it away from unnecessary headaches or rather keeping the headaches of "complementarity" and "uncertainty" separate) we think the complementarity principle should be restricted to its primary meaning (i.e. being about wave-particle) and should not be extended to include the uncertainty principle (and possibly other quantum mechanical formulations and features). Accordingly, the complementarity and uncertainty principles should not be associated with each other so that we avoid the confusion and mess caused by the looseness and lack of discipline in using terms like "complementary" and "complementarity". In fact, we even recommend avoiding the use of "complementary" to label the "incompatible" observables in the uncertainty principle. We also recommend relabeling the "complementarity" principle (if it proves to be meaningful and useful and can serve a purpose in the quantum theory that justifies its existence) so that the essence and content of this principle become consistent and compatible with its name and label. This, of course, requires at first reformulating this principle rigorously and technically so that we get rid of looseness in essence and content as well as looseness in name and label.[316]

---

[314] For example, it means (or suggests) "matching, corresponding, paired, harmonious, compatible, twin, supportive, interrelated, etc." as well as "opposite, contrary, etc.".

[315] I wonder if this brilliance should be attributed to Bohr or to Bell or to both (or even to Einstein and his "superhuman" nature which makes him different from the rest of us), although the main credit should belong to this extraordinary principle. In fact, this sort of nonsense and irrationality is not uncommon in the literature of modern physics.

[316] In fact, it may be better to get rid of this principle altogether. Yes, we can recognize a complementarity feature (as we did in § 5.9) which expands across many aspects and zones of the quantum theory and represents a loose characteristic or generic attribute rather than a (strict) principle.

6.8.1 Slit-Grid Experiment

This experiment,[317] which is a modified version of the double-slit experiment and is commonly known in the literature as Afshar's experiment, is supposed to refute the complementarity principle.[318] It is also claimed that it endorses certain interpretations (specifically the transactional school; see § 8.4) against other interpretations (particularly the Copenhagen and many-worlds schools; see § 8.1 and § 8.3). However, all these claims are questionable and can be challenged.

For example, the analysis of the experiment (which allegedly leads to the claimed conclusions) seems to assume the existence of definite and specific trajectories for the "particles" in the intermediate stages (i.e. between the emission at the source and the detection at the screen) and this can be challenged and questioned (at least by some schools of interpretation and in some stages of the experiment). Without experimental detection of the "particles" over their "trajectories" (which is almost impossible or at least has not been done in this experiment) it seems difficult (if not impossible) to substantiate some of the claims made by the supporters of the slit-grid experiment especially with regard to the existence of interference pattern where the existence of this pattern should be based on an implicit assumption of having definite and specific trajectories that cause the construction of this pattern.

It should be obvious that even if we can accept that the particles do not pass in their alleged trajectories through the minimums (i.e. where the grid is placed) it is not evident that the particles pass through the maximums in such a way that results in the construction of the interference pattern. In fact, these supporters seem not only to envisage and treat the quantum "particles" like classical particles (e.g. by having determined trajectories) which can be challenged and questioned (at least according to certain views and interpretations) but they gave themselves the right to determine these trajectories specifically by guess (to justify the existence of interference pattern for instance).

Also, the experiment was criticized as essentially being about the existence of waves and particles (as inferred by their characteristics) as if quantum mechanics is an ontological theory while the reality is that quantum mechanics is essentially about measurement and observation (i.e. it is basically a phenomenological theory). In the following points we outline some of our observations and assessments about this experiment and its value and significance:

1. As indicated, this experiment is challenged from various aspects (e.g. procedure, analysis and interpretation). This should diminish its value and cast shadows on its alleged implications and consequences. The misery of the slit-grid experiment is worsened by the fact that the complementarity principle itself is surrounded by many ambiguities, uncertainties, challenges, etc. (as indicated earlier in § 6.8) and this should not only diminish the value and significance of the principle itself but even the value and significance of any work or experiment related to it (i.e. the slit-grid experiment in our case).

2. As indicated, this experiment is questionable from various technical and theoretical aspects. However, regardless of these questionable aspects and if they affect the validity or/and significance of the experiment or not, we think this experiment (like the principle itself) is of little value. This is because the complementarity principle (at least in our view) is a rule of thumb or a feature (see § 4.7.3 and § 5.9) rather than a strict rule and hence its validity or invalidity (in particular cases and circumstances) should have no significant impact. Yes, the complementarity principle may have been used by some (as if it is a strict rule) in certain arguments and debates and hence only those users could be affected by the slit-grid experiment. Anyway, the complementarity principle is marginal (regardless of being valid or not and regardless of its status and details) and hence it has no impact on the formalism (and no one should be allowed to claim such an impact).

3. According to some interpretations of the complementarity principle (in the context of analyzing the slit-grid experiment), what violates complementarity is not the co-existence of the interference pat-

---

[317] The following discussion relies on the explanation of this experiment that is given in the appended section § 10.1 (as will be indicated later). So, those who are interested in this experiment should read and understand § 10.1 first, while those who are not interested should jump to the next section (i.e. § 6.9).

[318] In fact, it is a modified version of the "double-pinhole experiment" rather than "double-slit experiment". We should also note that there are other experiments which claim to demonstrate the breach of the complementarity principle (and hence this is not the only experiment in this regard although it is distinguished by receiving more publicity and attention).

tern (which is allegedly established in this experiment) but the co-observation (or co-measurement) of this pattern (and this co-observation is not achieved in this experiment although it is inferred, rather unjustifiably, by guess from the presence of the grid). Anyway, the significance of the slit-grid experiment in refuting the complementarity principle or not seems to depend on the interpretation of this principle. For example, if the principle is interpreted as denying the coexistence of interference pattern and particles then the experiment could be significant (although it should still be questionable from other aspects), while if it is interpreted as denying the direct observation of these then it may not be so. This similarly applies to the "which way" aspect of the experiment.

4. Regarding the alleged impact of this experiment on the interpretations of quantum mechanics, all the existing interpretations of quantum mechanics are questionable, and therefore even if we assume the validity and soundness of this experiment it should still have little (if any) impact on the interpretations because those interpretations which are supposedly endorsed by this experiment are still in need to address other challenges (which possibly could be more serious), while those interpretations which are supposedly refuted by this experiment are already facing other challenges (which also could be more serious) and hence the result of this experiment will not be decisive in determining their fate (apart from possibly worsening their misery).

So in brief, this experiment in our view is not only questionable but it is of little importance (like the complementarity principle itself which this experiment claims to challenge and refute). We therefore think it is better to save the time and effort for more serious and significant investigations. The readers who are interested in more detailed investigations and discussions about this experiment should refer to the literature. We also provided in the appended section § 10.1 basic explanation and illustration about the essence of this experiment for the interested readers to consult if they wish.

6.9 Quantum Tunneling and Scattering

Quantum tunneling is commonly seen as a characteristic feature of quantum mechanics with no classical explanation or justification (and this also applies to certain aspects and features of quantum scattering). What seems perplexing in quantum tunneling is the requirement of negative kinetic energy in rationalizing this phenomenon and making sense of it. However, we will see that this perplexity may not be too perplexing (or at least there is some exaggeration about it) and hence we may be able to rationalize this phenomenon at least partially. In fact, the confusion and mess about this phenomenon may be seen as another instance for the failure of the "particle-wave duality" approach (which characterizes the entire quantum mechanics) where "particle" and "wave" are treated as exact models (and hence we literally have "particles tunneling through barriers") rather than being imitates and approximates.

In the following points we investigate and assess quantum tunneling from different perspectives (with some assessment to quantum scattering as well):

1. According to the literature, the envisaged mechanism for quantum tunneling is either by surmounting the quantum barrier or by penetrating (or going through) this barrier. However, these quantum barriers are not real barriers (like walls) and hence these models and scenarios should be seen as mimics (of pedagogical nature) with no actual consequences (although we use them frequently in the terminology and visualization of this phenomenon).

2. The tunneled quantum object can be envisaged either as a particle or as a wave. So, from this perspective we actually have two main possibilities (or scenarios) for envisaging the phenomenon of quantum tunneling and these possibilities should have an effect on the assessment and interpretation of quantum tunneling (as will be seen next). However, as indicated above these should be seen as imitates rather than exact models although they should help in rationalizing the process of tunneling and facilitating its visualization.

3. There is no direct observation of negative kinetic energy. In fact, such observation seems impossible because actual detectors can detect (by design) only positive energy. So, the "negative energy" is actually in the theoretical models of tunneling processes (or in their rationalizations and justifications) rather than in the actual physical events and processes. To make sense of all this we can propose that although the theoretical models may be based on (or imply) negative energy, the actual physical

6.9 Quantum Tunneling and Scattering

process of tunneling occurs to quantum objects (e.g. alpha particles or electrons) whose actual energy is positive. In fact, being positive can be explained and justified statistically according to their energy distribution where some objects acquire sufficient (positive) energy to "jump over" or "penetrate through" the barrier. So in brief, even if we seemingly accept that in quantum tunneling we have "particles jumping over or penetrating through barrier" we can still (by means of the statistical proposal) justify quantum tunneling rather easily (and even classically) although this justification may not apply in all cases of tunneling.

4. As indicated earlier (see § 5.5), tunneling in itself is not a characteristic quantum phenomenon although tunneling of particle-like objects is characteristic to quantum mechanics. Now, if we note the particle-wave duality in quantum mechanics (assuming we accept this "particle"-"wave" duality literally) we can still explain tunneling "classically" by the wave nature of the tunneled objects although they are usually classified or seen as particles. In other words, these objects tunnel-through thanks to their wave nature, which is allowed and rationalized classically, and hence tunneling can be justified and interpreted (rather easily) by this classical feature. Accordingly, what is supposedly strange in quantum mechanics (or characteristic to it) is the particle-wave duality which allows particle-like objects to tunnel through barriers by means of their wave nature.

    As for quantum scattering, it may be explained similarly (i.e. by the wave nature of quantum particles) noting that classical waves that impact barriers (or steps of abrupt change) of finite height are partially scattered (or reflected) and partially passed (or transmitted). However, we should note that in general this dual behavior of classical waves occurs simultaneously unlike the behavior of individual quantum objects which either scatter or pass.[319] So, the analogy with the classical waves is not exact and this may be seen as another example for the failure of the classical concepts and analogies to provide a complete and exact picture of the quantum reality (i.e. the classical concepts and prototypes that we use in the description and modeling of the quantum world should be seen as approximate models).

5. To provide more clarification about negative energy (which we discussed rather briefly earlier) and if it is a necessary requirement for quantum tunneling through potential barriers (or it is possible to explain quantum tunneling without negative energy as we suggested earlier) we can say: the issue here is that negative energy is based on considering the (tunneled) quantum object as a localized particle and hence (according to our classical conception) it either "jumps" over the barrier which implies negative energy or "penetrates" through the barrier which seems inconsistent with localization (since it requires passing through the barrier; see § 1.4) as well as negative energy (since the penetration of the barrier requires, if possible, an amount of energy that is greater than the kinetic energy of the object due to the existence, for instance, of very strong repelling electrostatic forces which are supposed to exist if the "barrier" is really a barrier). However, if the object is behaving (during tunneling) as a wave (as suggested in the previous point) then the penetration requires none of these things (i.e. negative energy or penetration). This is because waves are not localized packets (of matter or energy) and hence in their propagation they can surmount (or jump over) barriers and can also penetrate through barriers and this sort of wave behavior (i.e. surmounting and penetration) is accepted even classically.

6. It is important to analyze and assess the role of the uncertainty principle in quantum tunneling through potential barriers. In fact, the uncertainty principle is used to justify the excess kinetic energy required for quantum tunneling (and this should imply the creation of the excess kinetic energy, which should be rejected). So, the uncertainty principle (through a violation of the conservation of energy) is exploited and misused again (see point 2 of § 6.6 as well as § 6.12) to avoid (seemingly) the taboo of negative kinetic energy (as if the taboo of violation of energy conservation does not exist or it is less severe). Anyway, the excess kinetic energy (to avoid negative energy) is not needed if the tunneled object behaves (while tunneling through the barrier) as a wave as explained in point 5 and hence the uncertainty principle should be unnecessary in the explanation of tunneling. Similarly, the uncertainty principle is not needed if the aforementioned statistical approach is used (where and when it is relevant and applicable) since no deficit in kinetic energy exists (and hence no excess kinetic energy is required to call for a violation of energy conservation through the uncertainty principle).

---

[319] In fact, this difference between classical and quantum behaviors should apply (to some extent) even to quantum tunneling.

6.10 Quantum Entanglement 146

7. We may note that the attenuation[320] suffered during the penetration process (as the penetrating object supposedly go through the barrier) may endorse the proposal that the penetration should be seen as a wave (rather than particle) phenomenon (i.e. the object penetrates the barrier by its wave nature) and hence quantum penetration can be easily justified (even classically as a classical phenomenon). In fact, this may be an instance of the complementarity principle (see § 4.7.3) in one of its interpretations where the object (during penetration) is actually a wave. However, this cannot explain the fact that matter waves are actually probability waves and the object (while emerging from the other side of the barrier as a particle) does not suffer any sort of tangible or physical attenuation (see point 7 of § 5.5). So, this should put again a question mark on the authenticity of the "wave" and "particle" models (as well as their duality) and if they really and exactly reflect the "reality" of the object or they are approximations and imitations. The result of all this is that the statistical distribution explanation (which we proposed above) may seem more sensible although this explanation can also be challenged by the fact that it does not apply to single-particle systems where there is no statistical distribution (although other sources for the surplus energy from the surrounding environment may be suggested).

So to sum up, we have two possible (classically-sensible) explanations and rationalizations for quantum tunneling: penetration as a (pure) wave, and penetration by getting sufficient energy according to the statistical distribution (or getting energy from the environment). However, both these explanations are not tight enough or sufficiently general to provide convincing (classically-sensible) rationalizations for quantum tunneling in all cases and circumstances. Moreover, none of the available models for the quantum objects (such as "particle", "wave", "dual particle-wave", "wavepacket" and "pilot wave") can explain all the aspects of quantum tunneling and scattering consistently, intuitively and thoroughly. This should endorse the view that all these explanations and models are approximations and imitations and hence they do not reflect the "reality" of quantum objects and phenomena accurately and honestly (although they are generally useful for providing intuitive insight and useful educational means within their limitations and restrictions). Also see § 5.5.

6.10 Quantum Entanglement

Quantum entanglement is problematic on two front lines: quantum front line and non-quantum front line. The former is related to the quantum theory itself (as a sensible physical theory) while the latter is related to other branches and theories of physics whose rules and principles may become threatened or violated by the implications of quantum entanglement. We investigate and assess these issues in the following subsections (and in § 6.10.2 in particular).

However, before we go through the details we need to understand why quantum entanglement can be a threat to any physical or scientific theory (whether quantum or non-quantum). In other words, what are the characteristic features of quantum entanglement that make it a potential threat? There are two main potentially-threatening features in quantum entanglement: possible violation of locality (see § 1.8) and possible violation of causality (see § 1.7).[321] These features make quantum entanglement especially important to science (with far reaching philosophical and epistemological consequences as well).

Regarding the violation of locality, quantum entanglement can violate certain (allegedly-required) restrictions on the speed of physical signals. This is because entanglement should imply communication between (quantum-mechanically entangled) physical objects with superluminal or even infinite speeds. Regarding the violation of causality, quantum entanglement can violate the principle of causality in one of its interpretations (i.e. if causality between two spatially-separated objects requires an exchange of

---

[320] Attenuation here is about the probability amplitude of the penetrating object (not the energy of the object; see point 7 of § 5.5).

[321] In fact, there is a third characteristic feature that is very important in this regard, i.e. the requirement of quantum entanglement (and indeed any non-local occurrence like instantaneous wavefunction collapse on a global scale) to frame-free sense of simultaneity which has very important implication on the issue of absolute frame (which is of special importance to special relativity). However, we did not consider this as a main issue in this context because this issue is related to non-locality (which we considered already); moreover it impacts primarily special relativity (which is not the primary subject of interest in this book) rather than quantum mechanics. However, we will deal with this issue (rather briefly) in § 6.10.2.

physical signals). This is because if spatial action at a distance (i.e. with no intervening signal) is allowed or required in the explanation and justification of quantum entanglement then this should incur a violation of the causality principle according to this interpretation.[322]

We should remark that experiments based on quantum entanglement (as well as related theorems and issues) can be used (and are actually used) for testing and assessing some interpretations of quantum mechanics. In fact, quantum entanglement is particularly important to the interpretation of quantum mechanics and to certain schools in particular (as highlighted by the Bell theorem and its relation to hidden-variable interpretations). However, for structural reasons we defer this investigation to later chapters (see for instance chapter 8 as well as § 9.3.2 and § 9.3.3). We should also remark that quantum entanglement is particularly important to some novel quantum mechanical branches and applications such as quantum communication and cryptography. In fact, what makes quantum entanglement exceptionally important is its exceptionally weird and non-classical nature. Also see § 5.6.

6.10.1 Threatening Features of Entanglement

Regarding the **first threatening feature** of entanglement, the central issue here is non-locality (or remote interaction) where two spatially-separated physical objects seem to interact (presumably) through a type of physical signal whose speed is seemingly infinite (the so-called "action at a distance" although this may be restricted to the upcoming case where there is no exchange of signals) either because it is really infinite or because it is apparently infinite (because the superluminal speed is too high that its finity cannot be detected). In this regard we may make the following remarks:

A. Regarding the issue of infinity and superluminality of the speed of physical signals (noting that if entanglement is established then it may necessitate and imply infinite or superluminal speed), it may be argued that there is nothing illogical or non-physical about superluminality. In fact, it may even be argued that there is nothing illogical (and possibly even non-physical) about having infinite physical speeds although from the perspective of our common and familiar scientific experiences we usually deal only with finite speeds. In this context, we should note that prior to the development of modern science (from Renaissance onward) the infinity of physical speeds (at least the speed of light) was acceptable (at least tentatively and as a possibility) in the scholar circles (as can be concluded, for instance, from the debate that surrounded Galileo's attempt to measure the speed of light). In fact, almost all the practical and conceptual practices of our daily life are based on the implicit assumption of the infinity of the speed of light (considering its magnitude on the scales that we are familiar with in our daily life). This may be regarded as an indication to the logicality (and possibly even physicality) of the infinity of physical speeds. Yes, our physical "intuition" or "common sense" (as modern scientists) seems against this infinity due to education and indoctrination (thanks mostly to special relativity). So in brief, we may claim that "infinite speed" is not impossible or at least its impossibility is not so evident that it can be used to refute opposite claims.[323] Therefore, infinite speed could be accepted (at least logically) either strictly (if we accept the possibility of infinite speed) or practically (since superluminal speed whose finity is undetectable is effectively infinite).

B. The hypothesis of the constancy of the speed of light (which is claimed to imply being the maximum possible speed of physical signals) is related to the role of light in the Lorentz transformations and their special relativistic interpretation. So, even if we accept this hypothesis within the framework of Lorentz mechanics, we may limit this to light only. In brief, there is no physical evidence for this alleged rule of the finity of all physical speeds or the impossibility of all superluminal speeds even if we accept Lorentz mechanics since the evidence for Lorentz mechanics is restricted to the speed of light (in its extended sense that includes all forms of electromagnetic signals). Yes, the establishment of the

---

[322] In fact, potential violation of the causality principle is a possibility even with temporal action at a distance according to certain opinions, possibilities and scenarios. We remind the reader that we distinguished early between "spatial" action at a distance where there is no exchange of signals and "temporal" action at a distance where there is exchange of superluminal signals.

[323] We note that the denial of logical inconsistency or non-physicality may clash with what have been argued earlier against creation and annihilation by the necessity of physical process to have a finite time for its occurrence; see § 2.5.4. However, as we will see next we may not need to claim the possibility of infinite speed strictly but only practically.

6.10.1 Threatening Features of Entanglement

infinity or superluminality of (some) physical speeds could abolish the special relativistic interpretation of Lorentz mechanics and confirm the invalidity of special relativity whose postulates (according to its supporters) are based on or lead to the finity and non-superluminality of all physical speeds. However, the invalidity of special relativity is already established by its logical inconsistencies (as discussed in our book "The Mechanics of Lorentz Transformations") and hence this evidence will not add much to our existing knowledge.[324]

C. The non-locality (which is seemingly demonstrated by quantum entanglement) in quantum mechanics is not restricted to quantum entanglement. For example, wavefunction collapse is also supposed to be non-local (see § 5.7). In fact, the non-locality of quantum entanglement should originate (partly) from the non-locality of collapse. However, it may be argued that wavefunction collapse is essentially an interpretative and controversial issue (and hence the collapse itself let alone its non-locality) can be disputed and rejected. However, the essence of collapse (regardless of its specific conceptualization and interpretation) should be a firm quantum feature. Moreover, quantum entanglement (in its general features and not in its details and interpretation) is an established quantum phenomenon (at least among the majority of quantum physicists) and hence it should lend credible support to non-locality (despite the question marks and controversies which no scientific issue is totally free of).

D. As indicated earlier (see point 6 of § 1.8), quantum entanglement (as well as related quantum phenomena and features) is not the only threat to the principle of locality (since there are non-quantum instances of seeming violation of locality). In fact, this should give more support to the violation of locality by quantum entanglement (and perhaps other instances of quantum violations) since other (potential) violations make non-locality look more real and credible.

Regarding the **second threatening feature** of entanglement, the central issue here is causality where two spatially-separated physical objects seem to interact without intervening signal that communicates this interaction (the so-called "action at a distance", i.e. in its spatial sense) which seems to violate the causality principle (or so claimed). In this regard we may make the following remarks:

1. Regarding the (potential) violation of the principle of causality, it may be argued that it is not obvious that spatially-separated causal interactions should occur through exchange of physical signals. In other words, such interactions with no intervening signals are not illogical or physically-impossible (although they are counter-intuitive). In fact, the counter-intuitivity of "spatial action at a distance" (i.e. with no intervening signal) may be justified by the failure of our patterns and models (which we acquired from our evolutionary past) to describe or envisage such phenomena. So, "action at a distance" is at least a possibility (although it may be a weak and counter-intuitive one) and hence "action at a distance" in itself cannot be used as a cause for rejection. This means that the aforementioned interpretation of the causality principle (i.e. causality between two spatially-separated objects requires an exchange of physical signals) can be challenged and hence we may accept causal interactions between spatially-separated events with no exchange of physical signals.

2. To explain and justify quantum entanglement, we do not actually need the assumption of "action at a distance" (i.e. with no intervening signal) which allegedly violates causality. What we need is a *pseudo* "action at a distance" which does not violate causality. In other words, we can assume that what appears to be an "action at a distance" (with no intervening signal) is actually an action through intervening signals (whether of finite or infinite speed is irrelevant from this perspective) that are not detectable. Yes, this could be challenged by being a kind of metaphysical explanation.

3. As indicated earlier and will be discussed later (see for example footnote [324] as well as § 6.10.2), the "action at a distance" (with infinite or superluminal intervening signal) may cause another type of (alleged) violation of causality (which we may call accidental violation) that is the violation claimed

---

[324] There are claims (mostly found in the literature of special relativity) that the infinity and superluminality of physical speeds should have illogical consequences such as the violation of causality. However, these claims are not well established and can be challenged. In fact, even if these claims are accepted they have no serious impact on our case because these alleged violations should negate only the infinite and superluminal physical speeds that cause such violations and hence infinite and superluminal physical speeds that do not cause such violations could be logical and even physical. We may even claim that the mere existence of such apparently-infinite or -superluminal physical speeds (as exemplified, for instance, by the instances of quantum entanglement) could be regarded as an indication to the existence (at least tentatively) of such infinite or superluminal physical speeds that have no illogical consequences.

by special relativists where superluminal speeds are alleged to lead to violation of causality relations between spacelike events. However, these claims are not well established and can be challenged (as discussed and explained earlier and later).

We should finally refer to the many experiments that have been conducted in the last decades to investigate and verify quantum entanglement.[325] The most famous of these experiments (and possibly the pioneering one) is the Aspect experiment which is investigated briefly in § 10.4. Almost all these experiments seem to confirm the non-local nature of the interaction between the quantum-mechanically entangled objects (and hence the aforementioned potential impacts on locality and causality seem real and credible).

6.10.2 Front Lines of Entanglement

Regarding the **quantum front line** of quantum entanglement, it should be obvious that the features and implications of entanglement should have a direct impact on quantum mechanics and if it is a sensible and rational theory and consistent with other theories and principles (or alleged principles) of science. So, if for example we firmly believe in the physical impossibility of superluminal speed while quantum entanglement seems to imply or require such a speed then quantum entanglement (and hence quantum mechanics) or at least its interpretation become questionable. Similarly, if we firmly believe that causal interactions between spatially-separated events should take place through exchange of physical signals while quantum entanglement seems to imply or require "action at a distance" (without exchange of signals) then quantum entanglement and quantum mechanics become questionable.

We should also note that quantum entanglement can be (or is claimed to be) a threat to some specific quantum aspects and features. For example, it may be possible (allegedly) to challenge the uncertainty principle by entanglement where complementary variables can be determined precisely and simultaneously (by measurements of the complementary variables on the entangled objects) as in some forms of the EPR argument (see § 9.3.1 as well as the proof of the Bell theorem in § 10.3.1).[326]

Regarding the **non-quantum front line**, it is obviously related to the relativity theories (and special relativity in particular) because these theories are based on the assumption of the existence of a speed limit (represented by the characteristic speed of light c) for any physical signal (whether the signal is light or something else). In this regard we may make the following remarks:

1. If quantum entanglement is just a threat to special relativity then we should have no worry about this because special relativity is a logically inconsistent theory (regardless of quantum entanglement) and hence any evidence (whether definite or tentative or alleged) against this theory will not add much to what we already have. Yes, quantum entanglement can be a threat to Lorentz mechanics which is a logically consistent and experimentally supported theory (although in our view it is not a final theory). However, this issue is out of scope since our book is about quantum theory.[327] Moreover,

---

[325] We are mainly referring to Bell's experiments (see § 9.3.3) which are primarily about Bell's theorem although they are also about quantum entanglement (noting the link and dependence between Bell's theorem and quantum entanglement).

[326] In brief, we can (allegedly) measure precisely one complementary variable on one entangled object and measure precisely the other complementary variable on the other entangled object and hence violate the uncertainty principle.

[327] In fact, all we need to worry about (if we accept that Lorentz mechanics is a "practically correct" theory) is the interpretation of the (supposedly correct) formalism; specifically if the speed of light is constant. However, even if we accept the constancy of the speed of light (supposedly as a requirement of the formalism of Lorentz mechanics) there is no implication of this formalism to the claimed assertion that the characteristic speed of light c is the ultimate speed for any physical signal. In other words, let the speed of light be constant (i.e. c) but it is still possible that there are other types of physical signal (i.e. non-light signals similar, for instance, to the presumed gravitational waves) that can exceed the speed of light (noting that the existence of physical signals whose speed is less than c is a physical fact). The claims that such physical signals can lead to contradictions and inconsistencies (such as violation of causal relations) can be challenged and revoked (e.g. the alleged violations of causal relations may be challenged by being based on the acceptance of the alleged speed limit). In fact, even the defensive argument that such signals cannot violate causality (as well as similar arguments like the alleged impossibility of sending useful information by such signals) are irrelevant and redundant in this context and therefore we do not need them and we should not bother to go through them (noting that some of these defensive arguments are wrong or questionable). We should finally note that so far we assumed the acceptance of the premise that causal influences are communicated by physical signals; otherwise if we accept a spatial form of "action at a distance" in which no exchange of signals is involved (as some seem to suggest although this may necessitate proposing a rational physical mechanism) then most of the above discussion becomes redundant because we

we have dealt with this type of issues in our book "The Mechanics of Lorentz Transformations" which the interested reader is advised to refer to (in fact some of these issues have been dealt with briefly early in the present book).

2. Like special relativity, quantum entanglement may threaten (although seemingly less seriously) general relativity. However, general relativity can also be challenged from certain aspects (i.e. it is not a confirmed and fully validated theory) and hence even if it is threatened by quantum entanglement we do not need to worry about this.[328]

3. It is useful to notice that there are claims in the literature that the non-local interaction of the type that arises by quantum entanglement should not threaten special relativity or lead to relativistic paradoxes where these claims are justified, for instance, by alleging the impossibility of sending information by such non-local interaction (or similar excuses).[329] However, in our view these claims (and their justifications) are unfounded and refutable. This is because what threatens special relativity is the non-local interaction itself regardless of being able to send information or not (and regardless of any similar excuses). Yes, these claims may be useful in refuting some challenges and paradoxes (e.g. those based on sending information) but they are irrelevant with regard to refuting the fundamental challenge to special relativity which originates from the speed postulate and its special relativistic implications and extensions that ban any non-local interaction. Anyway, such a challenge remains a possibility and hence special relativity becomes under suspicion which means that its validity cannot be guaranteed (as required for any reliable and acceptable theory).

4. If non-locality implies instantaneous interaction then special relativity should face another challenge. In brief, the notion of "instantaneous interaction" (and even instantaneous occurrence in general regardless of any other consideration like causality) is based on the notion of simultaneity which is supposedly abolished (in its frame-free sense) by special relativity. This should pose another challenge to special relativity by quantum entanglement which requires a well-defined (frame-free) type of simultaneity. Yes, if we assume the existence of a privileged (or absolute) frame then we should have no problem from this perspective, but unfortunately (to its adherents) special relativity denies such a frame. So, we can claim that quantum entanglement can represent a serious challenge to the very idea of "relativity" (which is based on the denial of any privileged or absolute frame).

6.11 Quantum Measurement and Collapse of Wavefunction

As indicated earlier, quantum measurement is one of the most problematic issues in quantum mechanics and its interpretation. In particular, wavefunction collapse (which is supposed to be the result of measurement according to the dominant interpretation) is the most troubling aspect of measurement (and hence of quantum mechanics). The paradigm of collapse is needed to explain the indeterministic nature of the outcome of measurement (as a result of superposition in wavefunction). In the following points we try to assess quantum measurement and collapse:[330]

1. One of the main sources of perplexity and troubles in quantum measurement is indeterminism (see § 5.2). Because of the indeterministic nature of quantum physics, the outcome of a measurement can (in general) be predicted only probabilistically and possibly with an intrinsic uncertainty (even if we assume that the initial state, in the form of wavefunction, is well-known and the measuring equipment is perfect). This should put a question mark on causality and cast a shadow on the nature and extent of the principle of causality (see § 1.7) in the quantum world and if this principle should be defined differently to what it is in the macroscopic world (see § 6.4).

---

do not need to worry about any signals let alone their speeds.

[328] Refer to our book "General Relativity Simplified & Assessed".

[329] An example of this type of arguments is the following: "Even though the change in one of the particles in the system affects the other particle instantly, special relativity is not violated because information does not propagate through the system".

[330] In fact, the assessment of measurement and collapse is a big subject and hence what we present here is not really an assessment but an outline of some aspects of this assessment (noting that some of these aspects have already been indicated or discussed before).

6.11 Quantum Measurement and Collapse of Wavefunction

2. Another main source of perplexity and troubles in quantum measurement is the (sudden) transformation from the indeterminate state of superposition (which is governed by the Schrodinger equation; see § 4.3 and § 5.2.1) to a determinate state (which is a sub-state of the superposition state). This is the issue of wavefunction collapse (or other similar or dissimilar conceptualization proposals about this transformation) which raises many questions and concerns, e.g. about the violation of the principle of locality or the potential violation of the principle of causality or the cause and physical mechanism of this transformation.[331]

3. Unlike classical measurement, quantum measurement is not (or at least "perhaps not" according to some theories and interpretations) such a passive process whose objective is to reveal a physical reality that is independent of the observer.[332] In fact, the very nature of quantum measurement makes it difficult or impossible to be a passive process (see § 1.3.3). This "non-passivity" is exacerbated by the proposals of some theories and interpretations about quantum measurement where they gave the observer and observation more role (in the measurement process and its outcome) than is required to explain the intrusive nature of this intervention. For example, in some interpretations a central role (which is physically-unjustifiable and unnecessary) is given to the consciousness of the observer while others went even further by claiming that the measurement and observation determine the reality and give it a specific shape and form. Also see point 7.

4. There are many controversies and opinions about the collapse of wavefunction (e.g. its nature, if it is objective or subjective, the role of consciousness in it, and so on). In fact, even its occurrence at all is not agreed upon although the common version of the formalism of quantum mechanics is generally conceptualized and phrased in terms of collapse (or reduction) which may give an impression that it is part of the formalism. In fact, collapse may not be a (necessary) part of the formalism although its essence exists in most schools of interpretation (even some of those who deny collapse in its explicit and common form). So, it seems that collapse in its explicit and common form belongs to certain schools of interpretation (and hence it is an interpretative aspect of quantum mechanics) but the "essence of collapse" seems to exist (in some shape and form) in most schools of interpretation, and hence from this perspective it may be considered as part of the formalism (or rather in the conceptualization of formalism).[333]

5. The collapse of wavefunction is supposed to be a global and instantaneous "occurrence" and hence it should be non-local. This should open the gate for many challenges and criticisms to those who believe in the principle of locality (see § 1.8) as discussed earlier in various locations. Also see § 5.6 and § 6.10.[334]

6. As indicated earlier (see § 5.7), there is no known and sensible physical mechanism for wavefunction collapse and this should put question marks on the essence of collapse and its role in measurement (as well as on its sensibility). In fact, collapse should be seen as an arbitrary hypothesis proposed to fill a gap in the conceptualization and rationalization of quantum mechanics and make sense of it.[335]

---

[331] It is worth noting that the collapse of wavefunction is commonly described by terms like "discontinuous, instantaneous, non-local, temporally-asymmetric" process and these labels should suggest many sources of trouble in this process and its conceptualization and rationalization.

[332] As indicated earlier, quantum measurement (at least according to some interpretations that adopt the concept of collapse) is an influence exerted by the observer on the observed "physical reality" and hence it is not really measurement (in its traditional sense). So, the measurement is not actually a process that exposes the already-existing, well-determined and well-defined reality but rather a process in which the reality is materialized or formed or created.

[333] We note that "collapse" is embedded in the commonly accepted version of the quantum postulates (i.e. represented by the projection or measurement postulate which is somewhat controversial and is not recognized by some schools), and these postulates should be considered part of the formal structure of quantum mechanics. In fact, the axiomatic framework (which these postulates represent) essentially contains all the main elements of the formalism of the quantum theory (see § 4.9).

[334] It should be noted that there are attempts in the literature to avoid the instantaneous nature of collapse (see for instance Quantum Mechanics by Bransden and Joachain in the References). However, the theoretical sensibility and rationality of these attempts (as well as their compliance with observations) are questionable.

[335] Some may question the "non-existence of physical mechanism" as a sufficient justification for the rejection of "collapse" because this may apply in one sense or another to many physical theories. However, it should be sufficient justification for rejecting an interpretation (based on such non-existence) because such a mechanism is required for rationalization which is a major role for interpretation.

6.11 Quantum Measurement and Collapse of Wavefunction

7. There are many conflicting views about what brings about the effect of measurement and leads to the collapse of wavefunction and if it is, for instance, the interaction between the macroscopic measuring device and quantum phenomenon or the consciousness of the observer or even both (where the observer could potentially be considered as part of the measuring device). In fact, the role of the observer and his consciousness is one of the most perplexing and controversial issues in the theory of quantum measurement and collapse. Due to its non-classical and counter-intuitive nature, the role of consciousness in measurement and collapse has been challenged vigorously from different aspects (despite its acceptance by some respected physicists). For example:
   - What is the physical mechanism that makes consciousness affect or influence the observed system?
   - What if there were more than one observer that simultaneously observe the system? Could we have, for instance, different (classical) results (as a consequence of having different quantum outcomes) and hence multiple classical realities?[336]
   - Why should consciousness be needed at all for measurement and collapse (as quantum events) whereas there are many spontaneously-occurring quantum events that happen with no need for any conscious intervention (noting that the denial of such spontaneously-occurring quantum events is nonsensical and totally unrealistic; see point 8 of § 5.7)?

   However, regardless of these challenges, as well as their challenges, it should be obvious that consciousness in itself is not sufficient (even if we accept that it is necessary) for the realization of measurement and collapse since some physical interaction between the measuring device and the quantum system is required for the measurement to take place, and hence consciousness (at best) should have a complementary (and possibly marginal) role rather than a central or primary role in the measurement and collapse. We should also note that the "consciousness theory" is intimately related to the collapse of wavefunction (or rather its essence) and hence consciousness may lose its alleged role (or at least it diminishes) if collapse is challenged or denied.

8. There are many paradoxical challenges to the paradigm of wavefunction collapse. One of these is the so-called "quantum Zeno paradox"[337] which claims (in one of its simple versions) that if observation (or measurement) leads to collapse then if we keep observing a quantum system (e.g. an atom in an excited state) then it should collapse continuously to its eigenstate[338] and hence it remains stationary (i.e. the atom will never decay as long as we are watching it, which is supposedly nonsensical and experimentally false). However, this paradox was challenged by rejecting this interpretation of "observation" and "collapse". In fact, it was also challenged by alleged experimental evidence (i.e. against this alleged paradox and in support of the commonly-accepted interpretation of collapse). Anyway, regardless of the correctness or not of this alleged paradox and its alleged refutation (noting that none of the arguments and alleged evidence, for and against, are conclusive due to many ambiguities and controversies), this highlights the difficulty of interpreting quantum mechanics rationally (see for instance § 7.15) and hence it supports our view that quantum mechanics is not an interpretable theory (or at least there is no known interpretation to quantum mechanics that is totally consistent and comprehensively rational).

9. There are attempts to get rid of the problematic aspects of quantum measurement and wavefunction collapse by trying to embed the issue of measurement within the main formalism of the quantum theory (i.e. instead of splitting the quantum theory between the deterministically-evolving Schrodinger equation and the indeterministically-characterized measurement postulate; see § 4.8). These attempts are represented by relatively-popular interpretations like the many-worlds school (see § 8.3).

---

[336] If we accept that quantum mechanics applies to macroscopic systems (as to microscopic systems) then we may be forced to accept multiple classical realities (and possibly infinite number of realities) which seems inline with the many-worlds interpretation (see § 8.3). However, as indicated above we may not even need to accept direct applicability of quantum mechanics to macroscopic systems for the implication of this multiplicity because the classical effects (as seen on the macroscopic measuring equipment which could include the consciousness of the observer) should reflect the possibly-different quantum outcomes. We note that multiple "quantum realities" seems already accepted and "rationalized" (in some shape and form, at least according to certain possibilities and interpretations) with no need even to these classical scenarios that involve consciousness or the many-worlds interpretation.

[337] This may also be called "quantum Zeno effect" or "watched pot effect" (as well as other names).

[338] This is due to the fact that such repeated measurements should reproduce the same state (like, for instance, measuring the z component of the spin of an electron time after time which should always produce the same value; see § 4.6).

6.11 Quantum Measurement and Collapse of Wavefunction

10. As indicated, the paradigm of "wavefunction collapse" is one of the most bewildering features of quantum theory. However, in our view there is a more bewildering feature which is what we call reverse- or anti-collapse where a system in an eigenstate is prepared to be in another (incompatible) eigenstate, e.g. a system which was initially prepared in an eigenstate of spin-up in the z direction is re-prepared to be in an eigenstate of spin-up (or spin-down) in the x direction and hence its previous state (i.e. spin-up in the z direction) is destroyed and therefore the spin state in the z direction becomes "superposed".[339] In the case of collapse of wavefunction we actually (or at least potentially) have "multiple superposed realities" that "branch" into a specific one of these superposed and "already-existing" realities (which seems somewhat rational). But in the case of reverse-collapse a specific reality (represented by spin-up in the z direction) becomes superposed which means that a new reality (represented by spin-down in the z direction) is created.[340]

11. In certain circumstances (e.g. in the double-slit experiment), the collapse of the wavefunction by measurement (according to the reduction postulate) may be seen as a transformation from diffuse wave (i.e. wavefunction which is a probability wave) to localized particle (of definite position). However, this may not represent the familiar (or traditional) concept of "collapse" (although it may be familiarized by its spatial significance and from this aspect specifically if measurement is supposed to take place by hitting the display screen).[341]

12. As explained earlier (see § 5.8), non-linearity was proposed as a solution to the problem of measurement in quantum physics and to avoid the need for the paradigm of wavefunction collapse in addressing this problem.

13. The idea of wavefunction collapse is needed for the description of individual quantum events. Therefore, if wavefunction is supposed to describe the evolution of ensembles of identically-prepared quantum systems (rather than individual quantum systems and events) then the collapse of wavefunction may become redundant or even inapplicable and meaningless (see points 3 and 4 of § 6.1). However, this may be challenged because the concept of collapse can (allegedly) apply sensibly even to the wavefunction of an ensemble although it may seem less physical and less sensible. But it should be noted that if the collapse (of an ensemble) is supposed to take place gradually and over a period of time then this should be inconsistent with the nature of collapse as an instantaneous process (see for instance point 5 of § 5.7). Accordingly, those physicists who ascribe the wavefunction to ensembles (rather than individual objects) should address this issue (noting that some physicists seem unaware of this inconsistency or potential inconsistency as they ascribe wavefunction to ensembles and adopt the paradigm of "collapse" at the same time). In fact, they may need to look for an alternative to the paradigm of "collapse" or modify it.

14. It should be noted that the condition of *instantaneous* collapse (which causes the violation of locality and which may some try to dismiss to avoid this violation) is needed to avoid, for instance, violation of conservation of spin (as a type of angular momentum) in entangled quantum systems. This is because if the collapse is communicated by a finite speed then during the time between the collapse (by observing one entangled particle) and the time of reaching this collapse to the other entangled particle the wavefunction of the other particle can collapse (if the other particle is observed) in a way that violates the conservation of spin.[342] For example, if A and B are two spin-1/2 entangled particles with zero total spin and particle A is observed to have spin-up and during the transmission of this collapse to particle B this particle is also observed (i.e. before the effect of collapse reaches it)

---

[339] An example of such preparations is a Stern-Gerlach type experiment; see § 4.6.1.

[340] We may say alternatively: an already existing reality (represented by spin-up in the z direction) is destroyed and replaced by a newly created reality (represented by superposed spin state in the z direction).

[341] In fact, this is just an example of the possibly-different nature of collapse depending on the observed quantum phenomenon and the type of quantum measurement. In other words, we should practice some flexibility when conceptualizing and interpreting the collapse in a given quantum measurement. However, this sort of flexibility may no be consistent (in some cases) with certain schools of interpretation. For example, if measurement is supposed to take place by hitting the display screen then where is the (alleged) role of consciousness according to the "consciousness theory" (although this can be challenged and answered by the "consciousness theory").

[342] The laxity of expressions like "the wavefunction of the other particle" should be obvious because the entangled system has a single wavefunction rather than two "entangled wavefunctions". So, it may be more appropriate to say something like: "the wavefunction with respect to the other particle". Anyway, the meaning is obvious and the issue is trivial.

then there is a possibility that particle B will also collapse to spin-up (noting that its wavefunction did not collapse yet and hence it has equal probabilities of collapsing to spin-up and spin-down) and this violates the conservation of spin. In fact, this should exclude the possibility of finite speed of communication (at least in such cases) because such a possibility (of violating the conservation of spin) stands as long as the speed of communication is finite (and as long as the conservation of spin is supposed to hold) even if the speed is assumed superluminal. Also see points 4 and 7 of § 1.8.

Also see § 5.7.

6.12 Conservation Principles in Quantum Mechanics

In this section we briefly investigate the role of the conservation principles in the quantum theory (especially the commonly-known ones such as the conservation principles of energy and momentum which are inherited from classical physics). It seems that the status of some of these principles is rather vague (although the conservation principles seem to be accepted in general by quantum physicists). So, while they play some roles in certain arguments they seemingly can be violated sometimes by the uncertainty principle for instance.[343] Anyway, the role of these principles in quantum mechanics is not as crucial and central as in classical mechanics and hence they are not mentioned and used very frequently in quantum mechanics (or not as much as in classical mechanics where these principles are dominant and present almost everywhere).

In brief, the commonly-known conservation principles (which originate from classical physics) do not have in quantum mechanics the central role and natural position which they have in classical mechanics. An inspection of the quantum theory should suggest (or give an impression) that these principles are imported from classical physics and used rather wishfully when they are needed and this could be seen as an example for the (potentially-unduly) influence of classical physics on quantum physics (see § 7.2). In fact, this can be understood and appreciated by noting that classical mechanics is centered on physical paradigms and ideas like force and energy (among which the concept of conservation takes a natural position) while quantum mechanics is centered on "mathematical" paradigms and ideas like wavefunction and probability.

We should also remind the readers of the aforementioned examples and instances (see for example § 6.6.2) where these principles (or some of them at least) are not strictly respected in the literature of quantum mechanics and hence they are violated occasionally by some quantum physicists (e.g. the alleged violation of the conservation of energy by the uncertainty principle; see § 6.6.2) if not by quantum physics.[344] In fact, being violable (even though "occasionally") should destroy their status as "principles" (or at least as fundamental and very useful principles). This "violability" seems to be caused (in part and sometimes) by the vagueness of the meaning of the uncertainties involved in (at least some forms of) the uncertainty principle. For example, Δt in the time-energy form of the uncertainty principle is rather vague since its meaning depends on the situation and context and hence it has been interpreted in many different meanings depending on the physical situation (see point 2 of § 6.6 as well as § 6.6.2 and § 6.6.3), and therefore it seems to be exploited as a "source of justifiable violations".

As indicated earlier, in quantum mechanics we have new conservation principles that have no parallel in classical physics. For example, we have a new principle called the conservation of probability principle (or law).[345] Although we do not have a particular interest in this principle here (and indeed elsewhere in the book), we note that this principle reflects the nature of quantum mechanics, i.e. while the conservation

---

[343] If not the formalism of quantum mechanics itself allows violation of some conservation principles, at least some quantum-mechanically based theories allow this (as well as some quantum mechanical interpretations and rationalizations). In fact, explicitly conflicting views about this issue can be found in the literature.

[344] It is important to note that some quantum physicists reject all these alleged violations (by excuses like the uncertainty principle) and they declare explicitly and bluntly that these conservation principles are (strictly) respected in quantum mechanics as in classical mechanics.

[345] The conservation of probability essentially means: as time goes on the certainty of getting one of the possible outcomes does not change (even though the individual probabilities of the individual outcomes may change). For example, the certainty of getting a given free particle somewhere in the space is constant in time (i.e. as time passes the probability of finding the particle somewhere remains unity).

principles of primary interest in classical physics are essentially of physical nature (e.g. about energy and momentum) the conservation principles of primary interest in quantum physics are about things like probability (which strongly suggests an observer-centered nature rather than physical nature). Another example of a new (quantum mechanical) conservation principle is the conservation of parity (which also reflects the quantum nature or flavor of these new principles).[346]

Now, if we note that the conservation principles of classical physics may not be respected in quantum physics (at least as a possibility and in part and according to certain views and interpretations) as we saw for instance in the possible violation of energy conservation by the uncertainty principle (refer for instance to § 6.6.2) and we note as well the existence of specifically quantum mechanical conservation principles, then we may conclude that the sets of conservation principles in classical physics and in quantum physics are generally not the same, i.e. they have something in common but neither is a proper subset of the other. As indicated, some conservation principles (such as the conservation of charge) are strictly respected by both physics and hence they are common to both.

Interestingly, some of the new (quantum mechanical) conservation principles seem to be as violable (by quantum physicists if not by quantum physics) as their old classical cousins. For example, the quantum mechanical formulation of the weak interaction seems to violate parity and time-reversal invariances (noting that invariances generally imply conservation of certain physical attributes; also see the next paragraph). This may confirm our impression that conservation principles (whether old or new) in quantum physics are not as respectable as in classical physics (or at least quantum physicists are not as respectful to their principles as classical physicists to their principles). Jokingly, this may be seen as another form of indeterminism and uncertainty in quantum physics (or quantum physicists)!

It is noteworthy that the conservation principles are generally based on (and "derived" from) symmetry and invariance considerations. For example, the conservation of momentum is a result of space-translation symmetry, while the conservation of energy is a result of time-translation symmetry. Similarly, the conservation of angular momentum is a result of space-rotation symmetry, and the conservation of parity is a result of space-reflection symmetry. In fact, these conservation principles can be obtained from operations on the Hamiltonian that leave the Hamiltonian unchanged. Anyway, these conservation principles (as well as their symmetry and invariance considerations) should in general facilitate the rationalization and interpretation of the quantum world and its phenomena (although some of these considerations are too abstract, technical and artificial to be of any use for this purpose).[347]

6.13 Characteristics of Double-Slit Experiment

There are certain technical aspects and features that characterize quantum theory (as well as quantum phenomena) but they are usually not expressed or reflected explicitly and directly in the formalism of the theory. In fact, these aspects and features are typically found in certain quantum mechanical experiments. One of these experiments is the double-slit experiment (or rather double-slit-type experiments) which is very rich in these features and aspects. As indicated earlier, this experiment (in its quantum version) is one of the most important experiments in quantum physics regarding its significance in demonstrating the characteristic and perplexing aspects of quantum phenomena and the impact of this on the formalism and interpretation of the quantum theory.

So in brief, the importance of the double-slit experiment is that it represents a prototype for this kind of typical quantum mechanical experiments and phenomena that demonstrate characteristic quantum mechanical features (such as particle-wave duality and self-interference) and hence the conducted analysis and the obtained results are not as restricted in validity and application as it might seem. Hence, when we refer to this experiment it should be remembered that this generally applies to many quantum experiments and systems that have common features with this experiment.[348] We therefore think it is useful to

---

[346] We note that the conservation of parity is also violable. We should also note that there are some more specific quantum mechanical conservation principles such as the conservation of lepton number and baryon number.

[347] We note that some of the aforementioned conservation principles belong to more advanced level of quantum theory than the basic level that we are supposed to investigate in this book.

[348] In this regard we can quote Feynman: "You remember the case of the experiment with the two holes? It's the same

reassess (in the following subsections) some of the main aspects and features of this experiment despite our previous investigation and (initial) assessment of this experiment in § 1.6.[349] However, because the double-slit experiment is widely and meticulously discussed in the literature we do not go through its details during our discussion and assessment (and hence we refer the readers to the literature for most of its details which most readers should be familiar with).

6.13.1 Dual Nature

Dual nature (or particle-wave duality) refers to the fact that quantum objects (e.g. electrons or photons) generally behave partly as particles and partly as waves. This is typically demonstrated by the double-slit experiment where the quantum objects are detected (individually) as particles and distributed (collectively) on the display screen as interfering waves. As indicated before, this is one of the most striking aspects of quantum physics and possibly its most distinctive feature. In fact, we do not exaggerate if we claim that this duality is what makes quantum physics quantum physics. It is also one of the perplexing characteristics of the quantum theory (as well as quantum phenomena) and the center of many controversies, opinions and interpretations. In the following points we try to assess the particle-wave duality from different aspects and viewpoints:

1. As indicated in § 5.3.1, the wavepacket and pilot wave models (which are the most dominant models) for particle-wave duality are untenable (i.e. as exact and generally-valid models). Both models can be challenged for instance by their failure to explain certain phenomena or rationalize some aspects and features. In brief, these models cannot provide thorough and comprehensive explanation and interpretation in all cases and instances and to all aspects and features and hence they cannot be considered as acceptable "interpretations" (see § 2.9). These models should therefore be considered as approximations and imitations that may help to rationalize and envisage some quantum phenomena in some situations and contexts and with respect to certain aspects. In fact, we could not find in the literature a single model that can explain and rationalize particle-wave duality consistently and comprehensively (despite the ingenuity and originality of many of the proposals and suggestions in this regard). This should endorse our view that "particle" and "wave" (as well as their alleged duality) are no more than approximations and imitations (of classical nature) and hence they do not provide a solid basis for deep understanding of the quantum world. In fact, they may even become (if their physical significance and authenticity are overestimated) a source of delusion and confusion and can mask the vision of the quantum world in a more realistic way. We note in this regard that the confusion in the interpretation of quantum mechanics could be seen as a sign for the failure of these models (in addition to other signs and indications as well as other causes and reasons).

2. According to the de Broglie hypothesis, any moving particle should have an associated matter wave. But this should require the particle to be moving and hence the wavelength, as well as the wave nature itself, becomes observer-dependent (or frame-dependent). In this case a single particle can have different waves at the same time); moreover it may not have wave at all.[350] In fact, the restriction of the "wave" property to "moving" objects and the variability of the wavelength (according to the observer-dependent de Broglie relation) indicates that the (matter) "wave" property is not an intrinsic nature of quantum objects but it is rather an extrinsic and observer-dependent (or frame-dependent) property. This is unlike (rest) mass or charge for example which are intrinsic properties. This is also unlike classical waves whose wave nature is intrinsic (although the wavelength may also depend on the observer to a certain extent). Also see § 3.2.

3. According to the complementarity principle (see § 4.7.3), no experiment can reveal both aspects (i.e. particle and wave) simultaneously.[351] This may cast a shadow on the claimed duality nature (or at

---

thing".

[349] In fact, we investigate here only two of these main aspects, namely "dual nature" and "effect of observation and knowledge". The other main aspects have been investigated sufficiently in the past.

[350] From Eq. 16 we can see that if $p = 0$ (i.e. the object is at rest) then $\lambda = \infty$ which effectively means no wave property.

[351] As indicated above, the quantum objects in the double-slit experiment are detected individually as particles and distributed on the display screen collectively as interfering waves. However, this is not simultaneous detection of the particle and wave natures in a strict sense (at least according to the claim and interpretation of the complementarity

least on its interpretation and significance) since this alleged duality seems to be an artifact of the observation process and its nature. In other words, the duality may actually belong to the observation (or the observer or the frame of observation) more than to the observed object. In particular (and in reference to point 2), the wave property of quantum objects is unlike the classical wave (e.g. electromagnetic waves) whose existence and properties are generally independent of the observer and observation although some of its properties may be dependent on the observer and observation. In other words, classical waves are intrinsically waves while quantum waves (or matter waves) are not so (i.e. they are waves extrinsically or observationally and they seem to belong to the ensemble rather than the individual objects).

4. Even if we accept that physical objects (at least at the quantum scale) have dual particle-wave nature, it seems that we still need the distinction between particles and waves (i.e. between objects which are primarily particles but they have wave aspect and objects which are primarily waves but they have particle aspect). For example, according to our physical intuition (which stems from our understanding of modern physics) electron and X-ray photons of the same kinetic energy or wavelength are different (e.g. in rest mass, charge, speed, etc). So, the (primary) classification of electron as particle and X-ray as wave is still required (at least in some situations and contexts), and this should imply that the particle-wave duality is an approximation or imitation model rather than being an "exact" model (as may be suggested or implied in the literature).

In fact, this limitation does not stop at this but the particle-wave duality model seems to be restricted (at least for practical reasons such as the impossibility of the current capabilities to reveal the wave nature of macroscopic objects) in validity and applicability to quantum objects (and perhaps only to some types of quantum objects). Accordingly, we can question the many generalizations that can be found in the literature about the issue of duality (as well as similar and related issues). In brief, we should be careful about the particle-wave duality model and its significance, implications, applications, etc.

In fact, there are other questionable aspects about the particle-wave duality model (which seem more fundamental and physical). For example, "wave" for some quantum objects like X-ray photons is classical (or "real") wave while for other quantum objects like electrons is "quantum" (or "probability") wave and this should cast a shadow on the entire duality model as an appropriate model (at least from an epistemological perspective) for describing and visualizing quantum phenomena (also see point 4 of § 7.11). So in brief, the entire particle-wave duality model needs assessment to avoid possible mess and confusion in the underlying physics (particularly in its epistemology) even if it is working perfectly within the existing formalism of quantum mechanics. In fact, it is possible that this duality model (or its "particle" and "wave" components) could be one reason (among other reasons) for the difficulty (and perhaps impossibility) of finding a consistent and rational interpretation for quantum mechanics (see § 7.15). Also see points 2 and 3.

5. As indicated earlier, the most troubling aspect of the double-slit experiment may not be the particle-wave duality (which allegedly can be explained and justified rather easily, e.g. by the pilot wave or wavepacket models) but other aspects and features like the "effect of observation and knowledge" and "self-interference". Accordingly, we may accept (for instance) that quantum objects are particles guided by pilot waves or they are wavepackets that travel as waves and are detected as particles, but it is difficult to imagine the "effect of observation and knowledge" or "self-interference". So, despite its perplexity duality may not be the "worst" of the quantum features of the double-slit-type experiments (as well as other experiments and phenomena that demonstrate this feature).

Also see § 5.3.1.

6.13.2 Effect of Observation and Knowledge

In this context, the "effect of observation and knowledge" should mean that the behavior of the quantum object (e.g. photon or electron) seems to be affected by the observation process (as if the object is aware

---

principle). We also note that the "particle" and "wave" belong to different entities and aspects (i.e. individual objects and ensemble of objects) which should also question the "duality" and its meaning and nature.

6.13.2 Effect of Observation and Knowledge

of this process) and it adjusts its behavior according to it even though this process is presumed completely passive, i.e. it does not impact the object or interact with it. This "effect" seems unavoidable in certain kinds of experiments (whether of double-slit type or not) such as in the case of delayed choice experiments (and even in the case of quantum entanglement experiments according to some interpretations; see § 1.3.3 and § 5.6). In fact, this aspect or feature of the quantum theory was investigated earlier in § 1.6.2 from the perspective of the double-slit experiment,[352] and we continue our investigation and assessment (although from a rather more general perspective) in the following points:

1. There is no known and sensible physical mechanism for the effect (or influence) of observation and knowledge on physical reality, and this may explain why we do not see this influence in any scientific theory prior to quantum mechanics. Yes, in the case of parapsychology (and its alike of scientific or allegedly-scientific branches) the very subject of investigation is the possible influence of observation (or knowledge or consciousness) on the physical phenomena in the outside world (see for instance § 1.11 as well as footnote [254] on page 114). Hence, this type of investigations and disciplines should be excluded from the discussion in the mainstream and conventional science (including quantum physics) and hence they should be examined and assessed within a different environment using different rules and methods.

2. The influence of observation and knowledge could affect (at least in some of its interpretations) the independence of reality (which is a potential attribute of the "existence of reality" principle) and this should compromise this attribute and hence the existence of reality principle (which could cause an epistemological problem; see § 2.4.1 and § 2.5.2).

3. At least some of the experiments (mostly of the double-slit type) that allegedly demonstrate the effect of observation and knowledge are questionable either from a procedural perspective or from an interpretative and epistemological perspective (or even from both). Therefore, this alleged effect may be challenged fundamentally and denied even in principle.

4. The effect of observation and knowledge could lead (in some of its alleged instances) to delayed cause which violates the principle of causality (see § 1.6.4 and § 1.7). This could pose another fundamental challenge to this alleged effect (since causality presumably should not be violated in a rational theory). The possibility of allowing violation of causality by these instances should be excluded because causality is a well established principle (due to its vital role in rationalization) while these instances are questionable from various aspects.

5. Regarding the so-called delayed choice experiments (which we indicated earlier),[353] we have many reservations on this type of experiments both in their procedure and in their interpretation. In fact, we can safely say that at least in some of the reported delayed choice experiments there is no delayed choice at all. However, the question marks on the alleged delayed choice experiments do not rule out the possibility of post-determination of the path (i.e. the determination of the path follows the passage of the particle from the slit).[354]

Anyway, the purpose of some of these alleged delayed choice experiments is legitimate and justifiable (such as challenging the complementarity principle which we reject as a principle) and hence their results could be accepted regardless of the legitimacy and soundness of the employed method and if it is really of delayed choice type or not.

We also note that any aspect of the double-slit experiment (or indeed any other experiment or formulation whether of delayed choice type or not) that cannot be explained and interpreted sensibly and consistently should imply that the concepts and patterns used in the explanation and interpretation

---

[352] We note that the issue of the effect of observation and knowledge is related (but not identical) to the issue of the effect of consciousness which we investigated earlier (see for instance § 1.11 and § 6.11).

[353] As explained before, in delayed choice experiments (say of the double-slit type) the path of the particle is determined after, not before, the particle has passed the double-slit and hence any interference pattern should (presumably) not be affected by this delayed determination.

[354] For example, the actual path (i.e. the exact slit from which the photon passed) could be determined (in principle) after the particle hits the screen by calculating the time-of-flight, i.e. the time between emission and hit (noting that if the particle did not hit the center line of the screen then it should have traveled different distances if it passed through the left or the right slit and hence the difference between the time of emission and the time of hit, which is proportional to the distance traveled, should determine which slit the particle passed from).

6.13.2 Effect of Observation and Knowledge

are not adequate (and in the case of formulation, the formulation itself may not be interpretable). So, the implications of the delayed choice experiments (e.g. influence by knowledge or back-in-time effects) as well as any other implications of other aspects (e.g. self-interference) that cannot be explained and rationalized should indicate, for instance, the inadequacy of the particle-wave duality model (and possibly its "particle" and "wave" components) to represent the physical phenomena demonstrated by the double-slit experiment (and possibly their inadequacy in the entire quantum world and not only in this experiment).

Also see § 1.6.2.

Chapter 7
General Assessment of Quantum Theory

In this chapter we assess the quantum theory from general perspectives (rather than from technical perspectives as in chapter 6). Again, the assessments of the individual aspects are largely epistemological in nature and try to evaluate the significance of various factors and elements (external as well as internal) on the quantum theory in general and on its epistemology and interpretation in particular. However, before we go through the details of this general assessment we should draw the attention to the following remarks:

1. As well as the general aspects (which are the subject of investigation in this chapter), there are many particular non-technical aspects which are specific to certain parts and features of the quantum theory that require evaluation and assessment.[355] Some of these particular aspects have been investigated (and will be investigated) in their proper positions and contexts and within the scope and extent of this book.
2. The assessment of the schools of interpretation of quantum mechanics (see chapter 8) may be related to the sort of assessment mentioned in the previous point (noting that these schools, or at least some of them, are inherent to the quantum theory and in a sense they are part of it). In fact, the assessment of the schools of interpretation also contains elements of technical and formal assessment of this type.
3. The assessment in this chapter (considering its nature) represents the opinion of the author which other people may question or oppose.[356] However, this should not diminish its value noting that this type of assessment is naturally subject to deliberation and disagreement. Moreover, almost every aspect of the quantum theory (excluding possibly the bare minimum of its formalism as well as its empirical success and epistemological failure or perplexity) is controversial and debatable. Anyway, we put the proposals in our assessment as issues worth of consideration and investigation and subject to discussion and reflection.

7.1 Historical Evolution

Quantum theory was developed as a collection of ideas created and improved by a number of physicists of different backgrounds and visions (and even from different generations) and over a rather long period of time. Hence, it was built and structured collectively and incrementally rather than by a single individual and in one go. This could raise some concerns about the consistency and homogeneity of the theory especially at the conceptual and theoretical levels which are critical to the epistemology and interpretation of any theory. This is reflected in the many debates and controversies between the founding fathers and pioneers of this theory where their contradicting views and attitudes and the effect of all this on the development of quantum theory and its epistemology can be easily seen and felt (in the scientific dimension as well as in the historical dimension). We believe that many of the problematic issues of the quantum theory originate from its historical development and how it was born and raised. Also see § 7.15.

We should also note that many of the "obvious" things and "common sense" issues in the quantum theory (such as the particle-wave duality of quantum objects or the probabilistic interpretation of wavefunction or the collapse of wavefunction) were not so obvious and common sense for the founders and pioneers of the quantum theory. In fact, these issues became obvious and common sense by education and indoctrination which the generations following the pioneers were subjected to. So, this factor (i.e. degrading the ability to criticize and assess the "obvious" and "common sense" aspects of the quantum theory due to this historical

---
[355] In fact, in this regard quantum theory is like almost any other theory.
[356] In fact, even other assessments in the book (e.g. those in chapter 6) generally represent (or contain elements of) our personal views (although this is generally more deep and intense in the present chapter).

element) should be considered as an important element in determining the value of any assessment since the taboos on touching these sacred aspects in the quantum theory could mask our vision and lead to distorted assessment.

7.2 Classic-Quantum Duality

We can claim that quantum physics is a semi-classical theory in the sense that it is built on the legacy of classical physics (or rather classical knowledge to be more general). First, historically it is developed as a continuation of classical physics as can be seen vividly, for instance, in the Bohr atomic model and in the so-called old quantum theory. Second (and more important), is that quantum physics rests on an extensive classical heritage of concepts and models. In fact, many of the "contradictory" and perplexing aspects in quantum physics originate from the "semi-classical" nature of quantum physics. For example, the use of "wave" and "particle" (which are classical concepts) in quantum physics resulted in the particle-wave duality conundrum. In other words, if quantum physics did not start from these concepts (as valid models for representing and describing the quantum world) and their associated mathematical formulations (like the classical wave equation which is behind the Schrodinger *wave* equation) then the entire issue of particle-wave duality and its problematic aspects could become irrelevant.

Similarly, the "clumsiness" of using the classical model of "angular momentum" to describe quantum "spin" (which may by totally different from the classical concept of spin) should be evident in the subsequent problem in developing a sensible model for elementary particles like electron. It should be obvious that the disposal of the classical paradigms of angular momentum and spin could lead to a totally different physical formulation from the current formulation of these paradigms and could avoid several problematic issues in the quantum theory. The conservation principles (see § 6.12) may be another example for the (potentially-unduly) classical influence on the quantum theory where these principles are imported from classical physics but they do not seem to fit well within the framework of quantum mechanics (and hence we see alleged violations and random use as well as rather marginal role).

In fact, the entire quantum physics is influenced by classical physics and this is evidently reflected and represented by general principles and rules of quantum physics (as well as seemingly less general aspects like the above) such as the correspondence principle (which shows or claims the ability of quantum physics to return to its classical roots in some cases and circumstances; see § 4.7.2 and § 6.7) or the rule of obtaining the quantum operators from the classical expressions of their corresponding observables according to certain substitution rules (see § 4.4). In fact, even the uncertainty principle (which is characteristically quantum mechanical) has classical roots (see for instance point 7 of § 6.6).

In brief, by the scale factor (see § 1.3.2) and similar considerations we should expect the quantum world to be different from the classical world and hence we should expect the description of the quantum world by classical concepts and models (as it is the case in quantum mechanics) to lead to epistemological dilemmas (and that is what we see in quantum mechanics). In the following points we address some issues related to the classic-quantum duality (or semi-classical nature) of the current quantum theory:

1. It may be claimed that there is no way for the development of a quantum physics other than by the reliance on our classical heritage because after all we are classical creatures. However, we think this reliance is not a necessity (i.e. we can start in our modeling of the quantum world from different principles and use different approach and methodology). Moreover, even if it is a necessity, we can reduce this reliance substantially. Anyway, this assessment is about diagnosis rather than treatment (so if the treatment is impossible, due to the above claim, then the problematic nature of quantum physics and its epistemology is a necessity that we should accept and live with).
2. It may also be claimed that a quantum theory that is free of the classical influence will not lead to a better understanding of the quantum world and could even lead to a worse understanding because our understanding is fundamentally based on our classical concepts and models (i.e. a purely-quantum theory should be based on a completely unfamiliar concepts and models and hence it should be less understandable). However, even if this is the case we should at least get rid of some of the contradictory and confusing aspects of quantum mechanics and have a more consistent theory. In fact, the mix of classical and quantum concepts and models in the current quantum theory aggravates the

situation and makes quantum theory confusing and less understandable (or even non-understandable as it is commonly claimed). In the absence of this mix we can at least try to develop a new type of understanding and interpretation (and even "intuition") based on a new epistemological framework and conceptual infrastructure rather than struggling to address contradictions and perplexities. Anyway, living with a homogeneous non-classical quantum theory that is known to be non-interpretable may be better than having a semi-classical quantum theory that raises false hopes of having interpretation.

3. The indicated problems of quantum theory due to its semi-classical nature could imply that the optimality criterion of "simplicity" (see § 2.7.1) cannot be applied to any quantum theory because being consistent with the classical heritage should make it inappropriate for describing the quantum world and hence no quantum theory can be optimal from the "simplicity" perspective.[357]

7.3 Validity Domain

An important issue in the assessment of quantum theory is whether its validity is restricted to the quantum domain or it extends (at least in principle) even to the classical domain and hence it supersedes classical mechanics. The (seemingly) general consensus is the presumption of general validity of quantum mechanics (see for instance § 3.7 and § 4.7.2), i.e. it is valid (in principle) to classical (or macroscopic) systems as well as to quantum (or microscopic) systems. Our view (which we explained elsewhere in the book) is that quantum mechanics does not apply (and it is not valid) to the macroscopic world, at least due to practical reasons and considerations. The obviousness of this (in our opinion) can be simply based on our daily life experiences. For example, we do not experience superposition of states or collapse of wavefunctions or probabilistic outcome of measurements (in the quantum sense) or uncertainty principle or particle-wave duality or ... etc. Also, it is practically impossible to apply the formalism of quantum mechanics (e.g. wavefunction and Schrodinger's equation) on macroscopic systems even if we assume that such quantum paradigms and formulations are valid in principle to macroscopic systems. So, for all practical purposes (at least) quantum mechanics is invalid theory for the macroscopic world, and hence the microscopic world is governed by quantum mechanics while the macroscopic world is governed by classical mechanics.

Our view may be challenged by several arguments, some of which are outlined in the following points (which also include other issues):

1. The main challenge to our view may be the alleged success of quantum mechanics to produce (part of) the formalism of classical physics, e.g. by the Ehrenfest theorem or by the correspondence principle (see for instance § 3.7, § 4.7.2 and § 6.7). However, these examples of the convergence of quantum formulations to classical formulations are not comprehensive (i.e. we cannot get all the classical formulations from quantum mechanics by this convergence) and hence quantum mechanics cannot take the role of classical physics in the entire validity domain of classical physics. Moreover, this convergence can even be challenged in (at least) some of the instances of alleged convergence due to the question marks on these arguments and their artificial nature (as well as limitations and approximations in their derivation and application).[358] Anyway, even if we accept these instances and arguments they do not establish the generality of this validity (as indicated already). In fact, the success of this convergence in these instances may be explained more naturally by the semi-classical nature of quantum theory (see § 7.2) rather than by the general validity of quantum theory. The important criterion for the general validity is the applicability of the quantum formalism to macroscopic systems directly and naturally (e.g. by finding wavefunctions for macroscopic systems through Schrodinger's equation and the collapse of macroscopic states) and no one can claim that this applicability is achievable.[359] Therefore, the

---

[357] Noting that we consider the reliance on our classical heritage as an optimality criterion for our scientific theories, if this reliance proved to be impossible or it clashes with other eligibility or/and optimality criteria (which could be more important) then obtaining a quantum theory that is optimal from this perspective could become impossible (or at least difficult and conflicting).

[358] As well as questioning aspects like the assumptions which these convergence instances are based on, we may also question the "conceptual convergence" in these instances (see point 2).

[359] We note that in comparison to the convergence of the formalism of Lorentz mechanics to the formalism of classical mechanics, the formalism of Lorentz mechanics can still be applied directly and naturally to classical systems (unlike the alleged convergence of the formalism of quantum mechanics to the formalism of classical mechanics where the

7.3 Validity Domain

claim of general validity should be rejected or at least questioned (and hence we should restrict the domain of validity of quantum mechanics to quantum systems until conclusive evidence in support of general validity can be established). Also see point 7 of § 6.7.

2. Referring to the previous point, it is important to note that the Ehrenfest theorem does not extend the validity domain of quantum mechanics from the microscopic world to the macroscopic world (at least in a strict sense) due to conceptual and theoretical difficulties (as well as other difficulties; some of which are indicated already).[360] For example, Newton's laws of motion cannot be obtained in their exact classical sense from quantum mechanics (i.e. by the Ehrenfest theorem) because Newton's laws are based on the existence of determinate trajectories which do not exist (generally) in quantum mechanics neither formally nor conceptually (or so claimed).[361] Moreover, the derivation of classical equations by the Ehrenfest theorem contains stretches and twists that cast shadows on its status as real derivation of the classical formalism from the quantum formalism (at least rigorously and reliably). Anyway, this alleged derivation does not extend the validity domain of quantum mechanics to the classical domain but confirms the validity of classical mechanics in its domain (i.e. by the legitimacy of quantum mechanics). In other words, quantum mechanics (as it is) is still inapplicable in the classical domain although it supposedly has the ability to produce classical mechanics (partially) in the classical domain (and hence classical mechanics, rather than quantum mechanics, is the valid theory in this domain although it is justified supposedly by quantum mechanics). Yes, quantum mechanics is valid in the classical domain if we can apply (directly and naturally) the quantum paradigms and formulations (e.g. wavefunction, collapse, Schrodinger's equation, etc.) in this domain (as explained earlier).[362] Also see point 7 of § 6.7.

3. As indicated earlier, macroscopic phenomena should in general be ultimately based on quantum phenomena. However, this does not necessarily mean that the formalism of classical mechanics can emerge from the formalism of quantum mechanics (which what the claim of extended validity, in one of its senses, needs).[363] This is because the formalism of any theory is man-made and hence it contains an element of invention and therefore a theory about a phenomenon that underlies another phenomenon does not necessarily underlie the theory of the other phenomenon.

4. As indicated before, the "partial validity" of quantum theory in the classical domain should also be compared to the "partial validity" of classical theory in the quantum domain (and hence the alleged general validity of the quantum theory based on examples of its partial validity in the classical domain could be challenged by corresponding partial validity of classical theory in the quantum domain). We note (as an example) in this regard that many of the early experiments (prior to the appearance of quantum mechanics) related to atomic physics such as the cathode ray experiments were totally analyzed by classical physics and they produced correct results. This should indicate that classical

---

formalism of quantum mechanics cannot be applied directly and naturally to classical systems, at least because of practical limitations and restrictions).

[360] This should also apply to some instances of the correspondence principle (and possibly to other quantum mechanical formulations and applications). However, the Ehrenfest theorem should be enough (as an example) for demonstrating this point.

[361] As indicated earlier and will be seen later, definite trajectories (within certain practical and theoretical limits) may exist in certain observations and experiments involving quantum particles, e.g. experiments involving use of certain types of collimators or optical fibers to guide certain types of quantum particles (like electrons or photons) along specific trajectories. In fact, such trajectories can even be seen by naked eyes in some experiments and devices (e.g. in cathode ray tubes or bubble chambers). However, these examples should not affect the conceptual absence of specific trajectories (in general) in the formulations of quantum mechanics and the conceptualization of quantum phenomena. This should be seen clearly when we compare, for instance, the paradigm of "trajectory" or "path" which is central to classical mechanics (or even the paradigm of "world line" which is central to Lorentz mechanics), with the paradigm of "probability" (i.e. of existence somewhere in space or space-time) which is central to quantum mechanics. Anyway, the aforementioned examples of definite trajectories of quantum particles can be seen as special cases and hence the existence of definite trajectory cannot be assumed in general (i.e. when it is not observed) due to its absence in formulation and conceptualization.

[362] We draw the attention in particular to footnote [359] where we distinguished between the convergence of Lorentz mechanics to classical mechanics and the alleged convergence of quantum mechanics to classical mechanics.

[363] In fact, even such an emergence should not be enough for the general validity of quantum mechanics as long as the formalism of quantum mechanics cannot be applied directly to the classical domain.

7.3 Validity Domain

physics is "partially valid" at the quantum level (and hence it is like the quantum theory which is "partially valid" at the classical level). The challenge that these experiments are essentially macroscopic does not affect the fact that they reveal the behavior of quantum objects (such as electrons) and hence they are based on the physics that govern the behavior of these quantum objects.

5. As indicated earlier, there are some non-formal types of extended validity of quantum physics or instances of its convergence to classical physics (see for instance points 3 and 4 of § 4.7.2). However, these types and instances are not significant with regard to the extended validity of quantum theory to the classical domain because they are not based on the extension and convergence of the formalism of quantum theory (noting that there are reservations even on the extension and convergence of the formalism in certain cases and forms and from certain aspects and perspectives, as explained earlier in the last part of point 2 for instance).

6. We should also note that the (alleged) emergence of some classical formulations and theories (e.g. statistical mechanics) from quantum considerations and formulations should not be seen as evidence for the general validity of quantum theory. This is due, for example, to the limited number of such cases (and hence not the entire classical physics can be obtained in this way) as well as to the nature of this emergence as not being really based on direct application of the quantum formulation but rather being based on the underlying physics of the employed formulation. Moreover, some of these emerging formulations and theories stem from quantum considerations and assumptions rather than from quantum formulations. We should also repeat our reservations which we expressed in the end of point 5 where we referred to the last part of point 2.

    Similarly (and more obviously), the emergence of classical phenomena from underlying quantum phenomena should not be seen as evidence for the general validity of quantum theory. This is because what we have in this case is the emergence of actual physics rather than the emergence of formulations and theories.

7. We may classify the physical world and its phenomena from the perspective of size to three main categories (or levels): microscopic (or quantum), macroscopic (or classical), and superscopic (or megascopic or astronomical or cosmological). So far, we dealt with only the microscopic and macroscopic levels where we claimed that (in general) quantum physics is valid only to the microscopic world (i.e. not to the macroscopic world) while classical physics is valid only for the macroscopic world (i.e. not to the microscopic world). As for the superscopic world,[364] it should be obvious (in our view) that quantum mechanics is not valid for the superscopic world. It should also be obvious (in our view) that classical mechanics may not be valid for the superscopic world, and hence any alleged validity and applicability of classical mechanics to the superscopic world should be endorsed and established by evidence. The (seemingly) common consensus is that the superscopic world generally follows the laws of classical mechanics (as seen for example in applying the macroscopic laws of gravity in astrophysics and cosmology) and can potentially even follow the laws of quantum mechanics (as seen for example in some suggestions and proposals of finding the wavefunction of the Universe where these proposals may be based on the alleged validity of quantum mechanics to the classical world which allegedly includes even the superscopic world). In fact, we do not find a clear and explicit distinction in the literature between the macroscopic and superscopic worlds, and hence they are generally treated as if they are the same (i.e. both are "classical" and subject to classical physics). This position and attitude is questionable and very likely to be problematic and hence it should be revised. In fact, some chronic problems in astronomy and cosmology may be solved or (at least diagnosed correctly) if this issue is attended and addressed. For example, the problem of dark matter (or dark energy) which originates from the failure of the existing gravitational models and formulations may indicate that these models and formulations could be valid only at the macroscopic scale, and hence the real problem may not be solvable by assuming (arbitrarily and metaphysically) the existence of illusive dark matter, but possibly by reformulating these models and formulations and adapting them for the superscopic scale (or even proposing novel models and formulations that are appropriate and applicable specifically to the superscopic world).

---

[364] In our view, the superscopic world is not classic since our classical concepts, models, formulations, ... etc. did not evolve for this huge cosmological scale and hence they should not be expected to represent and reflect the cosmological world.

In this context, we should also refer to the issue of quantum gravity and its troublesome nature. Noting that "quantum" belongs to the microscopic world while "gravity" belongs to the macroscopic world, it is likely that this incompatibility or inhomogeneity in scale could be behind the difficulty of finding consistent and rational models and formulations to quantum gravity. The awareness of this could lead to the conclusion and acceptance that quantum gravity may not be physically very sensible (at least under the umbrella of the existing physics) due to the incompatibility or inhomogeneity and hence it should be discarded altogether. Yes, if new physical paradigms, models, formulations, ... etc. are proposed and employed then the essence of this issue may become physically sensible and addressable (although this may not be under the "quantum gravity" banner). Also see § 9.2.2.

Also see § 3.7, § 4.7.2 and § 6.7.

7.4 Mathematical Apparatus

The nature of the mathematics used in the formulation of a scientific theory should give a good idea (or at least an impression) about the nature of the theory itself. On inspecting the mathematics of quantum mechanics we can easily see that this theory is not really about understanding (or at least it is about practicing more than about understanding) and this should complicate the interpretation of the theory. For example, the use of a complex (rather than real) mathematical device (i.e. the wavefunction) to represent the primary aspect of quantum systems (i.e. their quantum state) should give a strong impression that the quantum theory is too mathematical and abstract to be of much help in understanding or interpreting or explaining the physical world (see § 4.1). A "truly physical" theory which is a reflection of physical reality should use physical real paradigms in describing and modeling the world rather than "artificial" complex paradigms. Another example is the extensive and inherent use of operators (which are essentially recipes for doing certain things) to formulate the entire quantum theory and obtain the physical observables (see § 4.4). This sort of operator-based theory may be a good practical or empirical theory (as it is indeed) but it should not be expected to be ideal for providing good understanding and deep insight in the physical world and its intricate mechanisms. So in brief, the mathematical apparatus of the quantum theory strongly suggests or indicates that this theory is about empiricism more than about understanding or rationalizing or interpreting.

7.5 Logicality

As far as the formalism of quantum theory (i.e. quantum mechanics) is concerned, there is nothing inconsistent or illogical about it, although it does not seem to have a consistent and rational interpretation (or rather it does not have an interpretation that is consistent with our classical intuition or common sense since the theory contains many non-classical and counter-intuitive perplexities). We note in this regard that perplexity or counter-intuitivity does not mean contradiction or logical inconsistency (as we distinguish between common sense, for example, and logic; see point 5 of § 2.5.1).[365]

Yes, there could be logical inconsistencies in some of the interpretations of quantum mechanics, but this should have no impact on the theory itself whose logical consistency and legitimacy do not depend on any interpretation let alone on a particular interpretation (noting that if interpretation is needed at all then it may be provided, at least partially and tentatively, by other interpretations which do not contain logical inconsistencies). However, it is worth noting in this context that the literature of quantum theory (specifically with regard to its epistemology and interpretation) contains all sorts of nonsensical propositions, illusions, bizarre ideas, contradictions, and so on. In fact, no other branch of physics (except the relativity theories) is as messy in this regard as quantum physics.

---

[365] We note that some inconsistency may occur even at the level of formalism of quantum theory if we accept certain views and possibilities (that we can find in the literature). For example, if the uncertainty principle can cause violation of the principle of energy conservation (see for example § 6.6.2 and § 6.12) then we should have a clash (at the level of formalism and according to these views and possibilities) between the two principles (noting that both principles are commonly accepted by quantum physicists and there seems to be no condition or exception that allows such a violation to the energy conservation). So in brief, energy is conserved according to the principle of energy conservation while energy is not conserved according to the uncertainty principle (and this is obvious inconsistency).

7.6 Correctness

As indicated earlier and will be repeated later, we should accept that up to now quantum theory proved to be empirically correct and practically reliable. Nevertheless, we should not rule out the possibility that some failures of the quantum theory may be found in the future (and hence we should keep our eyes open for this possibility). So, if we adopt a purely pragmatic criterion for the definition of "correctness" then we should accept that quantum theory is correct (apart from the possibility of future failure). Otherwise, we should accept that quantum mechanics faces some real and serious theoretical challenges (which are essentially of epistemological rather than formal nature). So to sum up we can say: quantum mechanics is empirically correct but it may not be so theoretically (or conceptually or epistemologically).[366]

However, it is worth noting that the empirical success of quantum mechanics should be considered and evaluated within the limitations imposed by indeterminism because indeterminism is actually a deficit in the ability of quantum mechanics to describe and predict physical reality.[367] Hence, by the non-uniqueness of science (see § 2.4.3) it should be possible in principle to find a different theory that is more successful empirically (by lifting some or all limitations of indeterminism) as well as more successful epistemologically (by having a rational and consistent interpretation).[368] In the following points we discuss and assess the "correctness" of quantum mechanics from various perspectives:

1. As discussed earlier (see for instance § 2.8) and indicated above, empirical correctness is not a guarantee of epistemological (or interpretative) correctness even if the theory has a unique interpretation (i.e. it cannot be interpreted in anyway other than a specific way). For example, the formalism can be based on certain associations and coincidences that lead to empirical correctness although the underlying conceptualization and modeling are fundamentally wrong and hence acceptable interpretation may not be possible or available. In other words, the formalism can be inferred from realistic connections and correct correlations without an underlying rational conceptual framework and hence despite the empirical correctness (which stems from the correctness of correlations) the theory (as represented by its formalism) cannot provide an acceptable interpretation due to the lack of a proper conceptual framework and hence it is epistemologically nonsensical or incorrect. Also see § 2.13.

2. Despite the aforementioned empirical success of quantum physics (which quantum physicists are very proud of), this empirical success has also limitations (which will be discussed next). The main limitation on this success originates from the perspective of indeterminism (as indicated already).[369] Regardless of the nature of this indeterminism and its origin (e.g. if it is inherent to the quantum reality or due to deficiency in our formulations and theory), the lack of definiteness and certainty in the quantum descriptions and predictions (according to indeterminism) should be regarded as a limitation on this empirical success. Quantum theory would certainly be superior if it was entirely deterministic (regardless of the possibility or impossibility of this in principle or/and practice).[370] So, when we talk about empirical success we should take into account the limitations of this success by indetermin-

---

[366] Being theoretically (or conceptually or epistemologically) correct means being an honest and sensible reflection of reality (rather than being just a reliable computational tool or calculus). We should note here that quantum mechanics is not only empirically correct (which is an extrinsic criterion for correctness) but it is also logically correct (which is an intrinsic criterion for correctness), as we saw in § 7.5. Also see § 2.5.1.

[367] In fact, (potential) limitations due to factors other than indeterminism may also exist (some of these potential limitations are proposed in the following).

[368] In fact, empirical success and epistemological success are generally independent of each other (and hence the above should express our optimism).

[369] "Indeterminism" here is mainly about probabilisticity and uncertainty (which affect the definiteness of the predictions of the theory). However, even superposition can be seen as limitation (i.e. on the description or knowledge of the physical state).

[370] It may be argued that if quantum reality is inherently indeterminate then there is no prospect of determinacy in our physical description and prediction, and hence no deterministic empirical success should be expected from a new quantum theory (to replace the indeterministic empirical success of the current theory). However, we first note that we are concerned with the epistemological, rather than ontological, perspective which can still be deterministic. Moreover, even if we assume the impossibility of determinism this does not diminish the value of determinism (as superior to indeterminism). It should also be noted that although some of the (current) quantum formulations are intrinsically based on the hypothesis of indeterminism, this does not rule out the possibility (in principle at least and with consideration of the aforementioned factors) of alternative formulations that are not so (where the non-uniqueness of science should provide a theoretical, but potentially-tentative, basis for such possibility).

ism which should diminish this success at least from classical viewpoint and attitude. Accordingly, we should ideally look for a more deterministic quantum theory that can lift at least some forms of indeterminism in the current quantum theory.
3. It is important to note that part of the empirical success of quantum mechanics is due to its "flexibility", or rather the willingness of the experimentalists (as well as theorists) to adjust, stretch and reshape the theory (and possibly the experiments) within certain limits to make the theory consistent with the observations. In fact, there are examples of the "success" of this theory where the formulation which that "success" is based upon proved to be wrong or inaccurate or problematic (see for instance § 7.11). Those who have direct experience with science (both theoretical and experimental) should know very well the ability (or rather the "elasticity") of most scientific theories to yield to the will and willingness of scientists when the theory is required to fit a theoretical model or an experimental or observational result.
4. We should also note that large parts of the recent advances (e.g. in electronics) that are attributed to quantum mechanics are actually technological (rather than scientific) advances based, motivated and driven by empirical and pragmatic approaches and they come from direct experiences, trial and error, approximations, ... etc. rather than from the theory itself (although the theory is fitted or adapted somewhere and at some stage to the functioning model, possibly arbitrarily and dubiously).

So, although we do not underestimate the astonishing success of quantum theory, we should also consider this element of flexibility and the other (non-quantum) factors in this success. We should also consider that this also applies in general to other theories (and hence quantum mechanics is not alone in this "preferential" treatment).

7.7 Optimality

In this section we assess (very briefly) the optimality of quantum theory to see if this theory is ideal (within the current state of development of science and according to the proposed criteria of optimality; see § 2.7 and § 2.11) or not. Referring to our investigation in § 2.7, we can see that quantum theory meets some of the optimality criteria (e.g. predictivity although not perfectly) and fails to meet other optimality criteria (e.g. interpretativity).[371] So overall and in general, quantum theory is not an optimal or ideal scientific theory. We may also say (to be less harsh), quantum theory is not optimal from certain optimality aspects and according to some optimality criteria (although it is optimal from other aspects and criteria).

Anyway, it should be obvious that talking about the optimality of the interpretation of quantum theory (see § 2.11) is nonsensical considering the messy situation of the epistemology of this theory and the absence of any rational and consistent interpretation (especially if we consider the possibility of the impossibility of having such an interpretation, i.e. being non-interpretable). Also see § 7.15 as well as chapter 8.

7.8 Experiments in Quantum Physics

The history of quantum physics (since its appearance) is full of experiments of all types and for all purposes (noting that the emergence and development of quantum mechanics are largely credited to experimental work, which is a fact justified by the experimental nature and roots of quantum physics). In fact, numerous quantum mechanical experiments are carried out around the world every day (some original and some not). In the following points we briefly investigate some general aspects about the nature of quantum mechanical experiments (noting that a few of these experiments, which have a particular significance for the formalism and interpretation of quantum mechanics, are investigated specifically and in some detail in different parts of this book; see for instance § 1.6, § 4.6.1, § 6.8.1, § 9.3.3 and § 10.4):

---

[371] The fact that quantum theory is optimal (to some extent) in predictivity and non-optimal in interpretativity is inline with the fact that quantum theory provides reliable rules from a practical perspective, but it does not seem to work from a theoretical or epistemological perspective (as can be easily seen, for instance, from the many problematic issues in its interpretation; some of which were investigated earlier and others will be investigated later).

7.8 Experiments in Quantum Physics

1. As indicated earlier (see for instance § 2.10) physical experiments can be related to interpretation as well as to formalism and this applies in particular to quantum physics. Accordingly, some of the experiments in quantum physics are related essentially to the formalism of the theory, while other experiments are related essentially to the interpretation of the theory. An example of the first type is the Franck–Hertz experiment or Stern-Gerlach experiment or Davisson-Germer experiment, while an example of the second type is the Aspect experiment or some variants of the double-slit experiment. However, it should be noted that many experiments (e.g. some modern variations of the double-slit experiment) are related to both aspects of the theory (i.e. formalism and interpretation). In fact, we can claim that most of the quantum mechanical experiments have interpretative significance (in the general sense of interpretation) as well as formal significance although the primary focus and purpose are usually related to one of these aspects.[372]

2. Some of the experiments related to the interpretation (e.g. many types of experiments in atomic and particle physics or the double-slit experiment and its variants) belong to the interpretation of quantum theory in general (which may be called the common or standard interpretation) and its significance in comparison to classical theory, while other experiments related to the interpretation (e.g. Bell's experiments; see § 9.3.3) belong to specific schools of interpretation (see chapter 8). In fact, most (if not all) of the quantum mechanical experiments should have significance and value with regard to the common or standard interpretation (since they follow and demonstrate the formalism of quantum mechanics and hence they should demonstrate and reveal its general features and physical attributes).

3. Historically, quantum mechanical experiments in the early days of quantum theory were almost entirely about the development of the formalism of the theory (as well as developing basic understanding of the quantum phenomena). However, in the recent decades (i.e. since the beginning of 1970s approximately) substantial experimental work related and dedicated (specifically and exclusively) to the interpretation of quantum mechanics has emerged (as represented for example by the Bell experiments or the modern variations of the double-slit experiment). In fact, some modern quantum mechanical experiments may even be designed and conducted to target (i.e. to endorse or refute) specific schools of interpretation (see chapter 8).

4. In general, the analysis and interpretation of most quantum mechanical experiments are subject to many controversies and criticisms and hence we can rarely find a consensus about the significance and value of a quantum mechanical experiment. This should not be strange to quantum mechanics which is surrounded with controversies and uncertainties (especially in its interpretative and epistemological sides). Accordingly, it is difficult to propose or design or conduct conclusive experimental tests about quantum physics (i.e. with universal approval) although claims of conclusivity are not rare.

Finally, it should be obvious that we were talking here (as usually elsewhere) about real experiments, and not about the so-called "thought experiments". In fact, we do not even classify these "thought experiments" as experiments in any sensible way since they are entirely useless and senseless as "experiments" (noting that they have no experimental content or value at all despite the wrong suggestion of their name). Moreover, we believe that the label "thought experiments" which is attached to these (rather dodgy) theoretical practices could be misleading (especially to students and young physicists who may get the impression that they have the legitimacy and authority of experiments). Therefore, we propose replacing this label with another (more honest and reflective) label. As we will see in § 7.15, these "thought experiments" could be behind some of the troubles of quantum theory.[373]

---

[372] It should be obvious that any (supposedly correct) quantum mechanical experiment should comply with the rules of quantum mechanics and hence it should be an endorsement (in general and from certain perspectives) to the formalism of quantum theory (possibly implicitly and indirectly). It should also be obvious that physical experiments (including those of quantum physics) should normally have some interpretative value and significance (in the general sense of interpretation).

[373] We should note, however, that thought experiments can be useful in certain cases and for certain purposes such as using them as pedagogical and educational tools.

7.9 Tone

A very interesting aspect of the quantum theory and its formalism is that it embeds a collection of elements that do not belong to the physical reality of the observed phenomena, in sharp contrast with the classical theory. This is reflected in the tone and language used in the presentation of the quantum theory which is laced with terms foreign to this reality. So, while reading a theorem in classical physics gives a strong impression of an independently-existing reality that behaves autonomously regardless of any observer and the theorem expresses a physical pattern or a feature of this reality, reading a corresponding theorem (or part) of the quantum theory gives an impression of a tentative "reality" and a confused observer who are equally at the center of the theory and the function of the theory is to find a solution or recipe for this confused observer on how to deal with this "reality".

To clarify our point let have an example where we compare Newton's second law (as representative of classical physics) with a statement that summarizes the issue of measurement in quantum mechanics (as representative of quantum physics). According to Newton's second law (in its simple and compact form):

Force equals rate of change of momentum

As we see, the tone and language of this statement are completely objective and physical with no suggestion of any observer that has any role in this fact of reality. So, this is a totally objective statement synthesized entirely of physical terms[374] with no reference or suggestion of any observer or observation. On the other hand, the statement of quantum measurement reads like the following:

Measurement leads to the collapse of wavefunction to one of its eigenfunctions (or eigenstates) where the observed result is a corresponding eigenvalue (of the operator representing the observable) and the probability of a specific outcome is proportional to the modulus squared of the overlap between the wavefunction before measurement and the wavefunction after measurement

As we see, the tone and language of this statement do not give an impression of objectivity or physicality for instance. My personal impression when I read a statement like this is to imagine a puzzled physicist (or rather mathematician) who faces an ambiguous "reality" and he is trying to find a mathematical recipe that can solve his problem and crack the enigma. For example:
- We have terms like "measurement", "observable" ... etc. which suggest an observer (central to the theory itself and not just a casual bystander like the observer in classical physics) who is concerned with a problem and he is about to tackle it technically.
- We have terms like "wavefunction", "eigenfunction", "operator", "modulus squared" ... etc. which suggest mathematical machinery (or recipe) as part of the theory itself and central to it (and not just as a casual tool to deal with this problem). These terms (or some of them at least) are non-physical (or rather they are purely mathematical) and hence the issue (as well as its theorem) does not seem to be essentially physical.

In fact, we can even extract from the above statement (or at least from another version of it) other things (like indeterminacy of reality) that cannot be found in a typical statement of a classical theorem. Also see § 2.5.4 and § 2.7.5. The reader is also referred to the paragraph in § 6.12 about the principles of conservation of probability and parity which should also reveal and reflect the tone and spirit of the theory.

To sum up, the tone of quantum theory gives (at least occasionally if not commonly and frequently) an impression that it is a theory about our impression or notion of the outside world rather than about the outside world itself. The tone should also give a strong suggestion of its practical and empirical nature. In fact, this should explain (at least partly) the epistemological and interpretative deficits and failures of quantum theory, i.e. it is about practical objectives rather than theoretical objectives (see § 2.1).

---

[374] Even "equal" (let alone "rate of change") should be understood in its physical sense and context.

7.10 Formalism-Interpretation Confusion

An important aspect in the assessment of the quantum theory is the lack of obvious border between formalism and interpretation. Although this may be seen as a practical issue[375] (which is mostly related to the literature of the theory rather than to the theory itself) it still plays a crucial role in the theory and its evolution (and perhaps even its application since application is generally based on our understanding and appreciation of the theory). In fact, the origin of this problem is the special nature of the quantum theory (like the special nature of Lorentz mechanics) and being unusual in terms of dealing with unfamiliar objects and phenomena and hence such a confusion is not unexpected.[376]

Our view is that some parts of the quantum theory (which are traditionally and conventionally classified as part of the formalism like wavefunction collapse; see § 5.7 and point 2 of that section in particular) are actually foreign to the formalism despite the opposite claims. It seems as if some physicists try (deliberately or not) to dictate their opinions and personal choices and make them part of the theory of quantum mechanics. In fact, this sort of misrepresentation has a cost in terms of (mis)understanding, appreciation and evaluation of the quantum theory. For example, many physicists and philosophers of science talk about the philosophical consequences of "quantum mechanics" while the reality is that many of these alleged consequences belong to certain views and interpretations and not to quantum mechanics itself and this leads to serious confusion and delusion and exacerbates the mess of quantum theory and its interpretation in particular.

In this context, we should wonder that even the most basic form of the formalism (as represented by the axiomatic framework of quantum theory; see § 4.2) may not be the best to represent the theory and establish its theoretical structure because this form of the formalism is problematic from various aspects one of which is the (apparently unavoidable) mix of formalism and interpretation which (in part) is based on the ambiguity of some of its fundamental concepts (particularly "measurement") and the lack of rigorous scientific definitions which open the gate wide to all sorts of contemplations, opinions, etc. and thus lead to this unnecessary mess. We should also note that the very essence of the formalism of quantum mechanics (in its current state) contains a very strong element of interpretation represented by the standard interpretation of wavefunction as a probability amplitude (or Born's statistical interpretation). These issues will be investigated further later on (see for instance chapter 8). Also see § 2.8.[377]

7.11 Inconsistencies and Question Marks

The literature of quantum mechanics is full of inconsistencies and question marks (as well as controversies and differences in views and opinions which in many cases reflect and demonstrate these inconsistencies and question marks). Although these are generally not part of the formalism and they mostly have no direct impact on it, they represent demonstrations of how quantum mechanics is understood and "interpreted" in a rather natural and "intuitive" way, and hence if it is "interpreted by our intuition" inconsistently and badly then its "real interpretation" is very likely to be inconsistent and bad. This should support (tentatively) our view which we expressed and justified elsewhere (e.g. in § 7.15 and § 8.6) that quantum

---

[375] In fact, this is not entirely practical since there are some elements of interpretation which are inherent to the formalism and part of it.

[376] We refer the reader to earlier parts of the book about this issue (see for example § 1.3.2 and § 3.6). The reader is also referred to our book "The Mechanics of Lorentz Transformations" about this issue (see for instance § 1.6 of that book). The basic idea in this regard is that some theories of modern physics (particularly Lorentz mechanics and quantum mechanics) are centered on the paradigms of frame of reference and observation and hence the line between the physical phenomena and the observer is blurred which results in confusion and mixing of formalism and interpretation. This is seen obviously in Lorentz mechanics where the special relativistic interpretation is confused with the formalism of Lorentz transformations (i.e. Lorentz mechanics). It is also seen in quantum mechanics where the formal and interpretative elements are commonly confused and mixed (and hence what is part of the formalism according to certain view or school or formulation may be considered as part of the interpretation according to others).

[377] We should also refer to the strong influence of the Copenhagen school and its dominance (associated with its simplicity and attractivity) as a potential factor in this confusion between formalism and interpretation (see point 11 of § 8.1). This should justify that some elements of the Copenhagen school (e.g. collapse of wavefunction) crept and sneaked to the formalism, and hence they are commonly seen as part of the formalism of quantum mechanics.

7.11 Inconsistencies and Question Marks

mechanics is not interpretable.[378] In the following points we give some examples (or at least potential examples) of these inconsistencies and question marks:[379]

1. One of these examples is the explanation of the quantization of atomic orbits by the standing waves of electrons without admitting the classical view (as expressed in Bohr's atomic model for instance) of the electrons as being circulating around the nucleus (noting that without circulation the de Broglie matter wave cannot be explained; see Eq. 16). In fact, even circulation may not be able to explain and justify the standing waves picture consistently and entirely. It should be noted that without definite trajectories (which are presumably absent or denied in quantum mechanics or at least in some of its interpretations) there is no meaning to circulation and hence no standing wave.[380] Other sources of inconsistency can also be found in this explanation. We should also note that the circulation of electron (which is against the probabilistic interpretation and the absence of specific trajectories) can be found (sometimes more vividly and in a more central role) in other quantum situations and scenarios like in the explanation and justification of spin-orbit coupling.

2. Another example is the violation of some conservation principles (notably energy conservation) by the uncertainty principle (as claimed by some) while the conservation principles are supposedly accepted as respected and fundamental principles and are used (in this capacity) in many arguments in quantum physics and by quantum physicists (even by those who allow such violations). See § 6.12.

3. As a result of the confusion between physical waves and matter waves (see § 3.2), we can find many examples in the literature of inconsistencies and question marks in this regard (such as attributing some real physical properties to matter waves while considering or treating matter waves as an abstract mathematical device). In fact, regardless of any distinction between physical and matter waves, the mere attribution of some physical properties of waves (in their conventional and classical conceptualization) while denying or discarding other physical properties is inconsistent and questionable (see point 2 of § 3.2).

4. For photons we have a problem that is they (supposedly) should have real (classical) wave nature as electromagnetic waves and matter (quantum) wave nature (to account for the statistical distribution of photons in the interference experiments like the double-slit experiments involving photons). So, for example in the double-slit experiments involving electrons we have matter waves while in the double-slit experiments involving photons we should have classical waves as well as matter waves (see § 3.2).[381]

5. The alleged universality of the particle-wave duality (which supposedly originates from the de Broglie hypothesis) is problematic in many situations and cases and cannot be maintained consistently (at least for practical reasons). In fact, the paradigm of duality itself is problematic and leads to inconsistencies from various aspects. Some of these inconsistencies have been discussed or indicated earlier (see for instance § 1.6.1, § 5.3.1 and § 6.13.1). In fact, this causes many inconsistencies and arbitrariness (e.g. in analysis and interpretation). Our view (which we indicated repeatedly) is that "wave" and "particle" (let alone their alleged duality) are approximate models in the quantum world and thus they should be treated with caution to avoid these inconsistencies (or some of them).

6. The normalization of the wavefunction of free particle is problematic despite the various proposals

---

[378] With regard to quantum mechanics, "not interpretable" should mean that we can see no prospect of having a rational interpretation to this theory, but we cannot prove this or assert it firmly and hence having an interpretation in the future remains a possibility.

[379] In fact, there are many other examples but the following should be enough. We should note that some of these examples may be ridiculed by some thanks to indoctrination which makes such examples look consistent and unquestionable.

[380] In fact, the "standing wave" model should be seen as an approximate model (like other wave models used in quantum mechanics such as wavepacket and pilot wave models) and hence no serious physical implication should be derived from this model (and its alike). Also see § 3.2.

[381] This sort of inconsistency requires very sophisticated arguments and models (within the quantum field theory) to fix (allegedly). However, this does not make any difference to the fact that quantum mechanics itself contains inconsistencies that are difficult or impossible to explain and interpret (which makes the interpretation of quantum mechanics itself very difficult or impossible). In fact, these complications (introduced by the quantum field theory for instance) make the quantum theory more abstract and less understandable (and even non-interpretable). We should also note that the quantum field theory and its alike (of theories based on quantum mechanics) should be also difficult or impossible to interpret because they are based on quantum mechanics which is difficult or impossible to interpret. In fact, some of these theories are even more difficult (or "more impossible") to interpret than quantum mechanics because they contain additional complexities or/and inconsistencies.

7.11 Inconsistencies and Question Marks

suggested for formalizing and rationalizing it (see for instance footnote [167] on page 80). In fact, none of these proposals is totally convincing and problem-free. This should cast a shadow on a number of issues such as the statistical interpretation of the wavefunction as a probability amplitude (or at least put a question mark on the generality of this interpretation).

7. In some situations, we find the phase and group velocities of a wavepacket move in opposite directions (which is at least counter-intuitive).[382]

8. According to the model of "wavepacket" (which supposedly represents a material "particle-like" quantum object), the wavepacket is made of a superposition of matter waves which are not physical (see § 3.2) while the wavepacket itself (as it represents a "particle-like" quantum object) is physical (as can be seen for instance from having energy, momentum, etc.).

9. The concept of "quantum wave" is not entirely consistent (at least when compared to its classical cousin). For example, "quantum wave" supposedly possesses a well-defined wavelength but it does not seem to have other well-defined physical properties of waves such as speed or energy (e.g. according to relations like $v = \lambda \nu$ or $E = h\nu$; see for instance point 2 of § 3.2). The absurdity of this appears in several quantum mechanical situations and contexts such as wavefunction collapse and quantum entanglement where infinite speed (of communicating the collapse and interaction) seems a physical possibility (or at least the speed is not well-defined).[383] In fact, this could be seen as an evidence or indication that quantum waves are not really waves in the familiar and traditional sense. In other words, "quantum waves" is an approximate and imitate model rather than a real and exact model.

10. Some of the indistinguishable (or identical) quantum objects may not be really indistinguishable although they may be treated quantum mechanically as indistinguishable. For example two photons (which are supposedly indistinguishable bosons) can differ in their (supposedly) intrinsic attributes such as frequency (e.g. "blue" photon and "green" photon) or polarity (and this could cause some inconsistencies and complications).[384]

11. Quantum spin is commonly regarded as a relativistic effect, but following its "relativistic derivation and justification" it is applied to systems in which relativistic effects are negligible.[385]

12. There are many inconsistencies and arbitrariness in using incompatible physical models simultaneously and wishfully with no regard to their contradictory physical implications and consequences. For example, some formulations (possibly done by the same individual and in the same context) may be based on modeling the electron once as a "diffuse charge cloud" and once as a "wavepacket" and once as a "particle guided by a pilot wave". This chaotic and arbitrary use of models makes these models lose their physical significance and value and renders them to be no more than tools or devices (or even pretexts) for achieving whatever objective at hand with no consideration or regard to rationality or physical sensibility or consistency. Moreover, it could lead to conflicting physical consequences (e.g. one model could lead to having a dipole moment while another model does not). Similar examples of inconsistency and arbitrariness (e.g. with regard to localization/non-localization or physicality/non-physicality) are abundant in the literature.

13. There are also some examples of inconsistency or absurdity or question marks that we discussed or indicated before such as using metaphysical paradigms and entities (like creation and virtual particles) or proposing vague concepts and ill-formulated principles (like complementarity) or legitimizing and sanctioning thought experiments as if they are real experiments.

---

[382] In fact, counter-intuitivity is very common in the quantum theory as it is full of examples of this kind (and this should explain the difficulty of understanding and interpreting the theory). However, this example is more than being counter-intuitive.

[383] It is worth noting in this regard that the basic idea of "matter wave" lacks the dynamics of propagation (seen in classical waves) in space since matter waves seem to be frozen in space rather than being moving or propagating.

[384] In fact, some consider such "indistinguishable" objects as distinguishable unless they have identical attributes (and hence photons are indistinguishable only if they have the same frequency, polarity, etc.). They may also be treated as distinguishable/indistinguishable according to circumstances and conditions. Anyway, the issue of indistinguishability should be seen in the light of its effect (i.e. exchange interaction) and hence the objects should be treated as distinguishable/indistinguishable according to the significance of this effect (regardless of being "really indistinguishable" or not). The details of this issue are beyond the scope of this book and hence the interested readers should refer to the literature.

[385] We should note that quantum spin may also be derived non-relativistically (as indicated earlier).

14. Many quantum mechanical arguments and (alleged) proofs lack not only rigor and technicality but they lack sensibility and rationality as well.
15. Finally, the ultimate demonstration of inconsistency, absurdity and question marks is found in the schools of interpretation of quantum mechanics (see chapter 8) where most of these interpretations are either totally and fundamentally irrational or they contain strong elements of irrationality. For example, some of these schools are metaphysical (or contain elements of metaphysics), some contain ambiguous or/and nonsensical ideas, some are based on dubious formulations and dodgy presumptions, some lead to contradictory implications and consequences, ... etc. In fact, the real misery of quantum theory appears in these schools of interpretation where they show some of the most ugly facets of modern science.

We should finally note that the above inconsistencies (as well as other inconsistencies) are generally defended and justified by quantum physicists using various arguments and tricks (to make them look consistent and rational), but in most cases the proposed fixes are arbitrary and artificial. In fact, some of these fixes lack sensibility and consistency (and they could even be less rational and sensible than what they suppose to fix) and hence they themselves need sensible interpretations and justifications.

7.12 Problem with Causality

Referring to § 1.7 and § 2.4.2, the principle of causality is a scientific and epistemological (and possibly ontological) necessity although in essence it is not a scientific principle. In fact, causality is a requirement for rationality and sensibility (to make understandable correlations, establish predictability, avoid arbitrariness, provide consistent patterns, etc.). Accordingly, any scientific theory or interpretation that (totally) violates this principle should be rejected on the basis of violating rationality. As we saw earlier and will be investigated further, some interpretations of quantum mechanics may lead to such violations to the principle of causality and hence they should be rejected on this basis. Regarding quantum mechanics (in its bare formalism), we do not think it (in itself) leads to any violation of causality. After all, quantum mechanics as an empirical recipe (or mere calculus) should not have such philosophical or epistemological suggestions of causality or its alleged violation.

Yes, in the literature of quantum theory there is some confusion and blur between the formalism and interpretation of this theory (introduced mainly by its axiomatic framework as well as the dominance of some of its interpretations) where certain elements of interpretation are considered part of the theory (as represented mainly by its formalism; see for instance § 7.10).[386] Accordingly, some violations of causality may be implied by such confused elements of interpretation. However, as far as the theory itself (in its bare formalism and its application) is concerned no such implications should be considered seriously. In more simple terms, we should continue using this theory in practice (as long as it is working) with disregard to any such implications and if they could (or could not) be an indication to flaws in the theory and its formalism. In other words, on a purely practical and empirical basis there should be no concern about the validity of quantum mechanics from these alleged violations of causality (although legitimate concerns could be attached to its epistemology and interpretation and whether or not it is a good theory from these perspectives).

So far, we were talking (particularly in the previous paragraph) about the potential impact of the (alleged violations of the) principle of causality on the quantum theory (and on its formalism in particular). With regard to the potential impact of the quantum theory on the principle of causality we can say: if the quantum theory was the only possible quantum theory and there is a single valid interpretation that necessarily violates causality then we may need to amend the principle of causality to accommodate such alleged violations (refer to § 1.7 and § 2.4.2 for some proposals about such amendment). However, we do not need all this hassle and pain because we simply reject both premises. In fact, the premise of uniqueness (i.e. being the only possible quantum theory) should be rejected by the principle of non-uniqueness of science (see § 2.4.3), while the premise of existence of such an interpretation should be rejected first by the

---

[386] As indicated above, this situation was exacerbated by the dominance of the Copenhagen school (see § 8.1) over most of the history of quantum theory where (at least) some elements of this interpretation are considered (or treated like) parts of the theory itself.

non-existence (so far) of any valid interpretation to this theory (see for instance § 7.15 as well as chapter 8) and second by the existence of many interpretations some of which do not violate causality (if we have to accept the existing interpretations despite their flaws and failures).[387]

To sum up and conclude, quantum theory (both in its formalism and in its existing interpretations) cannot represent any serious threat to the principle of causality (at least in its basic form and essence), and hence all (or at least most) the arguments about the philosophical and epistemological challenges to causality by the current quantum theory should be rejected (at least if they represent threat to its basic form and essence). In fact, if we have to abandon (or even question) the principle of causality then we need to abandon (or question) the entire science (and indeed the entire rational knowledge) because no science or rational knowledge can be established without this principle (at least in its basic form and essence). Anyway, if we assume (arguably) that strict and strong form of causality at the quantum level cannot be maintained (for whatever reason or pretext) then we can accept a weak and mild form of causality at the quantum level (see § 2.4.2) while maintaining the strict and strong form of causality at the classical level. In other words, we compromise the principle of causality only minimally and only at the quantum level (i.e. no more than needed).

7.13 Problem with Locality

We explained earlier (see for instance § 1.8) our position about the alleged principle of locality. In summary, this alleged principle in its relativistic interpretation cannot be validated or approved and hence even if the quantum theory (whether in its formalism or in its epistemology and interpretation) leads to explicit violations of this alleged principle there should be no problem to the quantum theory.[388] Actually, such violations by the quantum theory (as well as by any other theory) should discredit this alleged principle further (rather than threaten the theory). In fact, such violations should discredit and refute even the relativistic roots of the locality principle, i.e. they should be a serious threat to the relativity theories (and special relativity in particular). Anyway, (potential) violations of the principle of locality could be a serious challenge to those who believe in this principle (who are seemingly the majority of physicists) and hence they should either reject or adjust quantum mechanics (or/and its interpretation) or modify or reinterpret the principle of locality (if they are not ready to discard it). Alternatively, they can look for a compromise (if this is possible) that keeps and respects both.

To conclude, it seems that non-locality cannot be avoided in the current quantum theory (noting that non-locality seems implicit even in the basic concept of wavefunction, which is intrinsically and essentially simple and global, regardless of any intervention or interaction such as reduction). Also see § 5.6 and § 6.10 as well as § 1.8 and § 9.2.1.

7.14 Problem with Realism

Generally speaking, quantum mechanics (or some of its interpretations) may form a potential threat to realism which (despite its state as an ontological choice) is an epistemological necessity (see § 1.10). However, in our view any potential clash between the quantum theory and realism should be decided in favor of realism and against the quantum theory. This is because realism (in its moderate form) is a necessity for rationality and hence no credible scientific or epistemological theory can be established without realism.

To be more explicit and comprehensive, the quantum theory in its formalism (i.e. quantum mechanics) cannot be a threat to realism because the formalism is purely about science and physics while realism is related to philosophy and epistemology and hence they are different species. Yes, quantum mechanics can be a threat to realism by its epistemology and interpretation. However, this should not be a serious and

---

[387] For example, some (non-local) hidden-variable interpretations (or formulations) may provide such causality-preserving explanations.

[388] There should be no problem as long as the violations are restricted to the locality principle, i.e. they do not lead to other violations (e.g. to causality). The consequences of other potential violations (associated with or caused by violations to locality) are discussed elsewhere (see for instance § 7.12).

7.14 Problem with Realism

imminent threat. **First**, because of the miserable state of quantum theory from the epistemological and interpretative perspective, quantum theory cannot make a serious challenge or threat to realism (or indeed to any well established principle or knowledge) from this perspective. As we explained earlier, there is no credible or reliable interpretation or epistemology for the current theory of quantum mechanics and hence the existing schools of interpretation (or rather those schools that supposedly threaten realism) cannot be a serious challenge. In fact, if we have to accept the existing schools of interpretation (despite their miserable state) then we can accept the ones that do not contradict or threaten realism (e.g. non-local hidden-variable interpretations and formulations).

Second, because quantum mechanics is not the only possible theory for the quantum world (at least in principle), it cannot demolish realism because realism can possibly be demolished if there is a necessity that any theory about the quantum world must have consequences (or interpretations or implications) against realism. As we repeated time after time, science is not unique and hence we should always be capable (at least in principle) to replace one theory by another (where both are empirically correct and equivalent).[389] So, if the current quantum theory is incompatible with realism we should keep realism until we can prove that no other quantum theory (i.e. theory about the quantum world) can be compatible with realism. Until (and if) we get this proof, we continue to use quantum mechanics as an empirical recipe (or mere calculus) for our quantum business and practices while keeping at the same time our epistemological belief or faith in realism.

In the following points we discuss and assess further the problem of quantum mechanics with realism from various aspects:

1. As indicated before, what is required (when we talk about embracing or keeping realism) is the compliance with realism in its moderate form, not in its extreme form (see § 1.10 and § 2.5.2 as well as point 4 of § 8.6).

2. The existence of realistic deterministic formulations (like Bohm's mechanics) to quantum mechanics despite the limitations of these formulations should be an indication that quantum phenomena can in principle be described by a realistic deterministic theory. In fact, if theories (like Bohm's) that can supposedly reproduce all the predictions of quantum mechanics do exist, then this in itself should be a proof for the possibility of the existence of other theories that are totally realistic and deterministic (although not necessarily local), noting that the (possible) violation of locality by these theories should not be a problem at all because we do not believe in the validity of the locality principle (at least in its temporal form; see § 7.13). In brief, the mere possibility (which no one can rule out and which is inline with the principle of non-uniqueness of science; see § 2.4.3) should justify maintaining the confidence (or at least hope) in realism even at the quantum level, while the existence of realistic formulations of quantum mechanics (e.g. Bohm's formulation) despite the fact that they are not ideal is sufficient to hold this possibility (and in fact making it an actuality). So, if realism at the quantum level is possible in principle (since there is no logical inconsistency in it) and even realizable (even though the current realizations are not ideal) then it should be possible unconditionally and even realizable in a more ideal form than the existing forms.[390]

3. If (for whatever reason) we were forced (by the implications of the current quantum theory) to compromise realism then we should limit any damage to realism to the bare minimum. Accordingly, any threat to realism by the quantum theory should not affect (classical) realism at the macroscopic level. Moreover, any damage to realism at the quantum level should be restricted and limited to the minimum amount required by the quantum theory. So in brief, realism is sacrificed if and when (and as much as) sacrification is absolutely necessary. However, as noted earlier, by the principle of non-uniqueness of science (endorsed by other indications) a totally-realistic novel quantum theory should be possible and hence we should not be forced to abandon realism even at the quantum level and even partially.

---

[389] It should be noted that we are not referring here to local hidden-variable type of quantum theories (which are supposedly refuted by Bell's theorem and its alike). In fact, we are not referring in this context even to the existing non-local hidden-variable type of quantum theories and formulations. But we rather refer to novel quantum theories that can be significantly different from the current quantum theory.

[390] In fact, this in our view is the main value of the realistic formulations of quantum mechanics even though they may not have a great value on their own and may not be ideal from various perspectives (at least because they just emulate quantum mechanics).

7.15 Difficulty of Interpretation

In this section we discuss the issue of the difficulty of finding a consistent and rational interpretation to quantum mechanics. In fact, the essence of this investigation is the problematic aspects in the quantum theory that make its interpretation difficult and perhaps impossible. In the following points we outline some potential reasons for this difficulty:[391]

1. One possible reason is the semi-classical nature of the quantum theory (see § 7.2), i.e. it is based on employing classical macroscopic concepts and patterns to describe the quantum microscopic world (e.g. using "particle" and "wave" to describe quantum objects or using classical spin as an angular momentum to describe the quantum "spin").[392] In fact, this sort of conceptual inhomogeneity (as well as the possible unsuitability of employing classical concepts and patterns) may not only make finding a consistent interpretation difficult, but it may make the existence of such an interpretation impossible.[393] In fact, a close inspection of the quantum theory and its historical development should reveal that considerable part of its theoretical structure and framework is based on classical paradigms and models (sometimes explicit and sometimes disguised) although this classical heritage was subject to dilution and "quantumization" later on. Also see § 7.1 and § 7.2.

2. Another possible reason is the non-physical nature of the main aspect in the theory (i.e. wavefunction) since it is essentially a mathematical device rather than a physical object or entity (see for instance § 4.1, § 6.1 and § 7.4). In fact, non-physicality[394] (to some extent) is a feature of the entire quantum theory (see for instance § 7.4 and § 7.9) and this could be a strong reason for the difficulty or impossibility of interpreting this theory physically (since interpretation is essentially a physical description of the physical reality).

3. A third potential reason is the blur between formalism and interpretation, i.e. the formalism of the theory is not really a pure formalism (see § 7.10) but it is based (partly) on a certain interpretation (represented mainly by the Born probabilistic interpretation and potentially by elements from the Copenhagen school).[395] In this context, we should remember the failure of Schrodinger himself to recognize the significance (i.e. interpret "correctly") the wavefunction in his equation and its physical meaning. This failure should demonstrate the interpretative nature of the probabilistic interpretation (as well as the blur between formalism and interpretation) since it is not part of or inherent to the formulation of Schrodinger. So, the subsequent development of quantum mechanics (which established the formalism in its mathematical framework) was based on an interpretation, i.e. the Born probabilistic interpretation (as well as other potential elements of interpretation). Although Born's interpretation seems now (thanks to education and indoctrination) an essential part of the formalism, it stays an interpretation of the formalism.[396] In fact, this blur between formalism and interpretation

---

[391] We should note that some of these reasons are tentative and they mostly apply to certain aspects of the theory and possibly according to certain views. We should also note that most of these reasons have been investigated earlier and hence this list (and indeed this section) builds on the previous investigations and represents a summary from this perspective.

[392] For example, "particle" and "wave" (as well as their duality) could be behind many of the quantum mechanical epistemological failures (some of which are indicated or discussed earlier). Similarly, it is possible that part of the confusion about the spin state (e.g. impossibility of simultaneous determination of more than one spin component) is because spin is treated like classical angular momentum.

[393] In fact, this sort of inhomogeneity is not restricted to quantum physics but it extends to science in general which in its current state is an inhomogeneous mixture of theories, ideas, methodologies, approaches, personal preferences, historical relics, prejudices, etc. Hence, we should not expect the current science to provide a completely consistent picture (neither as formalism nor as interpretation although the latter, which represents the epistemological and philosophical aspects, is generally the most affected by this chaos).

[394] "Non-physicality" here should mean that the center of attention of the theory is not on the physical phenomenon itself but on other things and perspectives like measurement and observation (or at least on the phenomenon from these perspectives).

[395] In fact, the interpretative aspects embedded within the formalism should be more than just the probabilistic interpretation as can be felt from inspecting the axiomatic framework of quantum mechanics (see § 4.2) and analyzing some of its aspects (like measurement and its meaning and physical significance, e.g. how we define "measurement" and what type of action realizes it and satisfies its conditions and requirements, as well as the collapse of wavefunction by measurement).

[396] We note that the interpretative nature of the Born probabilistic interpretation should be obvious (even from its name),

7.15 Difficulty of Interpretation

can complicate the building of a consistent interpretation because we need, for instance, to construct an interpretation on an interpretation and hence some (potentially logical) interpretations may be excluded while some (potentially illogical) interpretations may be included.

4. We should also note the nature of the mathematical and conceptual framework, i.e. we should consider the nature of the mathematics and concepts of quantum mechanics as a possible cause (or at least an indicator) of the difficulty of interpreting quantum mechanics (see for instance § 7.4 and § 7.9). For example, when a theory is intrinsically based on very abstract mathematical machinery and technical concepts like complex numbers, operators, eigenvalues, eigenvectors, probability, ... etc. it should be difficult to interpret since the essence of interpretation is to make the theory understandable and justifiable by using familiar physical concepts and language, and this should be hindered by the use of such complicated and non-intuitive machinery and concepts. In brief, the inherent and excessive use of abstract mathematics and subtle concepts makes ordinary interpretation (using familiar concepts and language) difficult if not impossible.

5. Another potential reason is the history of quantum mechanics, i.e. the theory was historically developed as a collection of ideas created and developed by people of different backgrounds where each one takes the theory a step further by fixing previous flaws and improving its theoretical structure and practical aspects (see § 7.1). This sort of theories developed by a group in a step-by-step fashion could be perfect from a practical and empirical perspective (where such a group effort over relatively-long period of time should be capable to fix all the defects and limitations from this perspective) but such theories are very likely to have conceptual and epistemological shortcomings (due to the inhomogeneity and lack of a unified theoretical background from which the theory emerges spontaneously and naturally). In short, the theory was a big and great renovation (rather than innovation) project where raw and crude ideas (like the primitive atomic model of Bohr) were created and subsequently improved and refurbished (gradually and collectively).[397]

   In fact, the effect of the history of quantum mechanics on the difficulty of interpretation extends beyond what have been said already. For example, we may recall in this context proposals like "the formalism of quantum physics preceded its interpretation" which indicates that this theory was developed with rather little insight (unlike many other theories which were preceded by and emerged from a deep insight and hence the formalism of these theories is based on a sort of insight that should lead to an interpretation unlike the situation in quantum theory).[398]

6. Another potential reason for the difficulty of interpretation is the heavy reliance during the historical development of quantum theory (which is still going on in some aspects) on thought experiments and hypothetical arguments some of which are completely detached from physical reality while most of which are affected by subjective factors that reflect the attitude of the developers and their "inner reality" more than the outside reality. This should complicate the theoretical and conceptual structure of the theory and make it difficult to interpret physically and realistically. This should also explain the fact that quantum mechanics (particularly in its epistemological aspects) is contaminated with subjective elements more than most of the other branches of science (especially classical branches).

7. We should also consider the possibility of negative impact by specific elements and factors in the theory that hinders its interpretation. For example, it is quite possible that the (rather clumsy) particle-wave duality model (or its "particle" and "wave" components or its conceptualization into the wavepacket model) is behind at least some of the difficulties of finding an interpretation to quantum mechanics

---

noting that it cannot be seen as a definition of the symbols of the Schrodinger equation. This should also apply (tentatively and to some extent) to other potential elements of interpretation which are embedded in the formalism and appear to be part of it.

[397] In this regard, we can even find historical "successful" attempts to use hydrodynamics and *continuum* mechanics to formulate quantum mechanics (which is supposedly about *discrete* "quantum" objects like electrons). However, this should not be very surprising because even the current quantum mechanics is essentially a *wave* mechanics (where wave is essentially a *continuous* phenomenon) which does not seem less weird.

[398] Although this proposal is true in general (or at least not far from reality) we should consider some limitations and reservations and hence it should be understood and interpreted in a proper sense and within an appropriate context. We should also note that this proposal (and its alike) should indicate the empirical and pragmatic nature of quantum theory (i.e. being about practice more than about understanding; see § 2.1).

(refer for instance to § 6.13.1). As indicated earlier, this model is questionable from various aspects and it seems to be unfit (at least epistemologically and partly) to be a basis (or ingredient) for a consistent and rational interpretation.[399] This could similarly apply to other problematic paradigms and models like measurement and wavefunction collapse.

8. Another potential factor for hindering the interpretation is the existence of uncertainties, suspicions and question marks about various aspects and parts of the quantum theory. An illuminating example is the uncertainty principle and its troubling and mysterious nature which was investigated earlier in § 6.6.3. Another example is the nature of spin and if it is really a form of angular momentum or even if it is appropriate to treat it conceptually as a form of angular momentum (see for instance § 4.6). A third example is the conservation principles and their (rather shaky) role in quantum mechanics (see § 6.12). A fourth example is the uncertainties about the domain of validity (at least from a practical viewpoint) of quantum mechanics and if it applies (at least in principle) to the macroscopic world.[400] So, even if we hypothetically found a consistent and rational interpretation, these uncertainties make such an interpretation tentative and subject to many questions and uncertainties.

9. We should also remember that the rather primitive nature of Schrodinger's equation (which represents the essence and heart of quantum theory) as well as its many limitations (see § 4.3 and § 6.3) could be a source of vagueness and uncertainty in interpretation and hence make the interpretation difficult and problematic. If the essence of a theory is surrounded by many sources of vagueness and limitation then we need to fill the gaps (especially the epistemological ones) by arbitrary speculation and wild guesswork, and this is unlikely to lead to complete, rational and consistent interpretation.

In summary, the mess of quantum mechanics created and exacerbated by the above factors (as well as other factors) does not only seem to make the interpretation of quantum mechanics difficult and challenging, but it may be a cause for confusion and a source of misunderstanding, and hence quantum mechanics may not be difficult to interpret but it could be impossible to interpret. In other words, quantum mechanics may be such a badly-conceptualized and poorly-structured theory (despite its empirical success within the previously-discussed limitations; see § 7.6) that it hinders (rather than help) any understanding, rationalizing and interpreting the behavior of the quantum world and its phenomena. In fact, we can find in the history of science many examples of such epistemologically-terrible and empirically-successful theories. For example, the Ptolemy model of planetary motion obscured the understanding of the solar system for many centuries despite its empirical success (although not at the level of success of quantum mechanics). Similarly, the phlogiston theory hindered the progress of chemistry in understanding chemical reactions for a while despite its empirical success (which should explain its existence in the first place and its survival for a considerable time). This may also apply (to some extent) to the Bohr atomic model and the old quantum theory.[401]

In fact, we may claim that such "epistemologically-wrong" (or "epistemologically-bad") and "empirically-correct" (or "empirically-good") theories are commonplace and this is one of the driving forces for the development and evolution of science where better new theories (even if only epistemologically) replace inferior old theories along the history of science. So in brief, "well-behaving empirically" does not mean or guarantee "well-behaving epistemologically", and hence quantum mechanics (which is obviously well-behaved empirically and badly- or poorly-behaved epistemologically) should be a candidate for replacement by a better theory which is well-behaved both empirically and epistemologically.

---

[399] Our concern here (and in similar contexts) is the treatment of classical macroscopic models (like "particle" and "wave") used in the formulation and interpretation of quantum theory as exact models rather than approximates and imitates. In fact, using such models within any theory about the microscopic world should be fine as long as we keep awareness of the fact that these models are (in principle) not valid as exact and true models (and therefore they should not participate, as exact models, in the formation of realistic and faithful epistemological and interpretative models). Accordingly, such models can play a legitimate role in the formalism and even in "approximate" interpretation which should be useful for educational and pedagogical purposes for instance (but we should keep remembering that these models may not reflect the quantum reality).

[400] There are many other examples but the above should be enough. In fact, we put many question marks on specific aspects of the quantum theory in our previous investigations.

[401] We may say: most (or at least many) "wrong" scientific theories (as we see them from a chronological perspective and within a historical context) should have achieved a reasonable level of empirical success for them to exist (in the first place) and to survive (for a while). Also see § 2.14.

7.15 Difficulty of Interpretation

It is important to note that although many of the perplexing aspects of quantum mechanics have rational explanations,[402] it is difficult to claim that all the perplexing aspects of quantum mechanics have rational explanations (e.g. it is difficult to explain self-interference or delayed cause or effect of knowledge and consciousness). Moreover, it is more difficult to propose a single interpretation that can fit and unify all these partial explanations within a single, thorough and universal explanation to the entire quantum theory. To the best of our knowledge and belief there is no such interpretation. Accordingly, in the absence of such a thorough (i.e. includes all aspects) and universal (i.e. fits all partial explanations) interpretation we can claim that quantum mechanics does not have an interpretation (at least until now). As indicated earlier, there is a strong possibility that quantum mechanics may not even be interpretable (because it is such a messily-conceptualized theory that it cannot be interpreted epistemologically although it is excellent empirically).[403] So, when Feynman says "I think I can safely say that nobody understands quantum mechanics" we should believe him (or almost) because quantum mechanics does not seem to be an interpretable (or "understandable") theory although we perfectly "understand" the formalism and know how it should be used as a *calculus* for the quantum world and quantum physics.

Finally, whether or not a theory (any theory) about the quantum world can be interpretable (considering that "interpretation" is essentially classical because we are classical creatures and hence we can interpret and understand the world only classically by using classical models and concepts while the reality of the quantum world is non-classical by its nature) is an open question that can be subject to debate and controversy. However, it should be possible (in principle at least) to develop a non-classical intuition (especially if we consider long-term evolution) that creates appropriate concepts and models for understanding and interpreting (as well as formalizing and predicting) the non-classical world and non-classical phenomena (whether quantum or not, e.g. the cosmological evolution of the Universe).

---

[402] We note that many of these explanations are discussed or indicated somewhere in this book (while many others can be found in the literature). However, the rationality of some of these explanations may be questionable or disputed and could be dependent on individual opinions and schools of interpretation.

[403] When we say about quantum mechanics things like "it is not interpretable" or "it cannot be interpreted" it should mean that we do not see (among the existing interpretations) any acceptable interpretation and hence it is very unlikely that we can find such an interpretation in the future (noting that if the huge amount of work conducted so far on the interpretation did not lead to such an interpretation then it is very unlikely that such an interpretation does "exist in reality" or can be found if it exists at all). So, "not interpretable" (and its alike) does not mean it is logically impossible to find such an interpretation in the future.

Chapter 8
Schools of Interpretation

In this chapter we investigate the proposed interpretations of the quantum theory focusing on the main and dominant schools. In fact, the investigated schools in this chapter are no more than a selected sample of the main interpretations of quantum mechanics (with some preliminary and brief analysis and assessment). However, before we go through the details of the investigation of these schools it is useful to be aware of the following remarks:

1. The investigation of schools of interpretation is like a trademark to quantum mechanics which characterizes it from all other branches of physics. So we may ask: why quantum mechanics requires interpretation unlike other (or at least more than other) branches of physics (and hence many chapters, papers and books, including our book, are dedicated to this investigation which is regarded as an essential part of any complete treatment of this theory)? In fact, the answer should be obvious that is: this requirement is due to the unusual nature of quantum mechanics in dealing with physical systems of unfamiliar size (or scale; see § 1.3.2) which leads to many perplexing and problematic issues some of which seem to threaten not only our classical science but even our philosophy about the nature of our Universe.[404] On the other hand, the need of classical physics (i.e. as opposite to modern physics) to interpretation is marginal because it is very usual and based on our daily experiences and concepts and hence it appears self-evident and easy to explain and justify (although a deeper insight is still required). In fact, very little (if any) attention is paid to the interpretation of classical physics. In brief, classical physics is consistent with intuition and "common sense" while quantum physics is not, and that it why they differ in their need for interpretation.

2. The interpretations of quantum theory may be classified generically into two main categories: epistemological and ontological where the focus in the former is on the observer while the focus in the latter is on the outside reality. In fact, any acceptable and successful interpretation should take both considerations into account, and hence such a division (if it is factual) should not occur. Yes, some interpretations could justifiably be more reflective of epistemological issues while other interpretations could justifiably be more reflective of ontological issues and hence such a division becomes justifiable and acceptable.

    Anyway, we do not put any limit or restriction on the nature of acceptable interpretation (i.e. whether it is epistemological or ontological or even a mix) as long as it supposedly provides rational and convincing explanation and justification to the theory. Accordingly, we will not bother about this classification (and hence we put all the investigated interpretations into a single list without worrying about their nature from this perspective). In fact, many other classifications can also be found in the literature but, as we indicated, we will not bother about these classifications and hence we will put all the investigated interpretations into a single list.

3. Many (or most if not all) of the proposed interpretations in the literature are not strictly scientific (and

---

[404] It is useful to note that this reason (i.e. the unusual nature due to scale) should also explain the exceptional need of Lorentz mechanics to interpretation and the many controversies about it although (for historical reasons) the interpretation of Lorenz mechanics was merged into its formalism (thanks to the dominance of special relativity) and hence the interpretation of Lorenz mechanics is commonly seen as marginal (or at least less important) compared to the interpretation of quantum mechanics. In fact, this dominance resulted even in a confusion (between formalism and interpretation) that led to the propagation of these controversies to the formalism of Lorenz mechanics itself since these interpretation-related controversies were seen as formalism-related controversies which, in its turn, led to the rejection of some to Lorenz mechanics altogether as if the problematic interpretation (i.e. special relativity) is a problematic formalism (i.e. Lorentz mechanics). The dominance and confusion also shut the door to any criticism to special relativity (since it is seen as criticism to the experimentally endorsed formalism) and hindered the effort to find a sensible and rational interpretation to Lorentz mechanics (and hence this dominance contributed badly to the development of modern science and philosophy).

hence they should be rejected as such) due to their failure to meet some of the eligibility criteria of scientific interpretation (see § 2.9) such as physicality and testability (see § 2.9.4 and § 2.9.5). Also, at least some of the proposed interpretations are partial and hence they are not acceptable (see § 2.9.7).

4. As indicated earlier, there are too many interpretations and hence we just present a sample of these interpretations which are the most important ones in our eye (e.g. by being more common or more acceptable or more interesting ... etc.). In fact, there are some very odd "interpretations" (like divine intervention or conspiracy theories) which are not worth of attention or consideration let alone investigation.

5. Some of the proposed interpretations are broad and generic and hence they embed a number of sub-interpretations. In fact, most (if not all) schools of interpretation come with different colors and flavors which reflect different preferences and tastes within the school (and hence it is legitimate to refer to a particular school as interpretation or as interpretations or even as schools).[405]

6. The majority of physicists adopt a general (or standard) interpretation about quantum mechanics which is based directly on its formalism regardless of the details of the specific schools of interpretation.[406] This general interpretation is based on generally accepted principles and rules (as well as definitions and conventions) based directly on the formalism (in its common version). The general interpretation is represented most distinctively and prominently by the interpretation of the wavefunction as a probability amplitude. It may also be represented by other (less distinctive and prominent) elements and features like the understanding of the uncertainty principle in its generic meaning and according to its direct implications. In brief, the general interpretation (which we may also call the standard or basic interpretation) is formalism-based (i.e. what the formalism supposedly suggests and means directly) without considering meticulous epistemological and philosophical issues. However, we should recognize that this general interpretation (or at least part of it) is commonly seen as part of the formalism. For instance, the interpretation of the wavefunction as a probability amplitude is stated formally and axiomatically (see § 4.2). It is also used systematically and technically (as an inherent part of the formalism) in calculating position probabilities and expectation values (as well as other things).

7. So far, no evidence (or at least no clear-cut evidence) *in support* of a particular interpretation do exist. Yes, there are recent claims that some experiments (e.g. the slit-grid experiment; see § 6.8.1) support certain interpretations (i.e. transactional; see § 8.4) and refute others (e.g. Copenhagen; see § 8.1). However, these claims are generally not substantiated and are challenged. So, the existing evidence generally belong (in their endorsement) to the formalism.

Yes, there are some types of experimental evidence (specifically those related to Bell's theorem like Aspect's experiment; see § 9.3.2, § 9.3.3 and § 10.4) that seem to negate some interpretations (like the local hidden-variable interpretations), and hence they may be regarded (if accepted) as evidence *against* some interpretations (and thus we may label them as negative evidence).

We should also remark (as noted earlier in § 7.8 for instance) that most of the quantum mechanical experiments should have interpretative significance and value with regard to the common or standard interpretation since they follow and demonstrate the formalism of quantum mechanics (as well as demonstrating and revealing its general features and physical characteristics). In fact, numerous quantum mechanical experiments are designed and conducted specifically to reveal, demonstrate and test these general features and characteristics of quantum mechanics and hence they categorically and unequivocally belong to this type of "evidence of general interpretative value" (noting that they do not

---

[405] In fact, some variants of some schools are so different that they contradict each other and could differ from each other (about certain issues) more fundamentally than with other schools.

[406] Terms like "standard interpretation" or "orthodox interpretation" are common in the literature of quantum mechanics. However, whether they refer to a particular interpretation (primarily the Copenhagen interpretation which is the dominant interpretation at least until recently) or they refer to a general interpretation that is common to all or most sensible interpretations (where this general interpretation is typically represented by Born's probabilistic interpretation) seems to depend on the view and terminology of individual physicists and authors. Accordingly, the readers should be aware of a potential difference in the terminology of each author. For this reason, we may use "basic interpretation" (in association with or instead of "standard interpretation" and its alike which are found in the literature) to refer to this general interpretation and distinguish it from particular schools of interpretation such as Copenhagen.

8 SCHOOLS OF INTERPRETATION

belong to a specific school of interpretation).

8. As is the case with the relativity theories, we meet in the quantum theory all sorts of fantasies, illusions, baseless claims ... etc. (although this time in its interpretation and epistemology in the first place).[407] In fact, the literature of quantum theory about the schools of its interpretation and how it should be "understood" and rationalized is full of nonsensical ideas and suggestions as well as many irrational proposals and theories.[408]

9. The literature of quantum theory is full of countless arguments with and against this interpretation and that interpretation where the adherents of each school try to "prove" their interpretation and rebut other interpretations. However, it should be noticed that many of these arguments are effectively invalid and cannot establish the underlying claims. One of the important reasons to this is the lack of common "language" and "conceptual platform", i.e. each group have (beside their illusions, fantasies, prejudices, etc.) their own conception and understanding to the theory and its elements and structure which are not exactly the same as those of other groups. So, in many cases they actually debate problems and issues that have common names but different content and significance.[409]

10. Before we go through the (investigated) schools of interpretation of quantum mechanics, it is useful to outline our view about the merit and worthiness of these schools. Although most of the proposed interpretations have their own merits and strengths and they succeed to fix certain problems and address some troubling issues, there is no single school that is totally satisfactory and consistent and is impermeable to criticism and challenge. So we can say: up to now there is no (totally-satisfactory) interpretation to quantum mechanics. Yes, there are many partially-satisfactory interpretations (which should not be acceptable according to our eligibility criteria; see § 2.9.7).[410]

In fact, the problem is more serious and deep than this because even if we accept that there are some (totally-satisfactory) interpretations, no one of these interpretations meet all the eligibility and validity criteria of scientific interpretation (see § 2.9 and § 2.10). For example, even if we accept the Copenhagen interpretation as a totally-satisfactory and consistent theory in itself, there is no evidence in support of this interpretation and its implications and consequences specifically. So in brief, there is neither eligible nor valid interpretation to quantum mechanics.

This should lead us to the view that although quantum theory (or rather its formalism which is quantum mechanics) is almost a perfect theory from an empirical (or practical) perspective, it is almost useless from an epistemological (or "theoretical") perspective. So in brief, quantum mechanics is a theory for dealing with the (quantum) world but not for understanding this world (see § 2.1). Some of the reasons for this epistemological failure have been discussed earlier (see for instance § 7.15).

11. We may criticize some schools of interpretation on a basis that is controversial or/and may not represent our view. So, the purpose of the criticism (or rather assessment) is generally to point out to the potential weaknesses and sources of challenge to those schools regardless of their overall value and

---

[407] We note that the illusions and fantasies of the relativity theories are mostly obtained from the theories and usually derived from them in a technical way (unlike the illusions and fantasies of quantum theory which are generally generated by the schools of interpretation). So, it seems that quantum theory is in a better position than the relativity theories in this regard and from this perspective.

[408] We make harsh and demeaning remarks like this not out of disrespect to others but to encourage the spirit and attitude of critique and to save the time and energy of many people (especially beginners) so that they avoid investing precious time and effort in hunting and growing worthless ideas. Most people hold the view that science as a whole is a perfect structure developed by extraordinary talented people while the reality is that science (like almost anything else in this world) contains as much pebbles as jewels, and scientists are not different (or at least not much different) from ordinary people (e.g. in making mistakes, holding prejudices, embracing superstitious beliefs, violating logic, etc.).

[409] This is inline with the fact that there is a considerable amount of ambiguity in the literature of quantum mechanical interpretations (e.g. lack of well-defined terminology, multiple-use terminology, expression of subjective mental and psychological experiences, difference in theoretical backgrounds and frameworks, etc.). Accordingly, it is natural that in many cases the discussions and deliberations are not based on a mutual understanding or they do not share a common ground.

[410] It should be noted that our view is supported by the fact that the Copenhagen school is the dominant interpretation (at least until recently) despite its weaknesses (see § 8.1). This sort of "vote" or "referendum" or "popularity test" should give an idea about the misery of the interpretation "market" (which makes the dubious Copenhagen school the most acceptable interpretation). In fact, the mere existence of too many schools of interpretation (some of which are exceptionally bizarre) should indicate and reveal the state of confusion and perplexity that dominates this market.

merit (or the value and merit of the criticism). For example, we may criticize some interpretations for being non-local even though non-locality can be accepted and justified in our view (see § 1.8 and § 7.13).[411]

12. It is useful to be aware that the focus of attention and the center of concern for many schools of interpretation (even though this may not be declared) is the issue of measurement and its role in quantum mechanics, and hence large parts of the literature of interpretation is related to this issue. In fact, some schools were constructed and developed specifically for this purpose (i.e. as a primary objective). This should be no surprise since measurement is seemingly the most problematic issue of quantum mechanics and it is the aspect that is mostly in need for interpretation and justification. Also, see § 5.7, § 6.11 and § 7.9.

13. Apart from the fact that many (if not most) of the circulating interpretations should be rejected on the basis of inconsistency or other reasons that we discussed earlier (see for instance § 2.9 and § 2.10), many of these rejected (as well as those remaining) are also susceptible to rejection on other bases (mostly of rather non-technical nature) such as aesthetic factors or harmony with other forms of knowledge or (scientific) common sense. After all, interpretation is not a precise science and hence its role is to provide a comforting explanation and justification and not just to provide something that satisfies the minimum requirement of rationality and compliance with the most basic rules of sanity. For example, the many-worlds interpretation (see § 8.3) may be acceptable at the very basic level of logic and rationality and may not lead to any form of inconsistency or absurdity but it should still be rejected on the basis of extreme oddity and counter-intuitivity.[412] This may also apply (but rather differently) to the interpretation (as well as formulation and conceptualization) of Bohm (and similar interpretations) due for instance to "artificiality" and counter-intuitivity.

14. It is important to note that part of the mess and confusion about and around the interpretation of quantum mechanics arises from the questionable assumption (or expectation) that the validity of quantum mechanics extends even to the macroscopic domain and hence quantum mechanics is expected to provide quantum mechanical explanations and interpretations even to the macroscopic phenomena (which are largely in the domain of classical physics; see for instance § 7.3; also see in particular point 7 of that section). This may be demonstrated by the Schrodinger's cat paradox (see § 9.1.2) which is based ultimately on the applicability of quantum mechanics to the macroscopic world and hence we need to deal, for instance, with "wavefunction of cat" (i.e. apply the quantum paradigms of "wavefunction" and "collapse" to the macroscopic object of "cat").

15. As seen earlier, the formalism of quantum physics itself (unlike other major scientific theories) is in need for an initial interpretation (represented by the basic or standard interpretation), e.g. the application and use of the formalism is based primarily on the probabilistic interpretation which cannot be understood or inferred directly and naturally from the formalism itself (as represented primarily, in this context, by the Schrodinger equation).[413] Accordingly, all the (known or mainstream) schools of interpretation depend on another (basic or standard) interpretation and this could be a reason for the difficulty of getting a consistent (school of) interpretation to quantum mechanics. For example, if the basic interpretation is wrong epistemologically (despite its empirical success within the formalism) then any school of interpretation based on this basic interpretation will be wrong epistemologically (or at least lead to wrong conclusions).

16. One of the (many) strange things about the interpretations of quantum mechanics is that some of these interpretations are more perplexing, ambiguous, dubious, irrational, ... etc. than the theory that

---

[411] As pointed out earlier, the criticism of non-locality, in this example, should apply specifically to the adherents of the relativity theories (who are the majority of physicists) due to the incompatibility of these theories (particularly special relativity) with non-locality.

[412] In fact, there are more fundamental (and even technical) reasons for rejecting this interpretation such as being metaphysical (at least in some of its variants). This could also apply to the Bohm interpretation (which we will refer to in the next sentence).

[413] We remind the reader of our distinction between the standard or basic interpretation of quantum mechanics (which is represented mainly by the probabilistic interpretation that is universally, or almost universally, accepted and is embedded in the formalism and seen as part of it) and the (epistemological) schools of interpretation of quantum mechanics (which represent the subject of investigation of the present chapter).

they are supposed to interpret (i.e. the theory which they are supposed to explain and justify and make understandable and rational). In fact, some of these alleged interpretations are more needy and desperate for an interpretation and rationalization than the theory itself. This does not apply to a few of these interpretations but to the majority (and possibly to all) of them (or at least to some aspects of all of them). This should reflect the oddity of this theory and its defiance to interpretation because it is simply not an interpretable theory. Also see § 7.15 and § 8.6.

8.1 Copenhagen Interpretation

This interpretation is generally seen as the conventional (or "standard" or "orthodox") interpretation of quantum mechanics and hence it seems to enjoy an overwhelming acceptance among physicists.[414] However, this may not be really the case noting that most physicists (and perhaps even quantum physicists) have little genuine interest in the interpretation of quantum mechanics, and hence only a minority of physicists hold definite views about these interpretations (including Copenhagen).[415] Moreover, in more recent years the attitude seems to have changed toward more tolerance and acceptance of other interpretations as well as more questioning and suspicion about the rationality and validity of the Copenhagen interpretation (and even if it is really and genuinely an interpretation; see footnote [420]). This is due to a number of reasons and pretexts such as:
• The proposal of new interpretations which are seemingly (or allegedly) more sensible and acceptable than the Copenhagen interpretation (or at least they are more capable than Copenhagen of addressing certain perplexities and difficulties).
• The emergence of more challenges and elaborate arguments against the Copenhagen interpretation and in favor of other interpretations.
• The emergence of new types of experiments that seem (or are alleged) to suggest refutation to Copenhagen and support to other interpretations.
These reasons (as well as other reasons) seem to have shifted the balance and changed the position about Copenhagen and other interpretations in the direction of reducing the support to Copenhagen in favor of other interpretations. So we may say: Copenhagen was dominant but (probably) not anymore.

In fact, the Copenhagen interpretation comes in different forms and flavors and hence we should be allowed to talk about "the Copenhagen interpretations" (i.e. plural). Anyway, in broad terms the main characteristic (and apparently common) feature of this interpretation is its probabilistic view[416] that is based on the paradigm of wavefunction collapse by the process of measurement, i.e. the state of the quantum system prior to measurement is a superposition (or mix) of sub-states and hence the effect of measurement is to reduce this mix of sub-states to a specific sub-state (which is the outcome of measurement) according to a certain probabilistic prediction. In fact, we may summarize the main features[417] of this interpretation in the following remarks (noting that some of these features may not be common to all the variants and sub-interpretations of the Copenhagen school):

A. Haziness of reality, i.e. the reality in itself (if it does exist at all) is a superposition of states (or substates) and hence prior to measurement reality is not well-defined and specifically-identified (assuming such characterizations are meaningful).[418] This seems to be incompatible (at least in part and

---

[414] As indicated earlier, it seems that expressions like "standard" and "orthodox" interpretation are not used consistently in the literature where some seem to use them to refer to a particular interpretation (mostly the Copenhagen school which is the dominant interpretation at least until recently) while others seem to use them to refer to the basic interpretation of the formalism (as represented mainly by considering the wavefunction as a probability amplitude) regardless of any specific school. In this book we generally use "standard interpretation" to refer to the latter.

[415] Yes, Copenhagen may enjoy (or have enjoyed) an overwhelming acceptance among those physicists (and quantum physicists in particular) who hold such views.

[416] To be more general we may say "indeterministic view" to include other forms of indeterminism (see § 5.2) that characterize this interpretation.

[417] In general, when we talk about features or characteristics (or things like these) of a school of interpretation they are not necessarily specific to that school unless it is stated explicitly or indicated evidently.

[418] Whether the Copenhagen school denies the existence of independent and definite reality or not (and hence it just denies the relevance of this sort of issues and knowledge to science), seems to be controversial and differ between the followers of this school. Anyway, the interpretation in itself can accommodate both (or all) views.

8.1 Copenhagen Interpretation

according to some variants of this school) with the principles of reality and truth (at least in their strong form which considers reality as a totally determined and independent entity), and hence Copenhagen seems to violate realism (at least in some aspects and from certain perspectives).

B. Collapse of wavefunction by measurement, i.e. the state of the quantum system (which prior to measurement is, in general, a mix of sub-states) collapses by observation to one of these sub-states and hence measurement (through collapse) can be seen as a cause for the emergence of a new and more specific reality from the old and hazy (or superposed) reality prior to measurement.[419]

C. Probabilisticity of outcome (which is a form of indeterminism; see § 5.2.2), i.e. the outcome of measurement is probabilistic in the sense that the collapse of wavefunction to any specific sub-state (as explained already) occurs according to a certain probabilistic expectation. This means that the outcome of quantum measurement is intrinsically probabilistic, i.e. its probabilistic and indeterminate nature is not because of limitations on the ability of the observer or restrictions on his knowledge or equipment. In other words, indeterminacy is an intrinsic and fundamental characteristic of reality and not a superficial and casual feature caused by our ignorance of some details or by limitations on our preparations and equipment.

D. Active role of measurement, i.e. measurement disturbs the measured system and changes its state from a superposition of states to a particular state. Accordingly, measurement is an active, not passive, process (see § 1.3.3). In fact, according to this interpretation measurement is *intrinsically* active (and not just *casually* active like some types of classical measurements; see point 7 of § 5.7). This is entirely opposite to the classical paradigm of measurement which is supposed (classically) to be a process whose objective is to reveal an already-existing, definite and independent reality (and hence classical measurement is intrinsically passive although it can be active casually and accidentally).

In fact, the Copenhagen interpretation also gives a central role to the measuring apparatus, i.e. in the process of measurement the appropriate attributes of the quantum system under measurement are determined by the measuring apparatus as well as by the system itself (and hence the proper basis of eigenfunctions that must be used to represent the state of the system should be selected accordingly). This seems consistent with the proposal (adopted by Copenhagen) that the wavefunction is a mathematical device used to represent the system and facilitate the description and prediction of its behavior (and hence it is natural to choose the basis that is appropriate for the setting and arrangement of measurement). It should also be consistent with the Copenhagen-type indeterminism which we indicated and outlined earlier.

E. We should finally note that the Copenhagen school is also characterized (not necessarily differently from other schools) by other features which may generally be seen as part of the formalism (or closely related to it) such as the adoption of the complementarity principle and the adoption of the uncertainty principle in a certain way (with a particular justification and a specific interpretation).

Regarding the assessment of the Copenhagen school, we may note the following:

1. Probabilisticity in its broad sense is not specific or distinctive to the Copenhagen school. In fact, probabilisticity emerges from the basic interpretation of the formalism of quantum mechanics (where the wavefunction is regarded as a probability amplitude). However, what distinguishes Copenhagen from some other schools (e.g. the hidden-variable and many-worlds schools) is the origin of this probabilisticity which (according to Copenhagen) is the indeterminacy of "reality" (assuming its existence) rather than being a matter of limited knowledge (like the hidden-variable interpretation) or multiplicity of reality (like the many-worlds interpretation).

2. As explained before (see for instance § 6.11), there is no rational physical explanation or known physical mechanism for the collapse (or reduction) of wavefunction.[420] In fact, some explanations and

---

[419] It is worth noting that the "collapse of wavefunction" is one of the most hated features and embarrassing aspects of the Copenhagen school and is one of the main factors for the migration of many physicists to other schools of interpretation (even though some of the other schools may have no less embarrassing, or even more embarrassing, oddities).

[420] Noting the nature of interpretation (as an attempt to explain, justify and rationalize), such an ambiguity should damage the status of the Copenhagen school as an interpretation (or at least as a complete interpretation). Therefore, such a gap should be filled for the interpretation to be complete and acceptable. It is useful in this context to quote the following (from Murray Gell-Mann) which expresses the reality about the deficits of the Copenhagen interpretation:

mechanisms (like "the interaction of the consciousness of the observer with the observed phenomenon" or "the perturbation of the measured system by the measurement equipment") have been suggested in the literature for the collapse of wavefunction. However, these alleged explanations and mechanisms are rather vague and hollow and lack physical substance and objectivity. In brief, they are not less vague, artificial and questionable than the "collapse" itself.[421]

3. The Copenhagen interpretation denies (explicitly or implicitly and at least in some of its variants) the existence of objective reality independent of the observer and the observation process since a quantum system, according to this interpretation, has no specific identity until it is observed (or its identity is not completely determined until the conduction of measurement). This denial is broadly in conflict with realism and the principles of reality and truth which are important for establishing rationality and avoiding inconsistencies and epistemologically-problematic consequences (see for instance § 1.10, § 2.4.1, § 2.5.2 and § 7.14).[422]

4. The Copenhagen interpretation may also lead to the negation (or at least limitation) of causality which is a fundamental philosophical and epistemological (and perhaps even scientific) paradigm that is needed to explain and justify what is going on consistently and rationally. In fact, this negation (or limitation) should be a consequence of the Copenhagen-type indeterminism which is based on the Copenhagen-type probabilisticity and collapse (as explained earlier). Also see § 1.7, § 2.4.2, § 2.5.3 and § 7.12.

5. Over its (rather long) history, the Copenhagen interpretation was continuously challenged theoretically by certain arguments and thought experiments such as the EPR argument (see § 9.3.1 and § 10.2) and the Schrodinger's cat paradox (see § 9.1.2). It was also challenged recently by more arguments and even real experiments (e.g. the slit-grid experiment which challenges the complementarity principle which is essentially a feature of Copenhagen; see § 6.8.1). So, the popularity of this school (which dominated the landscape of quantum mechanics for a considerable period of time) generally diminished over the last few decades (and this trend seems to be continuing).

6. There are many ambiguities about the "official" position of the Copenhagen school towards many issues (especially with regard to the details). In fact, there are many conflicting views and variants inside this school, and this vagueness should make this interpretation rather loose and short of theoretical clarity in various aspects (especially in comparison to some other schools). Also see point 2.

7. The Copenhagen interpretation may also be questioned and challenged in some of its generalizations (extensions, exaggerations, etc.) which can be inferred or understood from (or even declared by) some of its variants (or followers). For example, there is an apparent exaggeration by the Copenhagen school about indeterminism and its level and extent. The Copenhagen-type indeterminism my be challenged (i.e. in some of its generalizations and extensions) for instance by the fact that indeterminacy may be a characteristic of quantum systems but not necessarily of other types of physical systems, e.g. macroscopic systems are deterministic in their properties and behavior and this fact is reflected in the success of classical physics (which is deterministic) in its domain of validity.[423] In fact, some extreme forms of indeterminism found in the Copenhagen school (within certain variants and according to

---

"...Niels Bohr brainwashed a whole generation of theorists into thinking that the job was done 50 years ago" (where "the job" in this quote refers to the interpretation of quantum mechanics). Also see point 6.

[421] In fact, there seems to be an ambiguity about the position of the Copenhagen school even with regard to the nature of measurement (i.e. what makes quantum measurement "measurement"?). However, this seems to apply to the early days of this interpretation noting that a number of proposals (within the lines of this school) have been suggested since (although there seems to be no common acceptance or agreement within the followers of Copenhagen).

[422] A more moderate and rational view among the followers of Copenhagen is the dismissal of the issues of reality and realism (and their alike) by denying the relevance of such issues to science (and quantum mechanics specifically) and questioning the role and ability of science to address and answer such questions. So, they essentially deal with this problem by dismissing the issue of realism rather than by challenging and violating realism.

[423] It should be noted that the (potential) claim that classical physics is an approximate theory is essentially nonsensical in this context since being an approximation cannot be established or verified in any tangible and sensible way since it is beyond the capability and limits of experimental revelation and resolution of any physical measurement. For example, the application of superposition or probabilistic collapse or uncertainty principle on a one-kilogram object (let alone a star or a galaxy) is completely meaningless. Similarly, the trajectory of a one-kilogram object is entirely predictable (at the classical level and scale) with no probabilisticity or indeterminacy or intrinsic uncertainty.

certain views) are not needed at all (i.e. for the purpose of interpretation) even at the quantum level. Accordingly, indeterminacy (e.g. in the form of probabilisticity and intrinsic uncertainty) does not apply to some physical systems in any sensible meaning or observable form (even according to the criteria and standards of the Copenhagen school which effectively values only what can be observed and measured experimentally and usually distinguishes between classical and quantum). In fact, some extreme forms of Copenhagen's indeterminism are totally disposable and dispensable. So, at least some physical realities (whether classical or quantum) should or could be more deterministic (and definite and certain) than what is being alleged by the Copenhagen school and its followers.[424]

8. We should also note that there are many more specific questions and challenges to this interpretation (not all of which can be found or are stated explicitly in the literature). For example, what if the system is observed simultaneously by two (or more) observers? Is it possible that the wavefunction collapses differently for each observer? If no, what is the physical mechanism that ensures this to happen? If yes, we should then have different realities or different truths (which is against the principles of reality and truth; see § 2.4.1), and this could even lead to very odd implications and interpretations (like the many-worlds interpretation; see § 8.3).

9. As stated above, the Copenhagen school is historically the dominant interpretation although the tide has shifted (and is still shifting) with the appearance of other rival interpretations. As a consequence of this historical dominance, the conceptualization and terminology of the Copenhagen school can be felt even in the (commonly-accepted) formulation of quantum mechanics which is circulating today. In fact, part of this conceptualization and terminology (such as wavefunction collapse) is part of the technical language of quantum mechanics and hence it should not be given any weight more than this. In other words, using this language should not mean admission or following this interpretation. By priority, this language should not be seen as part of the formalism in its strict technical and empirical content and substance. Also see § 7.10.

10. Noting that the collapse of wavefunction is a global and instantaneous process, it should incur non-local effects in some situations (see for instance § 1.8, § 5.6 and § 6.1). Hence, the Copenhagen school should lead to violations of the principle of locality and this should bring it in collision with the relativity theories (and special relativity in particular; see for instance § 6.10 and 7.13). However, what seems strange is that the followers of the Copenhagen school generally accept the relativity theories (with no apparent concern about this clash let alone trying to address it and solve it).

11. The most attractive feature of the Copenhagen interpretation is its conceptual simplicity (i.e. it is seemingly the simplest among the popular interpretations). Also, despite its irrationalities (or ambiguous and counter-intuitive features) it may be (at least in some of its moderate variants) one of the most rational interpretations among other interpretations (possibly with the exclusion of the hidden-variable interpretation which is generally rejected due to the claimed evidence against it or rather against some of its variants). These factors (among other historical and non-historical factors) may account for the dominance of Copenhagen for rather long time. In fact, it still remains one of the most popular interpretations (if not the most popular).

We should also note that the Copenhagen school may be the most reflective of the essence and spirit of quantum mechanics since (in a sense) it is like a simple and plain description of the formalism. For example, the indeterministic nature of the predictions of quantum mechanics (as a formalism) is reflected in the indeterministic nature of the Copenhagen school (as an interpretation). Hence, in this sense the Copenhagen school may be seen as the most honest "interpretation" (or reflection) of the actual content of the quantum theory and its formalism. So, to some extent the Copenhagen school is like a translation (or "interpretation") of the formalism to ordinary language (with some simple and basic justifications and explanations). This should add more attractivity to this interpretation and provide more justification for its dominance. This should also explain its implicit presence in the

---

[424] It is worth noting that the above-mentioned questionable generalizations (which are exemplified by indeterminism) should also apply to other features and aspects of the quantum theory such as the principles of uncertainty, correspondence and complementarity (which are extended, exaggerated and misused by the Copenhagen followers in many cases and circumstances some of which have been indicated or discussed earlier within the context of our investigation and assessment of the quantum theory without a specific reference in most cases to the Copenhagen school).

axiomatic framework (or postulates; see § 4.2) which represents the essence of quantum mechanics (see § 4.9), since this reflectivity and conceptual compatibility (or homogeneity) should facilitate the conceptualization and axiomatization of the formal theory by the conceptions and terminology of this interpretation (and may facilitate even the confusion between formalism and interpretation; see for instance § 7.10).

12. Historically, the Copenhagen interpretation is primarily associated with Bohr and Heisenberg. However, it should also be associated with those who axiomatized and formulated quantum mechanics (following its emergence and during its early stage of development) in the language, spirit, style and concepts of the Copenhagen school (although some may not have declared their support to Copenhagen and may even had some differences with Copenhagen).

8.2 Hidden-Variable Interpretation

The main objective of this interpretation (which may also be called the "additional-variable" interpretation) is to restore determinism and realism which are sacrificed in some other interpretations (notably Copenhagen).[425] The essence of this interpretation (or most of its variants) is that the existing quantum theory is not complete as it ignores certain aspects and attributes of the quantum reality and hence a complete theory should include some variables that are missing (or "hidden") in the current quantum theory.[426] Accordingly, a complete quantum theory that includes all the relevant variables can be formulated and interpreted deterministically and realistically without need to the formal deficits and the epistemological gaps and embarrassments that the current ("incomplete") quantum theory leads to (such as quantum indeterminism or collapse of wavefunction or multiple reality which are adopted by other interpretations and may even be embedded in the formalism in some shape and form).

In fact, we can outline the distinctive features of this interpretation in the following remarks:

A. We may classify the hidden-variable theories into two main categories: in one category the hidden variables have an influence on the actual physical outcome (which implies that quantum mechanics is not complete formally), and in another category the hidden variables have influence on our knowledge but have no influence on the actual physical outcome (which implies that quantum mechanics is complete formally[427] but not epistemologically). In brief, according to the first category quantum mechanics is incomplete formally while according to the second category it is incomplete epistemologically (although this epistemological deficit should be ingrained somewhat in the current formalism and originate from it).

So, the first category should imply that the (current) formalism of quantum physics requires modification or replacement (so that it can make deterministic and realistic predictions), while the second category just tries to offer epistemological explanation to quantum mechanics within the limitations of its current results and predictions (although this epistemological explanation may require certain formal adjustments that do not change the current results and predictions).

In fact, these categories may be called (respectively) the "hidden-variable theories" and the "hidden-variable interpretations", but we will not follow this terminology (at least strictly) because in many cases and circumstances there is no reason to make this distinction. Moreover, sometimes it is difficult to put a sharp border line between the two. We also note that there is no convention or agreement on this terminology (if such a distinction is made or recognized in the first place).

It is noteworthy that (to the best of our knowledge) there is no successful theory of the first category so far, and hence in the following when we talk about hidden-variable schools and interpretations we

---

[425] Determinism here should include (in a sense) causality since (quantum-type) indeterminism could lead to violation of causality (see § 6.4) although causality may also be included (in some sense) in realism. It should also be noticed that determinism here is about reality (i.e. objective) rather than about observer (i.e. subjective) and hence the outcome of measurement could still be probabilistic or uncertain from the perspective of the observer.

[426] Depending on the nature of the presumed "hidden variables" and their role in the theory or interpretation (as well as for contextual considerations), it is more appropriate sometimes to use terms like "missing variables" or "additional variables" instead of "hidden variables".

[427] "Complete formally" considers the empirical aspect, and hence it could still be incomplete from the conceptual or epistemological aspect.

8.2 Hidden-Variable Interpretation

generally mean the second category (unless we refer to the first category specifically or indicate it evidently).

B. We may also classify the hidden-variable theories of the second category in point A (i.e. those whose objective is to explain, rather than lift, the empirical deficits in the current quantum theory) to two main sub-categories: those which treat the issue from a purely epistemological perspective (and hence they are purely interpretative), and those which try to modify the current formulation of quantum mechanics in such a way that it reproduces the predictions of quantum mechanics but in a conceptually deterministic and realistic fashion. An example of the former is the local hidden-variable interpretations (e.g. of EPR; see § 9.3.1 and § 10.2), while an example of the latter is the Bohm quantum theory.

C. As indicated above, the objective of the hidden-variable interpretation is to restore determinism and realism. However, to be more accurate we should have said: the objective is to negate the quantum mechanical form of indeterminism and non-realism (as represented typically by the Copenhagen school) since indeterminism (as represented by probabilisticity for instance) is part of the current (and empirically correct) formulation of quantum mechanics. So, the actual objective of this interpretation is to replace the "quantum indeterminism" by "classical indeterminism" and hence negate any form of non-realism.[428] In other words, the objective is to replace "indeterminacy in reality" (due to haziness in Nature) by indeterminacy in knowledge (due to ignorance), and hence the probabilisticity is a "probabilisticity of knowledge" and not a "probabilisticity of Nature".

We should also note that the purpose of the first category (of point A) is to establish or restore determinism actually and in practice by providing (or rather trying to provide) a formalism that is actually deterministic (by providing definite and certain predictions), while the purpose of the second category (whether with or without reformulation of quantum mechanics) is to establish determinism conceptually and in theory by providing re-conceptualization of the current quantum theory (possibly with some reformulation) so that the theory becomes conceptually deterministic.[429]

D. The singularity of "variable" in "hidden-variable" (which may also be reflected in the formulation of some hidden-variable theories and theorems by representing this "hidden-variable" by a single symbol which is usually λ; see for instance § 10.3.2) does not mean necessarily it is a single variable representing a single factor. Hence this "variable" can represent more than one variable representing more than one factor.

E. As indicated above, the hidden-variable theories are based on the claim that quantum mechanics is an incomplete description of reality (regardless of being formally/empirically incomplete or epistemologically incomplete and regardless of the nature of the supposed completion and whether it should improve predictability or not). However, none of the hidden-variable theories that proposed reformulation of quantum mechanics (whether of the first or second category of point A) to include the "hidden-variable" succeeded in filling this gap and providing a more complete description of reality than the description provided currently by quantum mechanics. For example, none of these theories (i.e. of the first category) could overcome the indeterministic and probabilistic nature of the quantum mechanical predictions by providing a more specific and deterministic description of reality (and hence getting more specific predictions at the empirical level). Similarly, no one of these theories (i.e. of the second category) could provide a better epistemological vision and understanding to the quantum indeterminism and probabilisticity (apart from providing tentative theoretical justifications). So, even those hidden-variable theories which proposed reformulation of quantum mechanics have essentially

---

[428] The focus here is basically on the second category.

[429] To clarify this further we can say: in any hidden-variable theory that replicates the probabilistic results of quantum mechanics (whether of formal nature like the Bohm theory or of purely interpretative nature like some local hidden-variable theories such as those of the EPR type), we may get determinism and realism but not predictability. In other words, it is not possible to predict the actual outcome of individual measurements in advance although the outcome is deterministic in reality (unlike, for instance, the "wavefunction before collapse" according to the Copenhagen school where the reality itself is not determinate). So, it is important to distinguish in this context between determinism-realism and predictability (or between determinism in reality and determinism in prior-knowledge). In essence, this type of hidden-variable theories effectively shifts the focus of attention in the theory from "observables" to "beables" (see point F) and hence they achieve "theoretical" determinism and realism (with no practical impact or improvement to predictability).

8.2 Hidden-Variable Interpretation

interpretative (or rather justifying or "apologizing") value only. In other words, they could be deterministic formally and hypothetically (in a subtle way) but not tangibly and conceptually (in an intuitive way) let alone actually and practically (in an empirical way).

F. An issue that is worth attention is the "hidden" label and its significance and justification. In fact, "hidden" indicates the absence of these alleged "hidden variables" in the theory (since there is no reference in the formalism to them) despite their existence in reality. So, their (supposed) existence in reality is conjectured rather than being "detected". In fact, we may even have no direct access to them (whether in reality or in theory). Such "hidden variables" may be suitably represented and typically exemplified by the "beables" of Bell.[430]

G. The main criticisms and challenges to the hidden-variable theories may be summarized in the following:[431]

 • Non-locality and hence clash with the relativity theories and (allegedly) potential violation of causality. This applies to some types of these theories which are based on non-locality or have non-local implications.

 • Clash with experiments based on the Bell theorem and inequality (e.g. Aspect-type experiments). This applies to the local versions of these theories. See § 9.3.2, § 9.3.3 and § 10.4.

 • Non-physicality, peculiarity, counter-intuitivity (and things like these) which lead to consequences and scenarios that are worse and more strange than what these theories are supposed to solve, explain and rationalize. This sort of criticism is directed to some peculiar versions or variants of the hidden-variable theories and interpretations (such as the Bohm quantum theory).

 • Violation of the principle of economy (see § 2.4.4). This criticism is also directed to the Bohm theory.[432]

Regarding the assessment of the hidden-variable school (or rather schools), we observe the following (noting that some of the following points may apply only to certain variants of this school):

1. Some variants or schools of this "interpretation" do not really offer an actual interpretation or explanation to the existing quantum theory (other than proposing an "apology" for the deficits of quantum mechanics). For example, if I want to explain the gravity law it is not sufficient (as a scientific theory or epistemological interpretation) to claim that there is something (which I do not know) behind gravity. In fact, this "interpretation" (in some of its forms and variants which lack real physical substance) is an excuse for the failure to provide a reasonable interpretation for the current quantum theory (rather than an actual interpretation to the theory). Hence, such "interpretations" should not be classified as (real) interpretations.

---

[430] In this context, it is illuminating to quote the following from Bell (see "Speakable and unspeakable in quantum mechanics" in the References):

In particular we will exclude the notion of 'observable' in favour of that of 'beable'. The beables of the theory are those elements which might correspond to elements of reality, to things which exist. Their existence does not depend on 'observation'. Indeed observation and observers must be made out of beables. (End of quote)

As we see, "hidden variables" can be seen as a sort of "beables" as opposite to "observables" with which quantum mechanics is conceptualized and formulated. So in brief, the absence or neglect of these (real) beables in the theory is what causes indeterminism and hence if these beables are included in the theory then the theory becomes deterministic (in its description and prediction) and totally realistic. Similarly, the experiments are indeterministic (in their outcome) because these (real) beables are not considered in the setting, procedure and (theory-based) analysis of these experiments.

[431] We note that there are many differences between the hidden-variable theories and hence these criticisms and challenges generally do not apply equally, simultaneously and necessarily to each type of these theories (as will be indicated). We also note that these criticisms and challenges reflect the characteristic features of these theories (and that is why this point is included in this list).

[432] We note that Bohm's theory faced many other criticisms and challenges which include non-locality, inconsistency and metaphysicality (and include even some seemingly-peculiar challenges like the violation of Newton's third law which may not be appropriate), and this should explain why it did not get enough support among physicists and philosophers of science despite its rational nature (as a deterministic and realistic theory) and its disposal of problematic quantum mechanical paradigms like wavefunction collapse. However, as we indicated earlier, the importance of Bohm-type theories is to show that quantum reality can be described (although possibly clumsily and in a trivial way) deterministically and realistically. So, despite their theoretical and practical triviality and loopholes, such theories keep alive the hope of formulating a new (deterministic and realistic) quantum theory that is hopefully more acceptable than these theories (noting that all the current schools and interpretation theories of quantum mechanics are questionable and challengeable and hence Bohm-type theories are not different in this regard or inferior to other theories from this perspective).

8.2 Hidden-Variable Interpretation

2. Referring to points A and B, an interesting feature of the hidden-variable schools is that they are not necessarily and entirely interpretations of the existing quantum mechanics, but some of these hidden-variable interpretations propose (at least hypothetically, or they try to propose) different or modified versions of "quantum mechanics" and hence the proposed "interpretation" affects the formalism although the predictions of the proposed formalism are usually identical (or supposedly so) to those of quantum mechanics.[433] This is unlike the Copenhagen interpretation, for instance, where the interpretation is entirely based on the existing quantum mechanics in its familiar formalism.
3. The hidden-variable interpretations have been challenged (and seemingly refuted in some of their forms) by theoretical arguments. For example, von Neumann argued that hidden variable interpretation of quantum mechanics is impossible (although his argument was challenged and even ridiculed). Also, Bell's theorem and inequality (see § 9.3.2) challenged local hidden-variable theories as an explanation to quantum mechanics and demonstrated that their predictions are incompatible with the predictions of quantum mechanics. In fact, some variants of the hidden-variable interpretation are (supposedly) refuted even by experimental tests (based on some of the aforementioned theoretical arguments) such as Aspect's experiment which is based on Bell's theorem (see § 9.3.2, § 9.3.3 and § 10.4).
4. Some types of hidden-variable theories and interpretations have non-local implications and consequences and hence they are susceptible to clash with the principle of locality and consequently with the relativity theories and special relativity in particular.
5. Based on what have been said we can conclude that the main problem with the hidden-variable theories is that they are either non-local (and hence they represent an embarrassment and challenge to those who believe in locality) or they fail to reproduce the verified predictions of quantum mechanics (and hence they are empirically untenable). This is typically and vividly demonstrated by quantum entanglement (see § 5.6 and § 6.10) and the challenges posed by it to these theories as represented typically by a number of theoretical arguments and experimental tests like the Bell theorem and the Aspect-type experiments (see § 9.3.2, § 9.3.3 and § 10.4).
6. The most attractive feature of the hidden-variable interpretation is its rationality (or intuitivity) due to its classical nature. In our view, it is the most rational interpretation among the existing schools. However, it is generally rejected due to its non-local implications or to the claimed experimental evidence against it (or rather against some of its variants) represented mainly by Bell's theorem and Aspect-type experiments.
7. As indicated earlier, the existence of realistic deterministic hidden-variable theories that replicate the predictions of quantum mechanics (such as the Bohm quantum theory) should endorse and consolidate the possibility of the existence of totally realistic deterministic quantum theories. In our view, this is the main value and contribution of such theories (noting that the currently-existing theories of this sort are generally trivial in theory and in practice and far from being ideal, e.g. because they are counter-intuitive or artificial or poor in physical content and credibility or non-economic or ... etc.).[434]
8. Historically, the hidden-variable interpretation is generally associated with de Broglie, Schrodinger, Einstein and Bohm (despite the difference in their motives, conceptualization, theoretical justification and approach, level of enthusiasm and certainty, historical rank, etc.). It should also be associated (more recently) with Bell (although his theorem and inequality led to the refutation of the local types of hidden-variable theories; see § 9.3.2). So in general, there was a good level of support (even among leading physicists) to this interpretation initially.

   However, following the experimental tests related to the Bell theorem (such as the Aspect experiment; see § 9.3.3 and § 10.4) there was a general tendency or trend to reject this interpretation. In fact, at one stage there was even a sense of triumphalism among the anti hidden-variable theories camp (especially among the Copenhagen followers), and it seemed as if the fate of the hidden-variable theories is sealed

---

[433] An example of this type of hidden-variable "interpretations" is Bohm's theory.
[434] Some of the currently-existing theories of this sort may also be criticized for being non-local, but from our perspective non-locality (in itself and at least in some of its forms) is not a problem. Some of them may also be criticized for being contextual, i.e. the outcome of a measurement on a quantum system depends on the type of measurement on another (independent) system (see point 2 of § 9.3.3). Also, some of them may be criticized for being metaphysical.

8.2 Hidden-Variable Interpretation

and they are doomed and destined to disappear from the quantum mechanical landscape. Nevertheless, in the more recent years there seems to be some return or acceptance to (or at least tolerance towards) these theories. This tolerance is caused, for instance, by questioning the significance and value of Bell's theorem or/and the associated experimental tests (e.g. of Aspect). It may also be caused by more tolerance among physicists towards non-locality especially when it (allegedly) does not contradict relativity (noting that Bell's theorem challenges only local hidden-variable theories and hence if we accept non-locality then there is no evidence, e.g. in the form of Aspect's experiment, against the hidden-variable theories, i.e. the non-local ones).

9. The existence of hidden variables may be justified by the claim that even at the quantum level the world is complicated and elaborate although we (according to our scale) cannot see many of the delicate details of this world.[435] If so, then the prospect of getting hidden-variable quantum theories of the first category (i.e. those which claim that the formalism is not complete; see point A) will be great. In fact, the impact of such possibility and prospect should be so fundamental that it could change the entire science, philosophy and epistemology that we currently have about the quantum world and produce a totally different understanding to the quantum reality.[436] This means that our current "quantum vision" is blurred and is expected to improve in the future (like the blurred "classical vision" at the early stages of Renaissance of modern science which was improved substantially in the following centuries).

10. Regarding the first category of this school (see point A), it may be argued (**against this category**) that this category does not offer an explanation or justification to the remarkable success of quantum theory. Noting that quantum mechanics is an extremely successful theory (within certain limitations; see § 7.6) we should explain and justify the theory and its predictivity within its given formalism which contains a specific number of *visible* variables (with no indication to any hidden variables). In other words, this category should explain why quantum theory (with only these visible variables) can produce these consistent results and astonishing successes. If there are really hidden variables (of the type that affects predictability and outcome) and the formalism is not complete then we should expect some failures to the formalism of the theory (which is applied and tested in a huge number of systems, situations, conditions, etc. and it is at the heart of many of the modern scientific and technological advances).

However, it may be argued (**against the above argument**) that the hidden variables have different natures and effects where some types of these variables may not have a direct impact on the outcome of measurement (and hence we should not expect the failure of the theory by these hidden or missing variables) although they have an impact on our knowledge or awareness of "elements of reality". So, this type of hidden variables (assuming their existence) should not have any impact on the correctness and even completeness of the theory itself (since the formalism can make correct and complete description and predictions of the concerned phenomena within the given limitations) although it may affect the completeness of the interpretation in the sense that an interpretation of such a theory cannot (or may not) be complete due to the missing "elements of reality". Moreover, even the other type of hidden variables (i.e. those which have a direct impact) may have an effect on predictability (through determinism) but not on correctness, and hence the inclusion of these hidden variables should improve predictability (and perhaps interpretativity as well) although their exclusion does not affect correctness. So in brief, either the hidden variables are purely of epistemological nature and value (with no impact on the outcome and thus predictability) and hence they should not affect the completeness of the theory although they may affect the completeness of its epistemology and interpretation, or these variables have an impact on the outcome of measurement and hence we may expect failure of the

---

[435] In fact, this may be justified by the existing limits on the resolution of our observation and measurement equipment which we expect to improve in the future and hence some of these limitations can be lifted or reduced. Accordingly, we will be able to observe and probe the quantum world more closely and intimately (possibly defying even some current taboos like those imposed by the uncertainty principle) and discover more delicate details and elaborations about the quantum reality.

[436] For example, indistinguishability and exchange symmetry (see § 3.1) which play a central role in many formulations and results of the current quantum theory could be affected or changed or even disappear. Similarly, the uncertainty principle could be pruned or abolished.

theory but this failure is not a necessity since the impact of their exclusion could be in the form of reduced predictability rather than failure.[437]

Anyway, although we do not claim that quantum theory is a perfect and ultimate theory, we should also note (for the sake of fairness) that its remarkable empirical success (despite some limitations) makes the possibility of the existence of physically-relevant hidden variables look less likely especially when they are expected to have an impact on predictability (by being of the type of the first category and hence having an impact on the outcome of measurement). Nevertheless, we need to keep looking for a different quantum theory that is superior empirically (with improved predictability) and epistemologically (with improved interpretation).

11. Should the hidden-variable proposal (in general) be classified as metaphysical (since it is based on presuming the existence of variables that are not seen or detected directly and specifically)? In our view, this question has no definite answer due, for instance, to the great variations between the hidden-variable theories (and even in the notion of "hidden variable") as well as the variations in positions and opinions towards the issues involved. However, from a general perspective we may say: if the claim of hidden variables is used to justify realistic aspects in the theory (possibly as a pretext for accepting the theory or proposing an alternative to it) then this could be regarded metaphysical (since such alleged variables play a real physical role in the theory although this may depend on the nature of these variables and their actual role or function), but if they are used to indicate potential deficiencies in the current quantum theory (i.e. put a question mark on the theory itself and hence the formalism itself could become questionable) then from this perspective this position is acceptable in principle (i.e. it should not be classified as metaphysical).[438] Accordingly, the theories in the first category (see point A) are more likely (but not necessarily) to be metaphysical.

8.3 Many-Worlds Interpretation

This school may also be called (among other possibilities and variations) the "parallel worlds" or "multiple worlds" or "multi universe" or "relative state" or "Everett" interpretation (although these labels or some of them may be associated with certain variants of this school). Despite its oddity it does not seem to be short of support or unpopular among physicists (which seems to be as odd as the interpretation itself). In fact, this interpretation enjoys an enthusiastic support even among prominent physicists. This should highlight the oddity of quantum theory (particularly in its epistemology) that makes such a bizarre interpretation look acceptable and even natural. This should also indicate the misery of modern physics and its illusory and hallucinating nature (which justifies all sorts of wild and hectic ideas under the banner and pretext of "theoretical physics").

The essence of this interpretation is that a measurement conducted on a quantum system in a superposition state leads to the branching of the universe into a number of separate non-interacting and non-communicating universes (or worlds or realities) where each one of these universes represents one of the superposed states. So, all the possible outcomes occur but in different worlds with only one of these worlds being actually observed (i.e. the "actual" outcome of measurement belongs to this "observed" world which the observer branched to; also see point 8). Accordingly, there is no collapse of wavefunction by observation but the wavefunction continues to evolve (irrespective of observation) since all the quantum possibilities are real although only one of these possibilities is actually observed.[439] Some of the distinctive features of this interpretation are summarized in the following remarks:

A. From philosophical and epistemological perspectives, the most important feature of this interpretation (in its common variant) is that it replaces the unique reality by multiple realities and this evidently contradicts the principles of reality and truth (i.e. uniqueness of reality; see § 2.4.1). However, as we

---

[437] We should note, however, that even if the current quantum theory is complete formally within the given restrictions and limitations, it can still be incomplete in comparison to a future quantum theory that can incorporate the missing variables (which supposedly represent missing "elements of reality") and hence become (more) complete (empirically or/and epistemologically).

[438] But then this should not be an interpretation in the strict sense (see point 1); moreover any suggestion of possible flaw in the theory (see point 10) may be challenged by the remarkable empirical success of the theory.

[439] For pedagogical reasons the presentation may not be very technical (although its alike can be found in the literature).

8.3 Many-Worlds Interpretation

will see, this despite its extreme counter-intuitivity and oddity may not violate rationality (which is the primary objective and ultimate justification of realism and the principles of reality and truth) from this perspective.

B. From a formal perspective, the most important feature of this interpretation is that it views and treats quantum mechanics as a simple and single theory with a single dynamics (see § 4.8). Accordingly, the quantum process is subject to a single dynamical protocol which is described by a single formalism (i.e. the Schrodinger equation) in all stages of quantum evolution (i.e. at measurement as well as between measurements). This means that we do not have a measurement theory or postulate that distinguishes the measurement process from the unitary evolution process of the Schrodinger equation.

C. As a consequence of point B, the many-worlds theory is not just an interpretation to quantum mechanics but it also introduces changes on its formalism in its commonly-accepted version (i.e. specifically on its axiomatic framework by discarding the measurement postulate and its consequences and implications; see § 4.2). In this respect the many-worlds school is like (some variants of) the hidden-variable school (see § 8.2 and point 2 of that section in particular) which also introduce changes on the (common version of) formalism.[440]

D. The most important "achievement" of this interpretation is that the problematic paradigm of "collapse of wavefunction" is removed and replaced by "branching" or "splitting" to different worlds or realities. However, the cost of this removal is very high and may not even solve the essence of the problem.

E. Another alleged achievement of this interpretation is that it removed (through unifying the dynamics and formalism; see point B) the sharp border between what is quantum and what is classical, and hence the evolution of the entire world (whether quantum or classical) is ruled by a single dynamical process and governed by a single formalism.[441]

F. This interpretation (unlike the Copenhagen school for instance) should seemingly be classified as "realistic" and deterministic in nature and essence since all the probable outcomes have reality and the wavefunction supposedly evolves deterministically in all stages (i.e. not only before measurement; see point B as well as § 4.8).

G. The above presentation of this interpretation represents (in general) one variant of this school (which is the variant that is more appropriate to be labeled with "many worlds"). There are less extreme and odd (or seemingly more sensible) versions of this interpretation.

Regarding the assessment of the many-worlds interpretation, we may note the following:

1. As indicated above, the cost of getting rid of the loathed and embarrassing paradigm of "collapse of wavefunction" is high and could be more embarrassing, and this should diminish the value of the alleged "achievement" of this interpretation. In fact, this interpretation is ontologically (and even epistemologically) very costly and troubling (as well as being against intuition and common sense which will be referred to next). Moreover, it seems to us that the culprit of the embarrassment of the "collapse of wavefunction" (e.g. lack of sensible or even known physical mechanism and justification) is still there. In brief, the "branching" of the world according to this interpretation is not more obvious or rational or definite (e.g. in its mechanism) than the "wavefunction collapse" which this interpretation intends to replace and get rid of its abnormalities and embarrassments.[442]

2. This interpretation is metaphysical in essence and nature. However, this should depend on the particular meaning of "many-worlds" (which has different meaning in different variants of this interpretation). So, if "many-worlds" means separate physical worlds (or universes or realities) then it is metaphysical (and hence it should not be suitable for interpreting a scientific theory like quantum theory).

3. As indicated already, this interpretation is extremely counter-intuitive and odd as well as metaphysical. Accordingly, if we are allowed to propose and embrace such an eccentric and metaphysical interpretation

---

[440] However, we should note that the nature of these formal changes is different between the hidden-variable and many-worlds schools. For example, the changes according to the many-worlds school are a necessity and they are essentially about the axiomatic framework (unlike the changes in the hidden-variable school in general).

[441] In fact, the authenticity and validity of this "achievement" may still require the assumption of general validity of quantum mechanics and its extension to the classical domain (see for instance § 3.7, § 4.7.2 and § 6.7).

[442] In our view, regardless of the ontological oddity and philosophical difficulties, "branching" is more problematic and less comprehensible and digestible than "collapse".

8.3 Many-Worlds Interpretation

then we should equally have the right to propose and embrace similar (but simpler) interpretations of this type (such as explaining the world by deities, angels, demons, imaginary forces and objects, nonsensical scenarios and mechanisms, and so on) and this should be the shortest and best way for relieving ourselves of all the pain and duties of science and avoiding its heavy cost. In fact, this interpretation, despite its scientific appearance, is highly unscientific.[443]

4. Although the idea of "multiple realities" contradicts the uniqueness of reality it does not lead to actual inconsistency (whose elimination is the objective of the principles of reality and truth; see § 2.4.1 and § 2.5.2) from this perspective. This is because the "active" reality is always unique (since no more than one reality can be observed) and hence no contradiction or inconsistency can occur. So in brief, although this interpretation is not compliant with the principles of realty and truth (i.e. uniqueness of reality), it should have no harmful consequences in itself and from this perspective.

5. As indicated in § 5.2.1 (and in footnote [297] on page 132), quantum superposition (i.e. of states prior to collapse) may suggest multiple realities since the superposed states should have some sort of reality (in the classical sense of reality). In fact, this may partly explain why the many-worlds interpretation is embraced by many respected physicists despite its odd nature (since this odd nature seems to be basically ingrained in the theory itself; also see point 19).

6. We may argue that this interpretation violates the principle of economy (see § 2.4.4) by introducing many (and possibly infinite) worlds which should complicate the situation unnecessarily and extravagantly. Although the compliance with this principle is not compulsory (and hence it can be ignored) the violation in this case seems unnecessary and does not offer any advantage to be justified (or at least the advantage is not enough to justify the cost).

    However, on the other hand we may also argue that the many-worlds interpretation is economic in its disposal of the "wavefunction collapse" and unifying the entire quantum process and formulation under the deterministic evolution of the Schrodinger equation which makes the measurement postulate and stage redundant (see points B and F as well as point 19).[444]

7. The aforementioned deterministic nature of this interpretation (see point F) may be challenged. The reason is that this "determinism" may be true ontologically (at least according to some variants of this interpretation) but not epistemologically because as far as the observer is concerned, the *expected* outcome is as much indeterministic as, for instance, in the Copenhagen interpretation. The reader should notice that this (i.e. epistemological indeterminism) does not apply to some other interpretations (e.g. the hidden-variable; see § 8.2) in which there is a unique definite reality even though the observer may not know (due to casual reasons) this reality deterministically and prior to measurement since in principle (at least) he could acquire such knowledge unlike many-worlds where the existence (let alone knowledge) of such a unique definite reality has no meaning prior to measurement. In fact, in this regard the many-worlds and Copenhagen interpretations are equal due to the similarity between "branching" and "collapse" from this perspective.

8. In our view, one of the inconsistencies (or ambiguities) in this interpretation is related to the role of observer. Referring to "the observer branched to" (which we stated in the preamble of this section) we have two possibilities:

    • The observer branches to a single world (i.e. the observed world) as stated in the preamble. In this case the observer will not be in the other worlds (i.e. the other branches) and this should lead to the separation of observers (i.e. each in his own world). So, with the passage of time the observers could end in isolation where each lives alone in his own world (which is a tragic consequence of this continuously evolving quantum process). In fact, this possibility should lead to other absurdities and oddities (which the readers should be able to infer and identify).

    • The observer branches to all worlds (i.e. to all branches). In this case the observer should observe all

---

[443] As indicated, this should apply to some variants of this school (which is seemingly the dominant and commonly-recognized variant of this school) but not necessarily to some other variants which are less eccentric than the dominant variant.

[444] In fact, the many-worlds interpretation violates the principle of economy (see § 2.4.4) from various aspects and perspectives and hence overall it is not economic which should be a factor against its acceptance (noting that although economy is generally not a necessity or obligation it is still worth of consideration when determining the virtues and vices of a theory).

8.3 Many-Worlds Interpretation

the possible outcomes (not only one) since there is no apparent priority that can justify the preferential treatment of a specific world and a particular outcome. This possibility should also lead to other absurdities and oddities (which the readers should also be able to infer and identify).

In fact, some physicists seem to be aware of this inconsistency and seem to have attempted to address it (rather indirectly). For example, according to one physicists (see Miller in the References):

Everett proposed in 1957 that there was no collapse of the wavefunction. Rather each possible outcome of a measurement actually exists, but in different "worlds". Performing a measurement causes reality to split into multiple branches or worlds, each corresponding to a different possible result. Multiple replicas of the observer then exist, one for each world, and in each world the observer believes a different outcome happened. In this approach, an observer can be a machine, and its main characteristic is that it writes down results (e.g., in a register of some kind). For each possible answer the machine might write down, there is a different world. An alternative version would have the observer have multiple different minds, one for each outcome, in which case it is known as a many-minds hypothesis. (End of quote)

As we see, this seems to be an attempt to address this issue (or a similar issue). However, in our view this does not address the root of this issue or address the perplexity of measurement because the observer (although may be considered part of the observation apparatus in some sense and from certain perspectives) cannot be treated as a dumb object since he is distinguished by having a consciousness. After all, he is the real center in all measurements and the ultimate justification of all scientific activities. In brief, the observer is not really an object (like any other object whether quantum or non-quantum) that we can copy and multiply because if we treat him like this then we may have a physical world (which includes this "object") but we will not have science, or quantum physics or observation or ... etc. So, the observer should be unique and distinct and we cannot have multiple copies of him if we have to maintain some sensibility and rationality in the observation process. The reference to "many-minds hypothesis" as a potential escape from this dilemma is not convincing noting that any sensibly-defined observer should have a single consciousness and mind.[445] If we allow this sort of nonsense then we should end in a complete epistemological disaster and sabotage the entire science and human knowledge. In fact, we may tolerate the ontological absurdities (or oddities) of these explanations and justifications but we cannot tolerate their epistemological absurdities since all our investigations and discussions (including the present one) are based on assuming a minimum level of epistemological rationality and sensibility which if lost there will be no subject to any of these investigations and discussions.[446]

9. It seems to us that although the many-worlds interpretation is ontologically very different (and even formally to some extent) from the Copenhagen interpretation, it is epistemologically very similar (and even equivalent or almost) to Copenhagen in some key aspects. This is because the physical situation (as well as the outcome) is identical in the two schools apart from the nature of the underlying reality at the roots of this situation (e.g. if it exists and is definite or not and if other realities exist or not). In other words, we cannot see any scientific experiment or physical test that can distinguish between the many-worlds and Copenhagen since the difference between them is entirely ontological and philosophical (as well as some formal difference) and of purely theoretical nature (e.g. whether measurement leads to the collapse of wavefunction as Copenhagen claims or it leads to branching and splitting of the "wavefunction" as many-worlds claims).[447] In fact, we may consider the many-worlds

---

[445] In our view, the "many-minds" is not only incapable of addressing this issue and providing an escape, but it has its own challenges which it cannot answer and solve satisfactorily. Also, if we make copies of the observer or his mind then we effectively destroyed him as an observer and destroyed the observation process.

[446] We should also note that the "many-minds" or the "many-consciousnesses" hypotheses face the same problem and challenge that we directed (in the beginning of this point) to the possibility that "the observer branches to all worlds", because we still need to explain why the observer with his multiple minds and consciousnesses should observe (in that particular world) by that specific mind but not by his other minds. In fact, there are many other challenges and question marks about these justifications and explanations, and hence if they are analyzed properly they should lose their persuasive power and their absurdity becomes evident.

[447] What may support this view is the use of "apparent collapse" in the context of conceptualizing "branching" in the literature of this interpretation which suggests (to me at least) that the "collapse" of Copenhagen is till there but in a

8.3 Many-Worlds Interpretation

interpretation as an attempt to rationalize the essence of Copenhagen and give it a new and more acceptable cladding (by keeping superposition in essence and repackaging collapse). Also see point 19.

10. Referring to point 9, the many-worlds interpretation should be seen as worse than the Copenhagen interpretation (even if we ignore the ontological oddity of the many-worlds). This is because while the many-worlds keeps the essence of superposition and collapse (as in the Copenhagen) it replaces the "collapse" (rather superficially) by "branching" which in our view is worse and less intuitive and natural. Yes, we should give the many-worlds the credit for the unification of dynamics and formalism (see point B) and hence it should be better than the Copenhagen in this aspect.

11. If the proposition of point 9 is accepted, then we may claim that this interpretation should face the same challenges as Copenhagen (i.e. mainly with regard to the issue of measurement), or at least similar challenges (and in most cases) in addition to its own challenges (see for instance point 16). However, noting that the many worlds has a better formalism (see point 10) it may avoid some of the Copenhagen challenges and difficulties.

12. Noting the similarity between "collapse" and "branching" (see points 9 and 10), "branching" should be (like "collapse") global and instantaneous and hence non-local effects should occur (at least according to some variants of this school) in some situations and circumstances (see for instance § 5.6 and § 6.10). Hence, the many-worlds school (like some other schools such as Copenhagen) should lead to violations of the principle of locality (see § 1.8) and clash with the relativity theories.

    However, we should remark that this represents our view (at least with regard to some variants of this school and as a possibility). The common view in the literature is apparently opposite to our view (see for instance Rae and Napolitano in the References). The opposite view may be justified by the fact (according to this interpretation) that the branches (i.e. worlds) are independent of each other and non-interacting or communicating. However, this may (and may not) be sufficient to remove non-locality ontologically but not necessarily epistemologically (which is the focus of science). We should also note that our view and the opposite view may apply to different variants of this interpretation (and possibly to different instances or different conceptualizations, e.g. of locality).[448]

13. The absence of "collapse of wavefunction" in this interpretation seems to make the wavefunction indestructible and hence it is an ever-existing (or everlasting) and ever-evolving and changing "object".

14. This interpretation is challenged theoretically by numerous arguments which cast doubt on its scientific eligibility (as well as being rejected for its oddity and counter-intuitivity which we indicated earlier). Moreover, (at least) some aspects of this interpretation are not physical and not testable (since no physical observation or experiment can reveal the multiple realities for instance) or unlikely to be so. Accordingly, this interpretation is not really scientific since it fails some eligibility criteria of scientific interpretation such as physicality and testability (see § 2.9.4 and § 2.9.5).

15. As indicated earlier, there is no known (or at least no reasonable) physical mechanism or justification for "branching" (i.e. how the measurement causes the "splitting" or "branching" of wavefunction to different worlds and why this should happen). There is also no known physical mechanism or sensible scenario and justification for the emergence of "our world" (or the actual world) from the many worlds (also see point 8). In fact, in this aspect "branching" (of the many-worlds interpretation) is like "collapse" (of the Copenhagen interpretation) which the many-worlds interpretation tries to fix or get rid of, and hence the many-worlds and the Copenhagen are equal in their failure to propose a convincing explanation and sensible mechanism for the process of measurement.

16. One of the major criticisms to the many-worlds interpretation (or at least to some of its variants) is that it is difficult to justify the probabilistic nature of the formalism of quantum mechanics (i.e. in a quantitative way) by this interpretation because all the possibilities do exist and there is no sensible sense of being "more existing" than others (i.e. there is no sensible meaning of having stronger/weaker forms of existence as probability demands). For instance, if the probability of the outcome in a spin experiment is 1/4 spin-up and 3/4 spin-down, then what 1/4 and 3/4 mean in relation to the existence of these states (which are supposed to be equally existing and real)?

---

disguised "branching" form.

[448] In fact, the threat of non-locality seems to apply to this school in general and not only to some of its variants (although some challenges to the opposite view may depend on the variant).

8.3 Many-Worlds Interpretation

In fact, we can find in the literature some proposed answers to this challenge. For example, according to one proposal, in any branching process the world actually splits to many (or infinitely-many) branches in proportion to the probabilistic expectation of the possible outcomes (i.e. the probability of a specific outcome is manifested in the number of worlds in which that outcome takes place). So, in the above spin experiment, we will have spin-up in 1/4 of the branches and spin-down in 3/4 of the branches. However, these answers are arbitrary, questionable and not convincing and they aggravate, rather than alleviate, the miserable state of this interpretation.[449] In fact, they are not less odd and dubious than the interpretation itself and hence they cannot be a remedy to this challenge.

17. There is an attempt to improve or moderate the many-worlds interpretation by suggesting that the many worlds are potentialities and not physical actualities (and this may be called "many-histories interpretation"). In fact, there are several other similar attempts to make the many-worlds more rational and sensible (e.g. by replacing "many-worlds" with "many-minds" or by including decoherence).[450] However, most of these allegedly-improved versions do not add much more (if any) rationality or sensibility or clarity or intuitivity to the original many-worlds interpretation. In fact, they are mostly as bad as the original version (or versions) of this school.

18. Historically, this interpretation was originally proposed (under the name "relative state") by Hugh Everett in his PhD dissertation under the supervision of John Wheeler (and hence the interpretation may also be attributed or linked to Wheeler who seems to accept it and even advocate it although he may also have criticized it and changed his mind later). It was later developed further and modified by Bryce DeWitt (who seemingly is the first to give it the "many-worlds" title). This interpretation is also attributed to Stephen Hawking and Steven Weinberg (as followers).[451]

19. There are several reasons for the relative popularity of the many-worlds interpretation despite its oddity and extravagance. For example:
    • It supposedly offers an escape from the problematic paradigm of "wavefunction collapse" which annoys quantum physicists.
    • It seems to offer simpler conceptualization and formulation to the quantum theory by unifying the dynamics of the quantum process and merging it into a single stage (see § 4.8) where this process becomes governed (according to this interpretation) by a unique and universal evolutionary dynamics ruled by the deterministic Schrodinger equation (and hence it disposes of the before-after measurement distinction and the classic-quantum divide).
    • It seems to restore some sort of determinism to quantum theory by subjecting the entire quantum process to the deterministic Schrodinger equation (as indicated in the previous bullet point).
    • Since its emergence, the many-worlds interpretation is associated with (and supported by) some prominent theoretical physicists (see point 18). This gave the interpretation some legitimacy, credibility and weight (whose effect is magnified by the celebrity culture as well as indoctrination).
    • As indicated already (see for instance point 5), the oddity of this interpretation seems to emerge naturally from the oddity of quantum mechanics itself, and hence if we accept quantum mechanics despite its oddity it does not seem strange to accept oddities (like the many-worlds) that emerge from it and may even be justified by it. Also see the next bullet point.
    • As indicated in point 9, the essence and spirit of the many-worlds interpretation seem similar to those of the Copenhagen interpretation (despite their difference in formalism and dynamics as well as ontology), and this should give the many-worlds some of the momentum and legitimacy of Copenhagen. In other words, the common acceptance and popularity (and even dominance) of Copenhagen make the many-worlds easy to accept and digest by the public of physicists (who are used to Copenhagen and adopt the Copenhagen-styled formalism) despite its oddity.
    • This interpretation is also attractive to some species of quantum physicists for special reasons due

---

[449] For example, if the observer branches to a specific world then how he senses the other worlds and their quantitative proportionality.

[450] In fact, including the decoherence may be seen as a necessity (because of interaction with environment) to "rationalize" the interpretation and make it consistent.

[451] In fact, there are many debates and discussions (and even controversies and contradictions) of historical nature (as well as non-historical nature) about this interpretation and its form, content, evolution, etc. The interested reader should refer to the literature for details although we do not recommend spending precious time on these trivial issues.

to their particular interests and priorities. For example, the many-worlds school seems to be favored by quantum cosmologists who like to see the entire Universe as a single quantum system evolving as a single entity (which this school is capable to offer; inline with its "world" nature as well as its unification of the quantum dynamics and the removal of the boundary between quantum and classic). This interpretation (in some of its variants) may also be favored by those (or some) physicists who wish to have a role for consciousness in the quantum process (especially in measurement) where the "branching" of the world may facilitate (or be facilitated by) a conscious interaction of an observer.[452]

• We should also note that the natural tendency of human (at the current stage of biological and cultural evolution) to superstition may help in accepting this type of interpretations (i.e. this tendency makes physicists, as humans, less immune against this sort of oddity and irrationality).

20. Finally, this bizarre interpretation reveals the degree of oddity and irrationality in modern physics and the level of illusion and detachment from reality that it can reach. In fact, when physicists (many of whom have exceptional talent and intellect) feel it is acceptable (if not necessary) to interpret quantum theory by such bizarre hallucinations then we should question the theory itself (i.e. epistemologically). In other words, when such bizarre interpretations look acceptable and rational (to many physicists including some brilliant physicists) then this should be an indication to the level of (epistemological) misery of the theory itself (which requires such ridiculous and absurd fantasies to digest) and this in its turn should ring alarm bells about the tendency and destination of modern physics.

However, we should acknowledge a positive contribution of this interpretation by exposing (through its absurdity) modern physics in its repulsive nudity to the naked eyes which some other interpretations may hide (or try to hide) behind opaque curtains of technical and aesthetic decorations (such as collapse, complementarity, decoherence, etc.). We should also recognize that although this interpretation may be the ugliest of all interpretations, it is not much different in its ugliness since all the known interpretations are ugly (or lead to ugly consequences).

It is noteworthy that although we reject the many-worlds interpretation, we cannot rule out the possibility of the existence of other worlds. However, we should note first that this possibility is not a physical possibility (but a philosophical and metaphysical one), and we should note second that the other worlds according to this possibility are not linked to "our world" through "branching" and "splitting" as claimed by this interpretation. So in brief, if other worlds do exist then they should be metaphysical (not physical as this interpretation explicitly or implicitly claims, at least by giving them a physical role in the interpretation of a physical theory) and they should not be linked physically in this many-worlds fashion (as this interpretation claims). Anyway, if such other worlds do exist, they should have no connection or relevance to quantum mechanics and its interpretation (as the many-worlds school tries to do).

8.4 Transactional Interpretation

This interpretation (which was proposed by John Cramer in the 1980s) considers the wavefunction ψ as a temporally retarded (or forward in time) wave and its conjugate ψ^* as a temporally advanced (or backward in time) wave. These waves form (in a handshake or transaction) a quantum interaction and thus proceed with the quantum event. Accordingly, a standing wave is formed in this interaction (or handshake) where it transfers energy and other conserved physical quantities and the event is realized.

We can summarize the main distinctive features of this interpretation by the following remarks:

A. The aforementioned handshake is supposed to explain and justify the "collapse" of wavefunction (which is a troubling issue in some other interpretations particularly the Copenhagen school).
B. Unlike the Copenhagen interpretation, observers and observations have no role in transactions, i.e. transactions with and without observers are indifferent.
C. Transactions are essentially non-local.

Regarding the assessment of the transactional interpretation, we may note the following:

---

[452] At least, this could be in spirit with the "many-worlds" if we have to avoid some bizarre ontological consequences of this school. Also note the "many-minds" in points 8 and 17.

1. This interpretation is counter-intuitive and difficult to digest. In fact, it is another example of the illusory nature of many (or even most) of the interpretations of quantum mechanics (although it may not be as extreme and fanciful as the many-worlds interpretation).
2. This interpretation is challenged theoretically by certain arguments and thought experiments. For example, the idea of waves traveling backwards in time is not only counter-intuitive but seems to imply violation of causality (in its common meaning and definition) through delayed cause (see for instance § 1.6.4 and § 1.7).
3. Noting that transactions are non-local, the transactional interpretation should lead to violations of the principle of locality (see § 1.8) and hence it may be challenged by those who believe in this principle.
4. This interpretation is not testable (at least positively and within the current capabilities of physics and from certain key aspects). Nevertheless, it is alleged that the transactional interpretation is supported by experimental evidence (see § 6.8.1).

8.5 Consistent Histories Interpretation

This interpretation (which is also known as decoherent histories) may be seen as a variant of the Copenhagen interpretation and this is justified by the similarities between the two. In fact, this interpretation can be considered as a reform to the Copenhagen interpretation by keeping the good features of Copenhagen while disposing the problematic features which brought much criticism to Copenhagen. The essence of this interpretation is centered on the idea of "decoherent histories" (where "decoherence" means *a process in which a macroscopic system evolves in a state where interference is practically banned* and "history" means *a series of quantum events*).[453] Accordingly, decoherent histories means in essence assigning classically-meaningful and consistent probabilities to various sets of histories of the system instead of relying on measurement (as in Copenhagen) to determine these probabilities. So, all time dependencies in quantum mechanics are probabilistic where the probabilities are determined according to the Born probabilistic interpretation.[454]

Some of the distinctive features of this interpretation are outlined in the following remarks:

A. This interpretation downgrades the role of measurement and measuring equipment (as well as the role of observer) which was given by the Copenhagen school (in a rather exaggerated manner). In fact, it downgrades even the role of the measurement theory in quantum mechanics.
B. It supposedly avoids the embarrassing "wavefunction collapse" (which is central to the Copenhagen school) in the description of quantum processes. However, it seems more appropriate that this interpretation tries to "explain" the collapse of wavefunction through proposing a mechanism for this collapse although the interpretation does not seem to be able to explain the collapse itself (i.e. it may be about how collapse occurs, when it occurs, rather than why collapse occurs at all). Anyway, the reader is referred to the literature for more clarity.
C. It allegedly respects the principle of locality (see § 1.8) and hence it is consistent with the relativity theories.

---

[453] It seems that different authors understand, conceptualize and define "decoherence" rather differently. In general we can say: in (quantum) superposition the superposed states are combined or merged indistinguishably (from each other) in such a way that they have specific phase relationships which makes the distinction of the individual states impossible, while in (classical or macroscopic) decoherence (or "superposition") these states are mixed with no specific phase relationships and hence they are distinguishable from each other (although they are mixed and hence they are "superposed" slightly with minor "interference" effects), and this should allow the emergence of "definite classic realities". However, we recommend that the specific conceptualization and terminology of each author should be consulted and analyzed carefully to avoid misunderstanding and confusion (where the understanding of one author may be projected wrongly on another author).

[454] We note that most of the references about this interpretation struggle to give a clear and tangible explanation to the substance of this school (although they pretend to have a clear idea about it). Moreover, there are contradictions between authors. It seems as if the authors (as well as the inventors of the idea of decoherence) have different "versions" of decoherence (and hence different ideas about its role and function in this interpretation). These ambiguities and conflicts make the presentation and assessment of this school difficult (and hence our presentation and assessment are generally tentative and should be received with caution).

D. It does not suggest any change on the commonly-accepted formalism of quantum mechanics and its axiomatic framework.

Regarding the assessment of the consistent histories interpretation, we may note the following:

1. This interpretation is not intuitive. In fact, it contains a number of subtleties and ambiguities and may not even be completely comprehensible or sensible.
2. This interpretation seems to have epistemological difficulties related to realism and the principles of reality and truth (see 2.4.1).
3. This interpretation seems to solve the problem of collapse artificially and apparently but not substantially and fundamentally. For example, the wavefunction still represents a superposition of states and no explanation is given to justify that only one component of this superposition becomes real. Also see point B.
4. Due to its affinity with the Copenhagen school, this interpretation could face similar challenges to those of Copenhagen (apart, possibly, from those Copenhagen's challenges which are addressed and targeted specifically by this interpretation).[455]
5. Overall (and despite the aforementioned ambiguities and conflicts which makes judgment more difficult), we do not think this interpretation is more worthy of attention and consideration (let alone acceptance and adoption) than the other interpretations.

8.6 Our Interpretation

Our "interpretation" of quantum mechanics is that it is not an interpretable theory. We can justify our view by a number of reasons and factors which generally have been investigated earlier (see for instance § 7.15). However, when we describe quantum mechanics as "not interpretable" it should mean it is very unlikely that an acceptable interpretation to this theory will be found in the future (although having such an interpretation remains a possibility). So, this view reflects our feeling and conviction considering the current state of quantum theory but without ruling out the possibility of finding an acceptable interpretation in the future (because we do not have evidence against this possibility).

Anyway, having no interpretation (according to our view) may relieve us from our duty and responsibility (as epistemologists) of providing or trying to find an interpretation but it does not relieve us from our duty to address some quantum mechanical issues (of epistemological nature) that seem to threaten our epistemological views and principles. For example, how can we keep our position towards realism or causality whereas quantum theory seems to have non-realistic and non-causal consequences? So, we need (as epistemologists but not necessarily as physicists) to explain and justify our epistemological views and positions which we adopted earlier (such as considering realism and causality as epistemological necessities) considering the epistemological consequences and implications of quantum mechanics that potentially contradict these views.[456] In the following points we discuss these issues rather briefly (by outlining our position in the light of our views which we expressed previously in various places and contexts):

A. Regarding **locality**, since we have no problem with violating the locality principle (because we do not believe in the validity of this principle at least in its relativistic justifications and implications; see § 1.8 and § 7.13), then we can generally (and as far as locality is concerned) accept the non-local consequences and interpretations of quantum mechanics (at least within certain conditions and justifications).
B. Regarding **causality**, we do not accept total violation of the principle of causality. However, no one can claim that quantum mechanics leads to violation of causality in this way (otherwise there will be total

---

[455] In fact, it may face challenges (possibly more difficult ones) even to its aspects and features that were introduced to replace problematic aspects of Copenhagen.
[456] As practitioner physicists, we can ignore the entire issue of the epistemology of quantum physics and the interpretation of quantum mechanics (as well as all its epistemological implications and consequences) and use quantum mechanics as a mere calculus or as a tool for doing our business, but as epistemologists (which what we are supposed to be in this book) we cannot do this.

8.6 Our Interpretation

chaos). So, all we need to do is to deal with (potential) partial violations due to indeterminism.[457] But as discussed earlier, such violations may not be a necessity noting that we have some interpretations that cause no such violations. Moreover, we believe in the possibility of creating new quantum theories that potentially respect causality totally and hence we can keep our position towards causality (while using quantum mechanics as a mere calculus) until we get evidence that creating such theories is impossible.[458] Anyway, if we are forced (for whatever reason or pretext) to abandon causality (i.e. partially) then we should restrict this to the quantum domain and limit it to the minimum required for explaining the situation and justifying the presumed violations (i.e. we maintain total causality in the classical domain and sacrifice causality minimally in the quantum domain).

C. Regarding **realism**, the situation is very similar to the situation of causality, which we discussed in point B.[459] Further discussion about realism from this perspective will be given later.

D. Referring to points B and C, we accepted (if necessary) a weak or partial form of causality and realism for the quantum world. A legitimate question then is: can we accept (if necessary) a weak or modified form of **logic** for the quantum world (or we may even accept a different logic if necessary)? In our view, the rules and principles of logic are more fundamental than any philosophical or epistemological or scientific principles, because they are essentially rules for ensuring consistency, sensibility and rationality intrinsically and at the most basic level, and hence they cannot be violated or modified. Yes, evolution may lead to biological (and hence "cultural") changes that may change the notion and rules of "intrinsic consistency" (which is the essence of the entire logic), but this is a different story and should be outside our concern as human species. Accordingly, logic is an intrinsic feature of our species and hence as long as we reserve our identity as humans it seems that logic cannot be replaced or amended (at least substantially).[460] So in brief, logic by nature is "classical" and strict and hence it cannot be "quantumized" or compromised. Also see § 2.5.1 and points 10 and 11 of that section in particular.

In the following points we list a few remarks related to the issues of the present chapter and section:

1. If we have to choose one of the available interpretations then our (personal) preference is as follows: the hidden-variable is the first (either local/non-local if possible or non-local if local is not possible), the Copenhagen (in its moderate variants) is the second and the many-worlds is the last.[461] The main criteria for our choice and preference are rationality and realism.

2. The possibility of having an alternative (fully causal/realistic) quantum theory according to the principle of non-uniqueness of science (see § 2.4.3) does not mean "actuality in the future" but it means a "possibility in principle but not necessarily". In fact, actuality does not only require the physical reality to be in a specific shape and form to be fully deterministic (or fully causal and realistic)[462] but it also requires certain coincidences, evolutionary paths, historical conditions, etc. related to human society, science, environment, etc.

Anyway, we should keep looking for such an alternative theory (ignoring the alleged impossibilities

---

[457] Regarding potential violations of causality due to non-locality (as claimed by relativists), we do not believe in relativity. Moreover, we addressed this issue previously in this book as well as in our book "The Mechanics of Lorentz Transformations" and hence we do not repeat. In fact, such alleged violations are more appropriate to investigate within the investigation of relativity rather than quantum physics (which is the subject of the present book).

[458] As indicated earlier, by the non-uniqueness of science (see § 2.4.3) it should be possible (at least in principle) to find a different quantum theory (and even theories) that is entirely and strictly causal and realistic (and hence the possibility of maintaining the strict form of causality and realism even at the quantum level cannot be eliminated). In fact, such a theory (if found) will be more successful empirically (by lifting some limitations of determinism) as well as more successful epistemologically (by having a strictly rational and consistent interpretation).

[459] As explained elsewhere, if we are forced by the implications of the quantum theory to abandon (epistemological) realism then we should abandon it as little as we can (and keep it as much as we can). This means that we should keep realism in the classical domain. Moreover, we should also keep it in the quantum domain where and when possible. So, we should always follow interpretations and justifications that are compliant with realism. We should also remind the reader that we adopt only a moderate form of realism and this could help in avoiding (or at least minimizing) any sacrifice in realism.

[460] In fact, we may have an analogy in mathematics and its tight rules and "everlasting" principles (noting that mathematics is ultimately based on logic and may be seen as a direct derivative and product of it).

[461] I may describe them as "the good, the bad and the ugly" (or "the good, the ugly and the bad").

[462] Or rather: to lend itself to be described and formalized in this way.

8.6 Our Interpretation

imposed by the current quantum theory or by some quantum physicists) hoping that we may find such a theory one day (if the reality allows the existence of such a theory or "truth" and if we are lucky and smart enough to find such a theory). In fact, if some alternative "realistic" formulations of quantum mechanics (like the Bohm theory) are identical in their predictions to quantum mechanics (as commonly claimed) then the possibility of finding theories of this type (some of which could be more perfect and intuitive than the current ones and could even be empirically superior) should be seen as an actuality.

3. Our view[463] (which we adopted as a last resort) about possible distinction between classical realism and quantum realism (i.e. by adopting, if necessary, total and strict realism at the classical level and partial and relaxed realism at the quantum level) may not look very rational and could be criticized or challenged (because if we have to adopt realism then we should adopt it equally and in the same sense to any reality whether classical or quantum). However, this view may be justified (epistemologically and possibly even ontologically) by claiming that non-quantum reality and quantum reality are generally different, i.e. the former is deterministic, definite and certain while the latter is not (or may be not). Anyway, in this case we need to make a distinction about the kind and level of applicability of the principles of reality and truth (or even possible minor modifications to these principles). So, we can continue with our original position that the principles of reality and truth should still apply to classical reality (or rather non-quantum reality or realities) because these principles are needed there (to avoid inconsistencies as explained in § 2.4.1, § 2.5.2 and § 2.9.2) and because there is no reason (such as indeterminacy) that necessitates the disposal or modification of these principles in the non-quantum domain. On the other hand, we modify these principles minimally (or rather how or how much they apply) to accommodate potential violation of strict realism (e.g. by adopting that reality exists and is unique within certain uncertainties which possibly could be purely epistemological rather than ontological and this may affect the truth although we may also compromise the truth only). In fact, if we have to accept that these principles could be violated or modified (to some extent) in the quantum domain then the success of quantum theory may provide justification for this since it means that the partial disposal of these principles in the quantum domain does not lead to inconsistencies (at least according to certain practical considerations and pragmatic leniency that maintain consistency and rationality in general).

Anyway, the difference in the state of the principles of reality and truth between the quantum and non-quantum domains may also be tolerated and justified from another perspective that is these principles are no more than useful conventions (see § 2.5.2) despite their crucial role for maintaining consistency and rationality and despite the possibility that this distinction may lead to adopting different epistemological rules (or epistemologies) towards different types of reality (i.e. quantum and non-quantum) which may not seem very sensible or rational. In fact, adopting different epistemologies for different types of reality means that the scale factor (see § 1.3.2) should play a role even in our epistemology and philosophy towards the Universe and hence the effect of scale factor is not restricted to science and other familiar types of knowledge (i.e. of physical nature).[464] Such a view is possibly more pragmatic than even the Copenhagen school (or some of its variants).

4. To summarize our position towards realism (which is the epicenter of most of the epistemological difficulties of quantum theory),[465] we can propose a compromise between indeterminacy (or dependency of reality on observer and observation) in quantum mechanics and realism. First, we consider the

---

[463] It is noteworthy that this view seems to exist among the followers of the Copenhagen school. We should also note that in the following we generally talk about realism in a rather extended sense which should include aspects like causality (so that we avoid complicating the text).

[464] This should be inline with our view that classical and quantum theories are two independent and equally-valid theories in their domain of applicability (i.e. macroscopic and quantum worlds) although they may share some common applicability (or they converge to similar predictions) at their borders where they meet (e.g. in quantum systems with large quantum numbers). In fact, this sharing should be seen as a healthy sign for the validity (as well as the consistency) of the two theories. So in brief, quantum theory is not more valid or more general than classical physics and hence the correspondence principle should be understood and appreciated accordingly (unlike what some statements or interpretations of this principle may suggest).

[465] In this regard we also refer the reader to point 5 of § 1.3.3 as well as to § 1.10 where we considered other factors related to realism and our position toward it.

8.6 Our Interpretation

moderate form of realism where reality is determined and independent but truth (as a reflection of reality) is subject to variations that consider the dependency of truth (to some extent) on the details of the observer and his equipment. This moderate form of realism may explain some types of indeterminacy/dependency (and hence realism is preserved at least in some cases). Second, we consider (if necessary and as a last resort noting that the moderate form of realism may not be able or sufficient to justify all types of indeterminacy/dependency and noting as well that other excuses that we proposed, such as non-uniqueness of science, may not be accepted or sufficient to maintain total realism) that quantum reality may not be as specific as classical reality and hence "quantum realism" could be partial (where some aspects of classical realism may not hold).[466] Accordingly, we allow ourselves to claim that realism in general (i.e. in its moderate, and possibly partial, form) is compliant with quantum mechanics. Also see footnote [63] on page 32.

---

[466] In fact, the distinction between quantum reality and classical reality should be essentially about our notion of these realities (since we are talking about realism and as epistemologists) and hence we can say that this distinction could be due to ontological reasons or to epistemological reasons or even to both.

Chapter 9
Challenges to Quantum Theory

In this chapter we investigate the challenges to the quantum theory. In fact, this theory is one of the most criticized and challenged scientific theories (at least in modern physics) despite its overwhelming acceptance at the practical level (thanks to its remarkable empirical success as well as to the absence of any credible competitor or substitute). Many of these criticisms and challenges have been investigated and assessed early in the book within our presentation and assessment of the theory (especially in chapters 6 and 7 which are dedicated to the assessment). However, the criticisms and challenges in the previous investigations were mostly specific and within the contexts of other investigations. In this chapter we want to present and discuss devotedly three general types or categories of the challenges to the quantum theory related to its contradictions (see § 9.1), its incompatibility with other theories (see § 9.2) and its incompleteness (see § 9.3).

In fact, the challenges to the quantum theory can be classified broadly considering its two main aspects: those related to the interpretation and those related to the formalism. **Regarding the challenges related to the interpretation**, we note that challenges and criticisms to the schools of interpretation are not of concern (or at least not of prime concern) to us in this investigation. This is because these challenges generally belong to these particular schools and hence they usually do not pose a challenge to the quantum theory itself (although they may have some implications and consequences of this nature).[467] Moreover, these schools have been investigated and assessed in chapter 8 where their challenges are outlined and dealt with there and hence we do not repeat. Yes, the interpretative and epistemological aspects and features of the theory (which may be seen as parts and ingredients of its basic or standard interpretation) should be within the topic of this chapter and hence they will be considered as potential subjects to the challenges of this chapter.

Regarding the challenges related to the formalism, we note first that any criticism or challenge to the theory that claims or implies empirical failure of the theory or casts doubts on its empirical success (within the previously-stated limitations) should be rejected (at least for the time being and as long as a solid proof on its failure is not established). This is because the correctness of quantum theory at the practical level is supported by overwhelming evidence and accepted by the bulk of scientific community (see § 7.6). Therefore, acceptable challenges should be generally restricted to its epistemological and interpretative aspects. Yes, challenges to some aspects and attributes of the formalism (such as its completeness or optimality) are acceptable in principle. Moreover, sometimes it may be necessary to deal with challenges to the formalism due to their connection to the challenges to the basic interpretation (or rather the interpretative and epistemological aspects) since the interpretation is generally based on the formalism. In fact, many of the challenges (including those investigated here) are mainly (and historically) proposed as challenges against the quantum theory itself (which is primarily represented by its formalism) but in the following they will be treated (when possible) as challenges to the interpretation (or rather the interpretative and epistemological aspects) of the theory since (as we stated already) they cannot pose a real threat to the theory itself (especially at this late stage in the life of quantum theory where the theory passed about a century of intensive examination and application; unlike in the early stages in its life when most of these challenges have been proposed initially and where the theory was not tested sufficiently).

---

[467] In fact, some challenges to the quantum theory are closely related to certain schools. For example, some paradoxes target quantum theory through its Copenhagen interpretation due to its dominance (historically) and its strong presence conceptually and terminologically in the theory, and hence it is seen as the official voice of the theory. Also, some challenges may be directed to the theory through a particular school of interpretation because this school exposed (intentionally or non-intentionally) a potential weakness or vulnerability in the theory.

9.1 Paradoxes

Since its early days, the quantum theory has been challenged by many claimed paradoxes that allegedly show inconsistencies and contradictions in this theory and its rational framework and logical structure and lead to false consequences or nonsensical implications or conflicting results or ... etc. A sample of the claimed paradoxes against quantum theory are outlined in the following subsections. However, before that we would like to outline in the following points a number of issues related to the subject of paradoxes in the quantum theory:

1. The claimed paradoxes against the quantum theory are of two main types: paradoxes against the theory itself (i.e. the formalism of quantum mechanics including its basic or standard interpretation and its interpretative and epistemological aspects and features) and paradoxes against its schools of interpretation (e.g. Copenhagen school; see § 8.1). As explained in the preamble of this chapter, we do not have interest here in the second type. However, we should remember that the second type may have relevance to the first type and hence it may be included marginally or referred to casually.

2. There are a number of typical and general sources (or reasons) for the paradoxes against quantum theory (whether these paradoxes are related to its formalism or to its basic interpretation and epistemology). However, most of these sources are dubious and challengeable. For example:[468]
    - Some paradoxes are based on ignoring the humanism factor or/and the scale factor (see § 1.3.1 and § 1.3.2) or based on the subjective nature of some aspects of the quantum theory.
    - Some paradoxes are based on thinking classically where the rules of classical physics are (or may be) invalid.
    - Some paradoxes are based on the implications of other (problematic) theories such as special relativity (whose questionable postulates and implications are treated as scientific facts) or dubious principles like locality (see § 1.8 and § 7.13).
    - Some paradoxes are based on making physical judgments on non-observable aspects (rather than restricting the attention to the observable aspects as it should be).
    - Some paradoxes are based on thought experiments[469] and their implications and results as if they are real experiments. In fact, many (and possibly most) of the alleged paradoxes are based on improvised thought experiments which are claimed to lead to inconsistencies and paradoxes in the quantum theory or some of its (formal or interpretative) consequences and implications.
    - Some paradoxes are based on the presumption of general validity of quantum mechanics and its applicability in the classical (or non-quantum) domain.
    - Some paradoxes are directed towards the quantum theory while they are actually and essentially related to certain schools (e.g. Copenhagen).

3. Different schools of interpretation generally address (and sometimes fail to address) these paradoxes differently and the debate about paradoxes and inconsistencies from this perspective takes a considerable part of the literature of this theory (especially in its early days). However, we do not present or assess the position of the different schools of interpretation towards these paradoxes due to the aforementioned restrictions (as well as restrictions on the book size, lack of originality in such investigation and minor benefit gained from it ... etc.). So, we generally investigate and assess these paradoxes from our perspective and viewpoint (rather than from the perspective and viewpoint of others) although we may refer to some of these interpretations when this is useful or relevant or necessary.

4. As indicated in § 7.5, there is nothing illogical in quantum mechanics although logical inconsistencies may exist in some schools of interpretation and potentially in certain interpretative and epistemological aspects that are supposedly based on the formalism. Accordingly, the paradoxes that allegedly refute or question quantum mechanics itself should be rejected without further ado (or at least they should be redirected to its epistemology and interpretation). So in brief, any alleged refutation of quantum mechanics by challenging its intrinsic consistency (i.e. its logical validity) or extrinsic consistency (i.e.

---

[468] We note that some of these reasons are proposed in the literature (possibly representing the position of certain schools) and may not represent our view. We should also note that we will meet instances of most of these examples in the following subsections.

[469] Typical examples of these thought experiments can be found in the famous Bohr-Einstein debates which contain loads of nonsensical arguments and funny thought experiments.

its empirical validity) should be rejected (also see the preamble of this chapter).
5. Many of the alleged paradoxes against quantum mechanics were proposed during the early days of development of quantum mechanics, and hence they are mostly characterized by their pedantic nature, poor quality and meager content. Fortunately (for science), the interest in such low scientific thinking and debate diminished (but did not vanish) in the more recent times. In our view, most of the discussion and literature about these paradoxes are mainly of historical value (and hence we do not recommend spending precious time on such trivial and rather silly topics).
6. Although we generally have a negative attitude towards paradoxes in science (including quantum mechanics which is our subject), we should accept that they may also have some beneficial outcome and positive impact (mainly on the assessment and scrutiny of scientific theories) as these paradoxes can draw the attention to problematic issues and potential loopholes in the theories. So, what we are actually against is the excessive obsession with these paradoxes which sucks and consumes considerable amounts of resources into Byzantine discussion and argumentation and may be accepted as a legitimate method for challenging valid science and replacing experimental validation by nonsensical argumentation.

9.1.1 Bubble Paradox

The essence of this paradox is that a spatially dispersed wavefunction (e.g. a bubble-like wavefunction of a spherically-expanding pulse of light) should collapse (and disappear) instantaneously at all spatial positions if it is observed (or detected) at a particular spatial position at a given instant of time and this requires superluminal (and possibly infinite) speed which violates the principle of locality (see § 1.8). So, this paradox is based on the principle of locality (which gets its legitimacy primarily from special relativity) and because this principle (as well as special relativity) is dubious and challengeable we should reject this paradox without further ado.[470] In fact, the bubble paradox (in its original form and as outlined and expressed above) can be challenged technically from other aspects but this is not needed here.

9.1.2 Schrodinger's Cat Paradox

This paradox (which is the most famous of all alleged paradoxes of quantum theory)[471] is a hypothetical (or thought) experiment whose purpose is to demonstrate some of the alleged nonsensical consequences and implications of quantum theory (and its Copenhagen interpretation in particular). In fact, this paradox is one of the hottest subjects in the quantum mechanical debate (especially in the early decades of quantum theory) where many arguments and theories (as well as huge amounts of nonsense) were developed about it and around it. This unduly excessive interest seems to be helped by its association with Schrodinger (who is a celebrity) and its emergence in the early days of quantum theory (noting that this paradox does not deserve even a fraction of this attention and interest).

In this "experiment" a cat is put in an execution chamber where the deadly mechanism (such as the release of a poisonous gas) that is supposed to kill the cat is designed to be triggered by a random quantum event (such as the decay of a radioactive atom) with a 50% chance of occurrence during the time of experiment (say one hour). Now, if this cat is actually a "quantum cat" then we can assume (according to the quantum theory or rather according to the claim of this paradox) that the quantum state of the cat at the start of the experiment is a superposition of an alive-state (represented by a state function ψ_a) and a dead-state (represented by a state function ψ_d) and hence we may pose the following challenging questions (among many other possible questions):
A. What is the wavefunction of the cat just before opening the chamber in the end of the one-hour experiment? Is it ψ_a or ψ_d or a combination of both (i.e. $\psi = \frac{1}{\sqrt{2}}\psi_a + \frac{1}{\sqrt{2}}\psi_d$)?

---

[470] See for instance § 1.8 and § 6.10 about the assessment of the principle of locality, and refer to our book "The Mechanics of Lorentz Transformations" about the assessment of special relativity.

[471] In fact, it should compete for the first place with the EPR paradox, but we classified the EPR paradox as an argument (see point 6 of § 9.3.1).

9.1.2 Schrodinger's Cat Paradox

B. What event causes the collapse of the wavefunction? Is it the trigger (which leads to the death of the cat assuming the trigger was activated) or when the chamber is opened in the end of the experiment?

C. If the trigger is the actual cause of the collapse of the wavefunction then what if the trigger was not activated (and hence the cat was still alive in the end of the experiment)? Is it in this case the collapse will be caused by opening the chamber? If so then why should the collapse be caused by different types of event in the two cases (i.e. by the actual trigger in one case and by the observation in the other case)?

For example, if we choose (in reply to question C) that the trigger is the actual cause of the collapse of the wavefunction then we contradict the commonly-accepted presumption (or interpretation) in quantum theory that the observation is the cause of the collapse of the wavefunction.[472] On the other hand, if we choose (in reply to question C) that the cause of the collapse is the observation (when we open the chamber) then we may be challenged that what actually killed the cat then is not the triggered deadly mechanism but the observation process. Similarly, if we choose that the cause of the collapse of the wavefunction is the trigger (if the trigger is activated) and the observation (if the trigger is not activated) then first we will violate the generality of the presumption that the cause of the collapse of the wavefunction is observation and second we need to justify the difference between the two types of events where in one case the collapse is caused by observation whereas in the other case the collapse is caused by an actual event (which belongs to an independent physical reality not to observation). In fact, we can pose many more questions and challenges but this will not add more to what we already have, that is this thought experiment can pose challenges to the interpretations of quantum theory (and the Copenhagen school in particular) and possibly even challenges to its formalism (or rather basic interpretative aspects of the formalism).

In our view, the confusion and mess in this alleged paradox arises mostly from using a classical system to represent a quantum system (i.e. using a cat which is a classical object to play the role of a quantum object) and hence mixing the "classical physics and epistemology" with the "quantum physics and epistemology". If we note that epistemology (like physics) is mainly about what we can observe and measure[473] and we note that what we can observe and measure quantum mechanically is not the same as what we can observe and measure classically we can appreciate the clumsiness of using a cat as a quantum object and trying to apply the rules and physics (as well as the epistemology and rationale) of the quantum world on it. In fact, the actual paradox is not in the quantum theory (and not even in its interpretation at least from this side) but in the mess and confusion created by this mix of classical and quantum things (i.e. physics, epistemology, rationales, arguments, etc.). By considering the scale factor we can say: what is valid and applicable to the classical world is not necessarily valid and applicable to the quantum world (and vice versa) although from a pragmatic perspective we may be justified and excused when we try to make the rules and principles of the classical and quantum worlds as close as possible because this approach ensures optimal adaptation to us with our environment (which is the physical world) and minimizes our effort to understand and conquer the physical world. So in brief, the culprit of this alleged paradox is the concept of "quantum cat" where "cat" is a classical object and hence it cannot be quantum or treated as quantum (at least from certain aspects and according to some considerations).

In the following points we inspect the Schrodinger's cat paradox further and assess its allegation of nonsensical consequences against quantum theory:

1. This paradox is presented in many shapes and forms and it has many modifications and elaborations (including replacing the cat itself by other types of creature and hence it is not a cat paradox anymore); see § 9.1.3 and § 9.1.4. However, the essence of all theses different versions and variants is to challenge the quantum theory (or some of its interpretations such as the Copenhagen school) by the intuitive sense of realism and determinism (which are embedded and demonstrated vividly in classical physics). The use of a "cat" with a macroscopic physical setting (which is a classical system) rather than an electron for example facilitates the rationale of the argument behind this paradox.
2. We should note (as explained in the preamble and indicated in point 1) that the major weakness in the

---

[472] In fact, this should also contradict the denial of objective independent reality (which is usually ascribed to the Copenhagen school and may even be ascribed to quantum mechanics).

[473] Our focus here is the epistemology of science (and quantum physics in particular).

9.1.2 Schrodinger's Cat Paradox

Schrodinger's cat paradox is the application of quantum theory to a macroscopic system in the form of a "quantum cat" which mixes the classical rationale with the quantum rationale and causes confusion and mess. Although this may seem a trivial or secondary issue, it can cause seemingly-paradoxical effects where we see one type of system being governed by the rules and rationale of another type of system. In fact, this paradox is implicitly based on the presumption of the applicability of quantum theory (in its formalism and rationale) to classical systems and this can be challenged (refer to § 3.7, § 4.7.2, § 6.7 and § 7.3). A deep analysis to this paradox should reveal that its paradoxical nature originates (partly at least) from this presumption. Also, the mix of a quantum trigger with a classical mechanism of killing (as well as a classical cat) should exacerbate the mess and intensify the confusion. So, a brief answer to this paradox is that quantum mechanics applies neither in formalism nor in interpretation to macroscopic objects like cat (see for instance § 3.7) and hence the entire paradox is baseless and meaningless.

3. Some of the proposed conceptualizations and interpretations of the Schrodinger's cat experiment are based on (or imply) delayed cause which violates the principle of causality (see § 1.6.4 and § 1.7) and hence they can lead to serious challenges to the quantum theory. Although these challenges can pose real threats to the quantum theory, they can be refuted simply by the baselessness and meaninglessness of the paradox itself (which we indicated in point 2). Other serious challenges (originating from conceptualizations and interpretations that lead to violation of established rules and principles) can also be refuted similarly.

4. There are many school-specific challenges and alleged remedies to this paradox (e.g. by employing decoherence or consciousness or non-linear term or ... etc.). However, as indicated already we generally have no interest here in this type of investigation and discussion.

5. An aspect of this paradox that does not seem right (or at least it is not clear to the author of this book) is the presumption of alive-dead superposition. This is because prior to putting the cat inside the box we should assume (for the sake of sensibility) that the cat is alive and hence its alleged wavefunction is ψ_a. So, how and at what point we got the superposition of alive-dead state? Is it by putting it inside the box (in which case what is the mechanism and justification of this "anti-collapse")? Or is it when the triggering mechanism is activated (which seems very nonsensical; moreover what if the triggering mechanism was not activated at all during the experiment)? Or is it at some instant in between (in which case when and why noting that all such instants are equal)? Or is it after the activation or even at the end of this experiment and prior to observation (in which case how and why)? In fact, there are many possibilities and potential remedies but they do not seem right or convincing. Also, some of these questions could similarly be posed (possibly more embarrassingly and perplexingly) if the deadly mechanism was not activated at all during the experiment. Also see point 10 of § 6.11.

6. According to some, this paradox can be resolved (partly) by taking into account that the wavefunction does not belong to an individual cat but to an ensemble of identically-prepared cats (see point 3 of § 6.1) and hence the uncertainty about life and death belongs to this ensemble (according to the probabilistic interpretation) which should pose no perplexity. However, this resolution does not address the issue of individual cats (even if we assume it makes sense for an ensemble of cats) noting that cats are classical objects and hence life and death can be (and should be) attributed to them individually (even if we tolerated the claim that life and death can be attributed to them as groups within an ensemble). In fact, this resolution should also fail to address other aspects and perspectives of this paradox. Moreover, the proposal of associating wavefunctions with ensembles (rather than with individuals) is problematic in itself (as explained earlier).

7. As indicated above, the Schrodinger's cat paradox is generally seen (especially in recent times) as a challenge to the (basic) interpretation of quantum theory (rather than its formalism). However, it may also be seen as a challenge to the formalism (especially in the early days of quantum theory) noting for instance the potential impact of this paradox on the axiomatic framework of quantum mechanics.[474]

---

[474] Not being a challenge to the formalism can be justified by the fact that the legitimacy and "correctness" of the formalism is based on its empirical success (which no one can deny). It is obvious that this thought experiment does not represent any challenge to this legitimacy since it has no empirical implications or consequences. Yes, the argument may affect the conceptual framework of quantum mechanics (e.g. the measurement postulate) but this framework (or this part of

9.1.2 Schrodinger's Cat Paradox

In our view, this paradox is (primarily and essentially) a challenge to certain schools of interpretation. Accordingly, if we have to deal with this paradox at all and consider it as worthy of investigation then it is more appropriate to deal with it on this basis and as part of the discussion and debate about the schools of interpretation (which we investigated in chapter 8). This should diminish the value of this paradox substantially and make the huge amount of literature written about it of very little use (or effectively redundant from our perspective).

8. As a marginal note, the Schrodinger's cat paradox highlights an issue which we discussed previously that is, the quantum indeterminacy can be reflected into a classical indeterminacy, i.e. the life/death of cat which is classical is determined by the indeterminate quantum event (i.e. the quantum trigger).

9. Another marginal note is about the issue (which may be debated in the literature occasionally and casually and could have epistemological roots and links) of potential moral and legal consequences. For example, if the wavefunction of the cat collapses by observation (rather than by the activation of the quantum trigger) then who should be considered the killer and held responsible for this crime: the one who set the device (which led to the activation of the lethal mechanism) or the observer? In this regard, we note that the moral and legal issues are generally determined and judged according to the rules of common sense and the cultural infrastructure of the society rather than by quantum mechanics. So, even if we assume, for instance, that quantum mechanically the cat is killed by observation the actual (moral and legal) responsibility should rest (according to my understanding to the currently dominant human culture) with whoever set the device that led to the activation of the lethal mechanism. After all, moral and legal issues are classical in nature and hence they should be judged classically (where a classical mind sees the cause of death is the activation of the lethal mechanism, as a result of setting the device, which should be inline with the classical realism where reality is independent of the observer and observation). In brief, quantum mechanics (including its epistemology and interpretation) has no role in determining moral and legal issues. In fact, this should remind us of similar arbitrary claims (about moral and legal implications) made with regard to the relativity theories (where similar refutations should also apply).

10. Finally, it is important to note that the paradox is based on an implicit assumption of classical realism at the quantum level. This is because when we talk about the "activation of the lethal mechanism" by a quantum event prior to observation we should assume the existence of determinate, definite and independent quantum reality (and this may be challenged by some schools and could be a challenge to the paradox itself).

To conclude, the Schrodinger's cat paradox does not really represent a challenge to quantum mechanics (although it may represent a challenge to some of its schools of interpretation and possibly even to some aspects of its basic interpretation). This should be the case even if we ignore the challenges and refutations to the paradox itself (some of which are indicated or outlined above such as the challenge to the implicit presumption of general validity of quantum mechanics) and consider the paradox as worthy of attention and consideration. So, this paradox is not a big problem (if it is a problem at all) noting that there is no reliable interpretation to quantum mechanics (regardless of this paradox) since all schools of interpretation (and even some aspects of the basic interpretation) are already challenged (and possibly refuted). Moreover, the basic interpretation as such (let alone the schools of interpretation) is not a necessity for the validity of the formalism at its basic empirical level. Accordingly, the above challenging questions and issues posed by this paradox (or some of them at least) may embarrass certain interpretative aspects and some schools of interpretation but they cannot embarrass quantum mechanics.[475]

However, to be fair we should also give some credit and merit to this paradox and its alike since they draw the attention to the perplexities of quantum mechanics and highlight the fact that it is difficult (and perhaps impossible) to interpret rationally and consistently. In fact, this should endorse our view

---

the framework) is of interpretative nature. So in brief, quantum mechanics (which is the formalism of quantum theory) is a well established and tested theory and hence it cannot be challenged or refuted by this type of paradox (or indeed by any paradox that challenges its logical or empirical validity as explained earlier).

[475] As indicated above, a huge amount of nonsense is produced in the literature about this trivial paradox and some people may take this seriously and treat it as real science and philosophy. So, my advice to the readers is to save their time and energy and avoid wasting their life in this sort of nonsensical pointless "physics" and childish paradoxes and arguments.

about the problematic nature of the epistemology of quantum mechanics and the possibility of having no interpretation. This should also provide a motive for trying to find a novel quantum theory that is epistemologically (and possibly even empirically) superior to the current quantum theory.

9.1.3 Wigner's Friend Paradox

This is a thought experiment proposed as an elaborate variant of the Schrodinger's cat paradox. In this hypothetical experiment a friend of Wigner performs the Schrodinger's cat experiment in the absence of Wigner who becomes aware of the result of the experiment after its completion.[476] Many challenging questions (similar to those of the Schrodinger's cat paradox) are posed and can be posed such as:

A. From the perspective of Wigner, what is the state of the cat during the time between completion and awareness? Is it alive-or-dead (according to the result of the experiment) or it is a superposition of alive-and-dead?
B. From the perspective of Wigner, when the measurement (or observation) took place? Is it in the end of the experiment (i.e. when his friend observed what happened to the cat) or when Wigner became aware of the result of the experiment (by observing the outcome himself or by asking his friend)?

For example, if we choose (in reply to question A) "alive-or-dead" then we may be challenged that this implies the existence of a reality independent of awareness (which seems contradictory to the Copenhagen interpretation for example), while if we choose "alive-and-dead" then we may be challenged that we then have more than one collapse of the cat wavefunction (i.e. one by the observation of the friend and one by the observation of Wigner). We may also be challenged (especially if we choose the latter) by the question: what is the role of awareness (or consciousness) in the collapse of the wavefunction?

In fact, the weaknesses of this paradox (and its variants) are similar to the weaknesses of the original paradox (i.e. the Schrodinger's cat paradox). Also, these questions and challenges (and their alike) may cast doubt on certain interpretative aspects or embarrass some schools of interpretation but they should not cast doubt or embarrass quantum mechanics (although they should highlight the fact that quantum mechanics is difficult or impossible to interpret). We also repeat our advice to avoid investing valuable time and effort in the heaps of nonsensical arguments and futile discussions generated by and about this paradox.

9.1.4 Suicide Paradox

This thought experiment is another elaborate variant of the Schrodinger's cat paradox (and rather more dramatic than the original paradox and its Wigner's friend variant) where the experimenter (or observer) plays the role of the cat in this "suicide experiment" (or "suicide mission"), i.e. the cat in the Schrodinger's cat experiment is replaced by the experimenter himself. Again, many challenging questions can be posed such as:

A. At the start of this experiment, what is the state of the experimenter? Is he alive or in a state of alive-and-dead superposition?
B. If (according to the answer of question A) the experimenter is "alive" then how can he be killed by the lethal mechanism (noting that there is no "dead" state in his wavefunction)? In other words, because there is no "dead" state in his wavefunction, should the effect of poison become nil and hence he stays alive?
C. If (according to the answer of question A) the experimenter is "alive-and-dead" and the lethal mechanism is activated then how his wavefunction collapses (noting that it may be argued that if he is killed by the mechanism then there is no observer or observation that causes the collapse of wavefunction). To put it differently, should he continue to be in a state of "alive-and-dead" even after the activation of the lethal mechanism?

---

[476] We note that "Wigner" in this experiment (or argument) represents the outside or secondary observer (noting that the "friend of Wigner" represents the inside or primary observer). We also note that the Wigner's friend paradox is reported in other forms and variants (e.g. the replacement of the cat by Wigner's friend although the result may not be as tragic as for the cat). However, these details are trivial and not worthy of attention or serious consideration. We refer the interested reader to the literature for further details.

For example, if we choose (in reply to question A) "alive" then we should have no collapse of wavefunction at any future event[477] which seems nonsensical (also see question B), while if we choose "alive-and-dead" then it is obviously nonsensical and contradictory (or at least it leads to nonsensical and contradictory consequences and implications) noting, for instance, that he is observing himself alive and hence he is actually alive (also see question C).

Again, the weaknesses of this paradox (and its variants) are similar to the weaknesses of the original paradox (i.e. the Schrodinger's cat paradox) and its Wigner's friend variant. In fact, there is another important weakness (or rather source of challenges and attacks) in this paradox that is the unification of the observer with the observed which introduces a fundamental difference on the paradigm of "observation" (and indeed on other paradigms of science and epistemology) noting that the observation process (by definition) is an interaction or relation between two separate and independent entities (even if we negate inertness and even if we accept dependency of reality to some extent and in some details on the observation, as discussed earlier). In other words, if the observer and the observed are the same then we need to revise all our principles and rules about observation and its epistemological status (as well as the role of consciousness) and this could lead to many challenges and complications.

We also repeat that the above questions and challenges (and their alike) may cast doubt on certain interpretative aspects or embarrass some schools of interpretation but they should not cast doubt or embarrass quantum mechanics (although they should highlight the fact that quantum mechanics is difficult or impossible to interpret). We also repeat our advice to avoid investing valuable time and effort in this trivial paradox and its alike.

9.1.5 Other Paradoxes

There are many other suggested (or can be suggested) paradoxes against quantum mechanics and its interpretations. In fact, we can synthesize a paradox (or rather paradoxes) from almost any problematic or perplexing aspect in the quantum theory (e.g. self-interference or wavefunction collapse or quantum tunneling or quantum entanglement). However, no one of these paradoxes can be a threat to quantum mechanics as an empirical theory and as a calculus for quantum physics (thanks to the undeniable empirical success of quantum mechanics). Yes, these paradoxes (assuming their validity) can (and actually do occasionally) threaten the schools of interpretation of quantum mechanics as well as some of its interpretative and epistemological aspects. However, this is not a big deal because no school of interpretation is safe from criticisms and challenges (regardless of these paradoxes) and hence these paradoxes may worsen the situation for these schools (by exposing their defects further) but they do not introduce a new effect (i.e. by changing the state of a school from being acceptable to questionable or non-acceptable). Moreover, even the basic interpretation can be subject to challenge and doubt from certain aspects without affecting the empirical validity of the theory. Nevertheless, all these paradoxes should highlight and endorse the fact that quantum mechanics is a difficult (and perhaps impossible) theory to interpret and make sense of, and therefore these paradoxes should support our proposal that quantum mechanics (despite its empirical excellence) is an epistemologically problematic and poor theory.

9.2 Incompatibility

Another potential challenge to the quantum theory is its incompatibility with other theories and facts (or alleged facts) of science. We can classify this incompatibility into two main types: clash with other theories/facts (i.e. by contradicting these theories/facts), and inconsistency with other theories (i.e. by not being able to mix or combine consistently and homogeneously with these theories to form new theories). These types will be investigated briefly in the following subsections.

---

[477] Or alternatively: we should have no possibility of collapse to a "dead" state.

9.2.1 Clash with Other Theories

Quantum theory may be challenged (at least epistemologically and interpretationally) by other scientific theories. The best known example of this kind of challenge is apparently its clash with the requirements (e.g. postulates, implications, consequences, ... etc.) of the relativity theories (and special relativity in particular) due to superluminality or non-locality or violation of relativity of time/simultaneity or requirement of global frame of reference or ... etc. Some of these clashes and conflicts have been explained or outlined or indicated earlier (see for instance § 1.8, § 5.6, § 6.10 and § 7.13), and hence we do not repeat.

However, as explained elsewhere (see for instance § 1.1 and refer to our book "The Mechanics of Lorentz Transformations") special relativity (as an interpretation to Lorentz mechanics) is logically-inconsistent and hence the quantum theory cannot be challenged by special relativity and its postulates or consequences. In fact, quantum mechanics could be used as evidence against special relativity (rather than being challenged and questioned by special relativity). For example, if quantum entanglement (which is supposed to violate locality and hence special relativity) is established and supported by experimental and observational evidence (as claimed and commonly accepted) then this could be an evidence against special relativity.

Similarly, general relativity is questionable from various aspects and perspectives and it lacks conclusive evidence in support of its validity (refer to our book "General Relativity Simplified & Assessed") and hence it cannot make a serious challenge or threat to quantum mechanics. Instead, quantum mechanics (which is overwhelmingly supported by experiment and observation) can be a challenge to general relativity. So in brief, any clash between quantum mechanics and the relativity theories should be decisively arbitrated on behalf of quantum mechanics. This should similarly apply to any clash or contradiction between quantum mechanics and other theories or propositions or principles whose validity is not established by a conclusive evidence. We should finally note that we are not aware of any clash or contradiction between quantum mechanics and any known and well-established fact of science (otherwise this could have been used as evidence against quantum mechanics).

9.2.2 Inconsistency with Other Theories

The known and prominent example is the supposed inconsistency with the gravity theory that prevents the merge into quantum gravity. However, if gravity is represented by general relativity then in our view we should favor quantum mechanics against general relativity (and possibly even put the blame in this failure on general relativity). Quantum mechanics (despite its epistemological weaknesses) is an empirically excellent theory and has passed countless tests (unlike general relativity which has problematic aspects and is not supported by conclusive evidence) and hence if we have to choose between the two theories then our choice should be quantum mechanics. So, the best option is to keep quantum mechanics and search for a replacement to general relativity (at least because of this failure and for the purpose of this merge). An even better option is to search for a new quantum theory which accommodates gravity (as well as being epistemologically and empirically superior to the current quantum theory). The best of all, of course, is to search for a new theory that considers all physical phenomena from the beginning and in one go (i.e. a theory that is genuinely a "theory of everything" rather than a "stitch-together theory"). Also see point 7 of § 7.3.

9.3 Incompleteness

This in our view is the main real challenge (or at least one of the main real challenges) to the current quantum theory. In brief, the incompleteness of quantum mechanics has two main aspects which no one can deny (regardless of whether this incompleteness is intrinsic to quantum physics and stems from the very nature of quantum reality or not, and regardless of being curable or not):
• **Empirical incompleteness** represented by indeterminism (mainly probabilisticity and uncertainty).
• **Epistemological incompleteness** represented by the absence of acceptable interpretation and the presence of many epistemological perplexities.

These issues have been investigated thoroughly throughout the book (see for instance § 7.6, § 7.15 and § 8.6) and hence we do not repeat.

So, in this section we investigate an argument (i.e. the EPR argument; see § 9.3.1) which historically played a central role in the debate and deliberation about the issue of incompleteness of quantum mechanics and the controversies about its interpretation and led subsequently to theoretical and experimental breakthroughs (represented mainly by Bell's theorem and Bell's experiments;[478] see § 9.3.2 and § 9.3.3) in the empirical and epistemological assessment of quantum theory. In fact, these breakthroughs opened the door to many novel areas and branches of research and development related to the quantum world and quantum physics (such as quantum information and quantum computing). They also raised the prospect of revolutionary technological advances some of which are (or were) considered to be on the verge of science fiction.

9.3.1 The EPR Argument

The EPR (i.e. Einstein-Podolsky-Rosen) argument can be a challenge to quantum mechanics itself (and this is the main reason for having this subsection here) and can be a challenge to some of its schools of interpretation. In fact, the EPR argument is certainly a challenge to quantum mechanics itself at least from the perspective of being complete (and possibly because of the EPR questioning of the uncertainty principle, as we will see for instance in point 4). This should be inline with the fact that we outlined in point 2 of § 8.2.

The literature about the nature of the EPR argument and its content and objective is rather conflicting from some aspects. In fact, the EPR argument has been understood and interpreted in the literature differently. For example, some people understood it as a challenge to quantum mechanics through a challenge to the uncertainty principle where a precise measurement of a complementary observable (e.g. position) on one of the entangled (or correlated) objects with a simultaneous precise measurement of the other complementary observable (i.e. momentum) on the other entangled object would violate the uncertainty principle. Other people understood it as a challenge to quantum mechanics through a challenge to the implied violation of locality (and possibly violation of causality) through interaction by superluminal signals. Other views and variants can also be found in the literature.

However, we can summarize the essence of this argument (according to the dominant view in the literature) by the claim that if quantum mechanics is correct and if the principle of locality is valid then quantum mechanics is an incomplete theory (in the sense that it ignores some aspects of physical reality). Accordingly, indeterminism (as represented mainly by probabilisticity and uncertainty) originates from this lack of completeness in the description of reality and therefore indeterminism can be lifted if this description is complete. In other words, an alternative theory (which we should look for) that includes all the relevant elements of reality in its description of the given phenomenon should be deterministic (as well as local).

In our view, this version of the EPR argument (which represents the dominant view in the literature) does not seem to be identical to the argument in the original EPR paper although it apparently reflects the essence of the original argument. The original argument of EPR (as presented in "A. Einstein, B. Podolsky, and N. Rosen Phys. Rev. 47, 777, 1935") includes a talk about certainty and hence it seems to target the uncertainty principle as well as incompleteness. Moreover, the argument is not presented and exemplified in the form and style in which this argument is usually presented in the recent literature. So, from a historical perspective the original EPR argument is not identical to the "modern EPR argument" although they are seemingly similar in essence and spirit.

Anyway, we have no primary interest in history (noting that documenting the history of science or quantum theory is not within the scope or objective of this book) and hence the readers who are interested in the original EPR argument are advised to refer to the literature (particularly the original paper of EPR which we cited already). However, for the sake of completeness (and to avoid disappointing some readers) we included a discussion to the original EPR argument in an appended section (see § 10.2) noting that

---

[478] We use "Bell's experiments" to mean experimental tests and verifications based on Bell's theorem and its quantum mechanical implications (such as the Aspect experiment; see § 9.3.3 and § 10.4).

9.3.1 The EPR Argument

the content of that appended section may not be very useful and may also be repetitive in part and hence it is not highly recommended for reading.

In the following points we discuss and assess the EPR argument (mainly from the perspective of its modern form):

1. Because the EPR argument is based on the assumption of locality, it can be easily rejected by those (like us) who do not accept (or at least they doubt) the principle of locality. So in brief, the EPR argument is a potential challenge only to those who believe in the validity of this principle.

2. As a local hidden-variable theory (or rather based implicitly on such a theory), the EPR argument is challenged (and should be refuted) by the Bell theorem (see § 9.3.2) which denies the possibility of replicating the predictions of quantum mechanics by any such theory (noting that refutation depends on accepting Bell's theorem).

3. From an experimental perspective, the EPR argument is (supposedly) refuted by the Bell experiments (see § 9.3.3) which demonstrate non-local effects (possibly through violation of Bell's inequality).

4. The EPR argument is (supposedly) about the completeness of quantum mechanics and not about its empirical validity (i.e. it disputes its completeness without claiming it is wrong experimentally or its statistical predictions are incorrect). Hence, it should be OK from this perspective (i.e. hypothetically). However, by considering the implications of the Bell theorem and Bell experiments (which we indicated already) it should be not (i.e. actually). Moreover, if the argument (in its original form) threatens the uncertainty principle then it should be a challenge to the formalism of the quantum theory (and possibly even its predictions), since this principle is part of its (current and commonly-recognized) formalism. In fact, threatening the uncertainty principle is not only about threatening indeterminism (by replacing indeterminism with determinism) but it can threaten some aspects of the current quantum physics (as represented by certain uncertainty-dependent formulations) which are based on the very concept and essence of uncertainty and which may (or should) lead to different predictions.[479]

5. Although we accept the incompleteness of quantum mechanics (according to our interpretation of "incompleteness" and for the sake of our own reasons and considerations as indicated in the preamble of § 9.3), we do not agree with most of the details and justifications about this incompleteness according to the EPR argument (even if we pretend to accept the EPR argument in principle). So in brief, we do not only disagree with the rationale of this argument (at least because of its presumption of locality), but we also disagree with (at least some of) its content and conclusions.

6. The EPR argument may also be classified in the literature as a paradox (considering some of its alleged consequences and implications) and hence it should belong to § 9.1 (but we did not include it there due to structural and non-structural reasons). This is inline with the fact that the majority of physicists reject the hidden-variable theories (especially the local ones) and their implication of incompleteness and hence the EPR argument can be classified aptly as a paradox according to their view.

7. From a historical perspective, the association of this argument with Einstein gave it an (unjustified) extra weight and significance and hence it dominated the following debates and conversations and oriented most of the research and deliberations about (almost all) the issues related to the interpretation of quantum mechanics towards certain (and sometimes disorientated) directions.

From another historical perspective, although the credit of this argument is attributed to Einstein and his collaborators, the idea of this argument (excluding some elaborations by EPR) can be traced back to an earlier date in the chronology of development of quantum mechanics (specifically to the time of the proposal of the probabilistic interpretation of quantum mechanics which is originated by Born and supported by Bohr and Heisenberg). We should also remember that the original EPR argument is not identical to the subsequent EPR argument (or rather arguments) which is generally more tidy and strong and hence the credit should not be attributed (at least in its entirety) to EPR.

---

[479] In fact, the rationale of the Bell argument should suggest that indeterminism in itself may have real physical consequences beyond what is expected from it as a form of ignorance or lack of definiteness.

9.3.2 Bell's Theorem

According to this theorem, it is impossible for local hidden-variable (LHV) theories to reproduce all the predictions of quantum mechanics, and hence no local hidden-variable interpretation of quantum mechanics is possible.[480] The essence of Bell's theorem is summarized or embedded in an inequality (called Bell's inequality) which is a classical-like relation (i.e. derived by an argument based partially on a local hidden-variable rationale that applies to physical systems behaving in a classical-like way), and hence the violation of this inequality refutes this rationale.[481] Accordingly, the violation can be taken as evidence against local hidden-variable theories.

In fact, the primary purpose of Bell's theorem was to address the EPR argument (see § 9.3.1) and its alike by proposing a mathematical relation (in the form of an inequality) that can be used as a basis for a decisive experimental test that demonstrates if quantum mechanics is a complete theory (as the proponents of quantum theory claim) or not (as the EPR argument and its alike claim). Accordingly, the violation of Bell's inequality supposedly implies that quantum mechanics is a complete theory (i.e. it does not miss or require hidden or additional variables that represent missing "elements of reality"), although honoring this inequality does not imply that quantum mechanics is not a complete theory. In other (and more precise) words, the violation of this inequality implies that the correlated or "entangled" system[482] (i.e. in the EPR argument and its alike) does not behave in an LHV way, and hence a system that behaves in an LHV way (i.e. it behaves "classically") must not violate this inequality (although non-violation in itself does not imply an LHV behavior or non-"LHV" behavior specifically since it is compatible with both).[483]

It is difficult to understand and appreciate Bell's theorem and inequality without going through the (messy and complicated) details of this theorem and its proof (or rather proofs and arguments). However, this is beyond the plan and scope of the book and above the average level of its intended readers. So, as a compromise between these conflicting factors we collected the technically-demanding and potentially-excessive or redundant material about Bell's theorem and put it in an appended section (see § 10.3). The interested reader can (and in fact is invited to) study this appended section if he wishes, while the non-interested reader can ignore it with no major loss or disturbance.

In fact, the difficulty of understanding and appreciating (as well as assessing) Bell's theorem and inequality is exacerbated by the fact that the subject of Bell's theorem is extensive and rich and hence we can find various forms of inequality and diverse types of proofs and arguments. So, if we have to understand and assess this subject properly we need to understand and assess (or at least be aware of) many of these forms and types (or at least a sufficient number of them). This, obviously, cannot be achieved in a section or chapter or appendix. So, as another compromise we put a sample of these proofs and arguments (some of which will lead to different forms of inequality) in our appendix.

In fact, our sample consists of two proofs (see § 10.3.1 and § 10.3.2) and two arguments (see § 10.3.3 and § 10.3.4) which we hope to be sufficient (for those who read them properly) to understand and appreciate

---

[480] According to Bell (see "J.S. Bell, Physics, 1(3), 195, 1964") a realistic theory that provides the statistical predictions of quantum theory must be non-local. In fact, Bell regards locality as the main reason behind the incompatibility between the statistical predictions of quantum mechanics and those of hidden-variable theories.

[481] To derive his inequality, Bell started by considering a quantum system made of a pair of spin-1/2 correlated particles where he assumed the existence of hidden variable(s) which determines the spin of the pair (i.e. whether up or down). He then showed that if we also assume locality then we should get an inequality relation (which is what we call Bell's inequality) that restricts the statistical predictions of this "local hidden-variable theory".

[482] In the context of discussing Bell's theorem (as well as similar subjects like the EPR argument) it is more appropriate to use "correlated" since entanglement (which suggests a quantum mechanical type of correlation) is the subject of investigation and is not confirmed since according to the LHV theories the objects in the system are not really entangled in a quantum mechanical sense but they are correlated (i.e. their properties are correlated).

[483] From the perspective of *logic*, Bell's relation is an inequality derived on the basis of LHV assumption and hence its violation should refute LHV and thus (supposedly) endorse quantum mechanics (i.e. in its non-"LHV" commonly-accepted basic or standard interpretation). However, since quantum mechanics (in itself and with disregard to the Bell theorem and inequality) is (or can be) compatible in principle with both LHV and non-"LHV", honoring the inequality refutes or endorses neither and hence both LHV and quantum mechanics (i.e. in its non-"LHV" standard interpretation) are possible in this case. The reader should, however, anticipate more clarification and examination to this *logic* in the upcoming points.

9.3.2 Bell's Theorem

this subject (as well as understanding and appreciating our assessment which will mostly be presented in the following remarks). However, the readers should note that some of our assessment and judgments are based on our general background knowledge and awareness, and hence they cannot be traced to what we provided in this book (and thus they could be difficult to understand and appreciate).

In the following points we present some important remarks about Bell's theorem and inequality:

1. Bell's theorem is (arguably) the most important development in the history of quantum theory since its "completion" in the 1930s. This is due for instance to:
   - Its extremely important scientific, interpretative, epistemological and philosophical implications especially with regard to hidden-variable theories and interpretations and the issues of realism and non-locality (which impact the relativity theories in particular).[484]
   - Its demonstration of the possibility of experimental refutation (and possibly even confirmation indirectly) of interpretations of scientific theories (and quantum mechanics in particular) and interpretative aspects that are commonly seen as (and believed to be) of purely philosophical and contemplative nature and beyond the reach of scientific test and experimental verification.
   - Its motivation, inspiration and direction of new types of quantum investigations and applications such as quantum information and technology.

2. Despite the undeniable merit of the Bell argument and theorem, the theorem is not as tight and firm as it might be depicted by some. In our view, it can be questioned and challenged through some of the implicit assumptions and suppositions (at least) in some of its proofs and arguments (see for instance point 2 of appended § 10.3.1). In fact, we can find a number of challenges to Bell's theorem in the literature. So, our overall judgment is that although the theorem is well established in general and the arguments and proofs in its support are sufficiently strong, it is not the final say or the last word in the debate about the validity of local hidden-variable theories and interpretations.

   We should also note that there are various aspects in Bell's theorem that require closer inspection and assessment. We feel that these aspects (or at least some of them) do not seem to be given sufficient attention and consideration in the literature. For example, Bell's theorem is based on a totally realistic and deterministic model for the hidden variables (which is inline with the EPR argument that was the main target of Bell) and hence certain restrictions may need to be imposed on the type of local hidden-variable theories and interpretations that can be eliminated by this theorem.[485] In fact, some (potentially) loose generalizations and vague aspects in the theorem and its implications require clarification and interrogation. For example, we feel there is a level of ambiguity (or lack of sufficient clarity) about the nature and role of the hidden variable(s) in the theorem and how it is modeled and embedded in the formalism (by referring to it rather ambiguously by just a symbol λ with no detail about its nature or role).

3. As there are some uncertainties about Bell's theorem as well as Bell's experiments (see for instance point 2 and § 9.3.3) some of which we cannot reach a conclusion about (because for instance they are related to experimental details which we cannot access or to unclear assumptions), we cannot reach a definite conclusion about the significance of Bell's theorem (and Bell's experiments) for disproving local hidden-variable theories definitely and unequivocally or endorsing quantum mechanics (i.e. in its non-"LHV" interpretation) specifically (through some specific results of Bell's experiments) although we should admit that Bell's theorem and experiments pose a serious (and possibly decisive) challenge to the local hidden-variable theories and interpretations (or at least some types of them). Accordingly, this issue needs further investigation and verification, and till then we should not rule out the possibility of local hidden-variable theories and interpretations completely, i.e. we continue treating them as potential (though not strong) candidates for replacing or interpreting quantum mechanics until we get a definite proof that they cannot be so.

---

[484] In fact, because of its impact on the relativity theories (as well as other important impacts) it should be regarded as one of the most important developments in modern physics in the last decades.

[485] We note that some of these possibilities (e.g. local non-deterministic hidden variable theories) are discussed in the literature where they are usually classified as being subject to Bell's theorem (and hence they are supposedly refuted by it). However, the state of these possibilities (as well as many other possibilities) is not clear cut and hence further investigation is required to determine their state conclusively (hopefully). We should also note that some of these issues could depend on the type of proof or argument used to establish Bell's theorem.

9.3.2 Bell's Theorem

4. We can classify the existing hidden-variable theories into two main types (noting that other classifications and types do exist): non-local hidden-variable theories (which supposedly replicates the predictions of quantum mechanics such as Bohm's theory), and local hidden-variable theories (which contradict the predictions of quantum mechanics according to Bell's theorem and related experimental tests). It should be obvious that non-local hidden-variable theories (like Bohm's theory) are not included in the Bell argument and inequality and hence they cannot be revoked by Bell's theorem and experiments. Accordingly, those who do not believe in the principle of locality (like us) should not be affected much by the Bell's theorem and experiments (i.e. from the perspective of hidden-variable). Also, some misleading generalizations in the literature should be understood in this context.

5. Bell's inequality depends on a number of factors (e.g. the type of the quantum system under consideration) and hence there are various forms of this inequality.[486] So, what we actually have is "Bell's inequalities" (or Bell-type inequalities which take similar forms and are based on similar rationales to that of the original Bell inequality). We also note that there are several (and possibly many) proofs and supporting arguments to Bell's theorem and inequality(s).[487] So in brief, Bell's inequality and proof come in various shapes and forms and have different flavors and justifications.

   Accordingly, the significance of Bell's theorem and its implications may vary according to the nature of the inequality and proof used. Also, some (potential) criticisms and challenges to the Bell theorem may depend on the particular variant of the inequality and its particular proof (and this should be an element of strength in favor of Bell's theorem since it can escape certain variant-dependent criticisms by switching to other variants). This sort of uncertainty (which is based on this diversity) should also be projected on Bell's experiments and hence the significance and conclusivity of some experiments may depend on the form and proof of the inequality used in the construction and analysis of that experiment (and this could be an element of weakness for Bell's theorem and its experimental tests since any challenge to that form and proof of the inequality could be a challenge to the experimental results based on that form). Also see point 3 of § 9.3.3.

6. Most of the systems considered in the literature (both theoretically and experimentally) as instances for the Bell argument and inequality are spin-1/2 particles (like electrons) or polarized photons. However, the argument and inequality are not restricted to these cases but they can be extended to other types of correlated systems. In fact, even the states (of the correlated or "entangled" systems used) are not required to be binary or dual (e.g. spin-up or spin-down) and hence the argument and inequality (and the experiments which are based on them) can be extended to multi-state (non-binary) systems.[488]

7. As indicated earlier and will be investigated further later on, many experimental investigations (of different types and levels of sophistication and elaboration) have been performed (since the appearance of Bell's theorem in 1964 until these days) to test Bell-type inequalities for various types of "entangled" quantum systems. The results of all (or almost all) these experiments show a violation of these inequalities and hence they generally endorse (with different levels of certainty) quantum mechanics against local hidden-variable theories (and against the EPR argument in particular).[489] In fact, experimental testing of Bell-type inequalities became in the recent years a standard and familiar educational practice in undergraduate labs.

8. As indicated earlier, the significance of the Bell theorem and inequality is that they provide a theoretical basis for experimental tests of local hidden-variable theories and hence these theories can be tested and possibly refuted scientifically and objectively instead of assessing and judging them philosophically and subjectively (e.g. by arguments based on thought experiments).[490] So in brief, Bell's theorem

---

[486] In fact, Bell's argument and inequality have been modified and expanded (since their appearance) time after time.

[487] As indicated earlier, a sample of these proofs and arguments are given in appended § 10.3.

[488] In this context we should mention the generalization and extension of Clauser, Horne, Shimony, and Holt (commonly abbreviated as CHSH and may be called Bell-CHSH inequality) which is widely used in the formulation of the Bell-type arguments and related experiments.

[489] Despite their general acceptance, Bell's theorem and inequality (as well as their implications) have been challenged and criticized. Moreover, the experimental tests related to Bell's theorem and inequality have also faced challenges and questioning and their value for proving the invalidity of local hidden-variable theories has been questioned.

[490] As we will see later (refer for instance to point 3 of § 10.3.1), specific experimental tests for the Bell inequality may not be necessary (at least in some cases) for refuting the LHV theories, since theoretical comparisons (in association with

9.3.2 Bell's Theorem

allows experimental tests of local hidden-variable theories and hence it refuted the previous belief that these theories can be accepted or rejected according to the philosophical and epistemological choice and preference and cannot be tested experimentally. In fact, Bell's theorem and inequality should also demonstrate that interpretations of scientific theories can be tested experimentally.

9. Bell's theorem and inequality are essentially related to the interpretation of quantum mechanics.[491] Yes, if we assume that Bell's inequality (associated with Bell's experiments) can falsify quantum mechanics (which we dispute at least in some cases as explained in footnote [490]) then Bell's theorem and inequality can be related to the formalism as well and can (in principle) pose a challenge to it.

10. It is important to analyze the implications of Bell's theorem and inequality correctly to avoid confusion and misunderstanding. For instance, we find in the literature claims like this: *violation of Bell's inequality refutes local hidden-variable theories and endorses quantum mechanics, while non-violation proves or refutes neither*.[492] Although such a claim is correct in general it could be misleading (if not associated with proper explanations and clarifications) from various aspects. For example:

    • This could be misleading because local hidden-variable theories are essentially interpretations to quantum mechanics and are supposed to explain it and hence (strictly speaking) they cannot be contrasted in validity with quantum mechanics (as this claim suggests).[493] So instead of this we may have to say something like this: *violation of Bell's inequality refutes local hidden-variable interpretations while non-violation means that these interpretations are possible*.

    • In the case of violation, there is no endorsement to any non-"local hidden-variable" interpretations. For example, these non-"local hidden-variable" interpretations could be as invalid as local hidden-variable interpretations due for instance to the non-existence of (eligible or valid) interpretation (whether local hidden-variable or otherwise).[494] It should be obvious that non-"local hidden-variable" interpretations include non-local hidden-variable type as well as other types (which could be totally different from the hidden-variable type).

    • In the case of non-violation, these local hidden-variable interpretations (as well as those which are not of this type) are possible in principle but this should not guarantee the existence (let alone the endorsement) of any interpretation of any type (whether local hidden-variable or else).

    So in brief, the only valid implication is the invalidity of local hidden-variable interpretations in the case of violation. However, we should note that some details in this point are largely based on the presumed (unconditional) validity of quantum mechanics (see for instance footnote [490]). Also see point 12.

11. There is some exaggeration in the literature about the value of Bell's theorem and its implications (as well as the value and implications of the experiments related to it) with regard to the endorsement of quantum mechanics against theories and interpretations that do not follow its common formalism and interpretation. As we said in point 10 the only (supposedly) valid implication in this regard is that local hidden-variable interpretations of quantum mechanics are invalid (i.e. if the inequality is violated).[495] So, even if we ignore possible challenges to this theorem (see point 2) and accept the theorem and its proofs and arguments (and even accept the claimed experimental endorsement of quantum mechanics by experiments based on this theorem; see § 9.3.3 and § 10.4) we cannot accept some explicit and implicit assumptions that can be found in the literature. For example:[496]

    • It does not rule out the possibility of alternative (novel) quantum theory that can replace quantum

---

the already verified predictions of quantum mechanics) can be sufficient for achieving this objective. Also see point 12.

[491] We should also remember (as explained elsewhere) that they also provide motivation and guidance to novel areas of research and application, and hence their value exceeds their role in interpretation (and regardless of their potential role in the formalism which we will discuss next).

[492] In fact, we made (for simplicity) claims similar to this earlier and later.

[493] In fact, we are implicitly assuming the unconditional validity of quantum mechanics (and hence any valid LHV theory even if it is based on different formalism should replicate the predictions of quantum mechanics).

[494] Yes, the theorem should imply (in the case of violation) that if an interpretation of quantum mechanics does exist then it should be non-"local hidden-variable".

[495] In fact, we are assuming the validity of quantum mechanics; otherwise the only valid conclusion is that local hidden-variable theories cannot be a valid interpretation to quantum mechanics.

[496] We remind the reader that some of the implications of Bell's theorem and inequality may depend on the argument and proof used to prove the theorem and derive the inequality.

9.3.2 Bell's Theorem

mechanics altogether. In other words, it does not finalize the say about the physics of the quantum world and has no capability to issue a judgment about the merit of quantum mechanics as an ultimate quantum theory (even if we do not believe in the non-uniqueness of science). For instance, a completely deterministic quantum theory (that agrees with the statistical predictions of quantum mechanics) remains a possibility from this perspective.

- It does not endorse (i.e. necessarily) quantum mechanics, neither in its formalism nor in its interpretation (let alone endorsing any specific school of non-"local hidden-variable" interpretations).[497] In fact, it does not even endorse (i.e. necessarily) a non-"local hidden-variable" interpretation as such and in its generality. Also see point 12.
- In our view, it cannot rule out (i.e. categorically and entirely) even the possibility of local hidden-variable quantum theories at least because of the uncertainties about Bell's theorem and experiments and their implications (see for instance points 2 and 3). For example, if these theories are totally novel and they are of a specific type (e.g. local non-deterministic hidden variable theories) then they may not meet the requirements and conditions (explicit or implicit) of Bell's theorem (or at least some of its proofs and arguments as well as experiments) and hence they may escape the challenge of Bell.

12. Some types of proofs and arguments in support of Bell's theorem do not (or rather they do not need to) refer to quantum mechanics or make comparison with it (and some may not even produce a specific mathematical relation or inequality).[498] Regarding this type of proofs and arguments we note the following:

- Some of these proofs (i.e. those which do not produce a specific mathematical relation) may have only theoretical value and hence they cannot be tested experimentally. Yes, the ones that produce a specific relation can be tested. In fact, experimental test may be a necessity for some of the latter type (i.e. for establishing Bell's theorem) since it does not refer to quantum mechanics and hence it may not be able to rely on the existing experimental support of quantum mechanics (see footnote [490] on page 218 as well as point 3 of § 10.3.1).
- This type of proofs cannot endorse quantum mechanics (since it does not refer to quantum mechanics) and hence its value (assuming its validity) should be restricted to ruling out LHV theories (as potential interpretations or absolutely; also see points 10 and 11). So, from this perspective we may classify the proofs and arguments of Bell's theorem to those which only rule out LHV and those which also have the potential of endorsing quantum mechanics in some way (either because they can directly appeal to the verified predictions of quantum mechanics or because they can be used as a basis for Bell's experiments).

13. We note that some of the proofs and arguments of Bell's theorem seem to be based on an implicit assumption of the validity of quantum mechanics (in its verified statistical predictions) as the only possible quantum theory, i.e. the proof is basically based on the uniqueness of science.[499] Hence, if we doubt the uniqueness of science or embrace the non-uniqueness of science (see § 2.4.3) then we may be able (from the perspective of these proofs) to question Bell's theorem and its implications.

Finally, regardless of the validity and invalidity of the Bell argument and theorem and the rigor of its proofs from a technical perspective, we may claim that the content of this theorem (i.e. local hidden-variable

---

[497] Yes, the violation of the Bell inequality in a certain manner that agrees with the predictions of quantum mechanics should endorse quantum mechanics (i.e. the formalism in its basic interpretation) like any other quantum mechanical experiment that follows the rules and predictions of quantum mechanics. It is worth noting that because Bell's relation is an inequality (not equality) its violation, as well as its non-violation, in itself cannot prove a specific thing. Yes, violation and non-violation in a certain pattern and a specific manner may prove something if it agrees with the specific predictions of a theory. We should also note that some arguments do not provide specific predictions (i.e. they do not produce specific mathematical relations) and hence all they can do is to show that the predictions of quantum mechanics and those of local hidden-variable theories are not identical.

[498] See for example our argument in § 10.3.3 (noting that the references to quantum mechanics in that argument are not essential to the argument). This is because the essence of our argument is that LHV theories do not necessarily provide identical predictions to the predictions of non-"LHV". The impossibility proof of von Neumann against hidden variables (see point 3 of § 8.2) may also be of this type (regardless of the correctness of the proof and some of its details and regardless of locality as well as its obvious independence of Bell's theorem).

[499] In fact, this could be inherited from the EPR argument itself which also seems to be based (implicitly) on the assumption of uniqueness of science (see point 3 of § 10.2.3).

theories *cannot* reproduce all the predictions of quantum mechanics, i.e. in its standard interpretation) is rather intuitive.[500] In this regard, we refer the reader to the appended subsection § 10.3.3 for a qualitative argument that we propose in support of the Bell theorem (although we may need to replace "*cannot*" by "do not necessarily" because our argument in support of Bell's theorem may not be strong enough to establish the strong "*cannot*" conclusion). We should also note that even if quantum mechanics is complete (in the proposed quantum mechanical sense that negates hidden-variable theories) there is still a possibility (by the principle of non-uniqueness of science) of the existence of a more complete theory (as explained earlier).

9.3.3 Bell's Experiments

As indicated earlier, we use "Bell's experiments" to mean experimental tests and verifications based on Bell's theorem and its quantum mechanical implications. The best known example of these experiments is the Aspect experiment (see § 10.4) which is widely regarded as the pioneering experiment in this field. In fact, the first experimental attempts to use Bell's theorem to test the predictions of quantum mechanics against the predictions of local hidden-variable theories dates back to the early 1970s.[501] However, the most serious, elaborate and technically sound of these early attempts was conducted in 1982 by Alain Aspect and his team. Since then many other experiments of this type have been designed and conducted where subsequent experiments usually try to improve their predecessors or confirm their results or address some of their limitations and gaps or ... etc.

The results of almost all these experiments are claimed to endorse (with various degrees of certainty and clarity) the predictions of quantum mechanics (in its basic interpretation) against the predictions of local hidden-variable theories, and hence they are generally regarded as evidence against local hidden-variable theories and as justification for their rejection. However, there are many controversies about these experiments and their significance especially their ability to determine the fate of local hidden-variable theories decisively and definitely and hence declare (allegedly) the victory of non-"local hidden-variable" interpretations against the local hidden-variable interpretations.

Anyway, in our view Bell's experiments generally have limitations some of which can pose a serious challenge to this type of experiments and their value and significance. Although subsequent (improved) formulations and experiments generally address (or attempt to address) the limitations in their predecessors (where the majority of the results are claimed to endorse quantum mechanics against local hidden-variable theories), there are still question marks and disputes about certain aspects of this work (i.e. the theoretical and experimental work that is based on the Bell theorem and whose objective is to test the validity of local hidden-variable theories).

At this stage, no one can claim that all the problematic aspects are addressed and resolved decisively and undisputedly, and hence further work is still required to justify the claimed victory. In fact, in the early stage of this investigation it looked as if this work resulted in a decisive and generally-accepted conclusion in favor of quantum mechanics and against local hidden-variable theories. However, in the more recent years there seems to emerge more cautious approach and wary attitude about the decisive nature of this conclusion (as more objections and suspicions, theoretical as well as experimental, were raised about the validity and conclusivity of the results of these experiments). Hence, currently the "decisive conclusion" that have been reached in the early stage of this work does not look as decisive as it was thought previously.

However, for the sake of fairness we should also note that some of these objections and suspicions are of interpretative nature (where they rest on school-dependent views and analyses) and hence their value (for providing a solid ground for a decisive judgment and conclusion) is not very great. We may also question some of these objections and suspicions for other reasons such as being counter-intuitive or being very unlikely and so on. Nevertheless, they are generally part of the overall "quantum mechanical mess" and hence they cannot be ignored or discriminated against since they do not differ much from the rest.

---

[500] In fact, what we claim to be intuitive is not exactly the statement of Bell's theorem (which we summarized above) but rather the proposition that LHV theories/interpretations may not be able to reproduce all the predictions of non-"LHV" theories/interpretations even if they are seemingly based on (or belong to) the same (or similar) formalism (with no specific reference to quantum mechanics).

[501] See for instance "S.J. Freedman and J.F. Clauser, Phys. Rev. Lett. 28(14), 938, 1972".

9.3.3 Bell's Experiments

So, our overall conclusion (considering all these messy issues and controversial details and the ambiguity about some aspects of this work noting, for instance, that certain aspects of some experiments are accessible only to the experimenters and hence they are totally based on trust) is that Bell's experiments generally support or favor quantum mechanics (in its basic interpretation and statistical predictions) against local hidden-variable theories although this result should be kept under scrutiny and revision in the future (since opposite indications or causes for suspicion and concern could emerge in the future and hence demolish or diminish this result).

In the following remarks we highlight some issues about Bell's experiments:

1. As indicated earlier, Bell's experiments were criticized and challenged by various general arguments as well as specific question marks and for various reasons such as the impracticality of such tests or by having experimental loopholes (even if these tests are assumed to be theoretically sound which may also be controverted).

   • An example of these criticisms is the so-called "detection efficiency loophole" which puts a question mark on the significance of the results of these experiments when some of the correlated pairs (i.e. the supposedly entangled objects) are not detected (due to limited efficiency of photo detectors for instance).[502]

   • Another example is related to locality and if the techniques and timing constraints imposed by the experimental setting and procedure were sufficient to test and establish the violation of locality.

   • A third example is about questioning the randomization process of measurements (which is used to avoid potential local communication), e.g. questioning the random nature of the shifting of switches which changes the orientation of the polarizers in the Aspect experiment.

   However, some of the criticisms and challenges to Bell's experiments (e.g. some of the alleged sources of impracticality and loopholes) are rather illusory and counter-intuitive and hence they do not represent serious challenge.

   Anyway, even if the individual Bell experiments may not be conclusive on their own (due to these criticisms and challenges), the totality of these experiments could be conclusive (or almost), i.e. they endorse each other and hence reduce the possibility of being fortuitous coincidences (especially if we note their abundance and diversity).

2. Bell's experiments may also need to address the issue of contextuality where (for instance) the choice of the spin direction in the measurement of one correlated object should determine the spin direction (and hence the spin eigenstate) of the other (correlated) object. In fact, non-locality (if allowed as we do) may be able to explain contextuality (at least in this sort of experiments) because if we accept the existence of non-local interactions then a measurement on a given correlated object (as part of an entangled system) is also an actual measurement on the other correlated object in the system. For example, if A and B are a pair of entangled electrons then when we measure the spin component of A in the z direction then (by the supposed non-local interaction) we actually and concurrently also measure the spin component of B in the z direction, and hence we do not need (at least in this sort of experiments) more than non-locality to explain contextuality. However, we should admit that in some cases non-locality in itself may not be able to explain contextuality and hence we may be forced to accept contextuality as well as non-locality (which could be seen as an extra burden and additional cost).

3. As indicated in point 5 of § 9.3.2, the significance and conclusivity of some of Bell's experiments may (and generally do) depend on the form and proof of the inequality used in the construction and analysis of that experiment.

Finally, it is noteworthy that the diversity of Bell's experiments (which originates partly from the diversity of the Bell inequality and its rationale and proof) is remarkable and reflects the ingenuity and sophistication of (some) theoretical and experimental quantum physicists.[503] Some examples of this diversity are:

• Use of different entangled quantum systems such as spin-1/2 particles and polarized photons.

---

[502] As discussed elsewhere (see for instance point 4 of § 10.3.1) the importance and criticality of detection efficiency generally depend on the proof used to establish Bell's theorem and derive Bell's inequality.

[503] For fairness, we should admit that this is one of the bright facets of modern physics (which we usually criticize for other reasons and from different aspects).

9.3.3 Bell's Experiments

- Use of multi-objects entangled quantum systems which consist of more than two entangled objects.
- Use of various innovative methods for randomizing the selection of measurement (e.g. how to choose the direction of spin or polarization).

Also refer to the appended section § 10.4.

Chapter 10
Appendix

In this appendix we accumulate some technically-demanding topics and parts as well as some non-essential material. This is to ease the reading and reduce complication and confusion while studying the main parts of the book. It should also enable some readers who have no sufficient background or interest to avoid going through details which they cannot digest or do not prioritize.

10.1 Slit-Grid Experiment

The essence of the slit-grid experiment can be summarized (and simplified) in the following points (or rather sequential steps) where the reader is referred to Figure 1 for illustration of the basic setting:

1. Let have a standard double-slit experiment (using a light source with both slits being open) where we observe a diffraction pattern on a screen S. This should show the wave nature of light (or photons).
2. Remove S and put a lens L just behind its position. An image of the two slits will appear on a display screen D behind the lens. This should show the particle nature of photons (since the photons at D are making localized images of the slits rather than an interference pattern).
3. Close one of the slits and you will see the image of the other slit unaffected (i.e. in shape and intensity). This confirms the particle nature of photons (since the photon "particles" that contribute to the image of each slit are different, i.e. blocking the photon particles that make the image of the closed slit will not affect the photon particles that make the image of the open slit).[504]
4. Put a grid G (of wires) at the minimums of the interference pattern that we saw earlier at the location of S (refer to point 1). The image of the slit will change (i.e. in shape and intensity) which indicates that there is no interference pattern (since the grid will cause diffraction and block some of the photons and hence change the image of the open slit).
5. Open the closed slit and you will see that the image of the open slit in point 3 is recovered (in shape and intensity) which means that the interference pattern is there and hence the grid does not diffract or block any photons (since the grid is at the minimums where no photons pass through).[505]
6. This means that we have (simultaneously) an interference pattern of waves (since the grid did not affect the image in the last step) and the photons behave like particles (as implied by the images of the slits). This simultaneous "observation" of particle and wave behaviors (supposedly) represents a violation of the (alleged) principle of complementarity.

10.2 The Original EPR Argument

We use in this investigation the following paper "A. Einstein, B. Podolsky, and N. Rosen, Phys. Rev. 47, 777, 1935" in which the original EPR argument is published and to which we refer as "the EPR paper". In fact, this paper is badly written, technically poor and even contains a number of technical mistakes and confusions which reflect lack of rigor or/and poor understanding of quantum mechanics (see for instance equations 5-8 in the EPR paper and the explanations related to them).[506] We should note that our investigation in this appended section is not thorough, i.e. it represents a sample of what we observe from reading the EPR paper. We also note that in this section we usually use the quotation marks " " for quoting from the EPR paper with no explicit reference to it. In the following subsections we will try to

---

[504] This supposedly indicates and demonstrates the "which way" aspect.
[505] In fact, we should expect some diffraction due to the existence of the grid (even though it is at the minimums) although this diffraction can be minimized and could possibly be eliminated-practically.
[506] This could be a reason for the different understandings of this paper and the variation of the interpretations of the original EPR argument.

10.2.1 The Essence of the Original EPR Argument

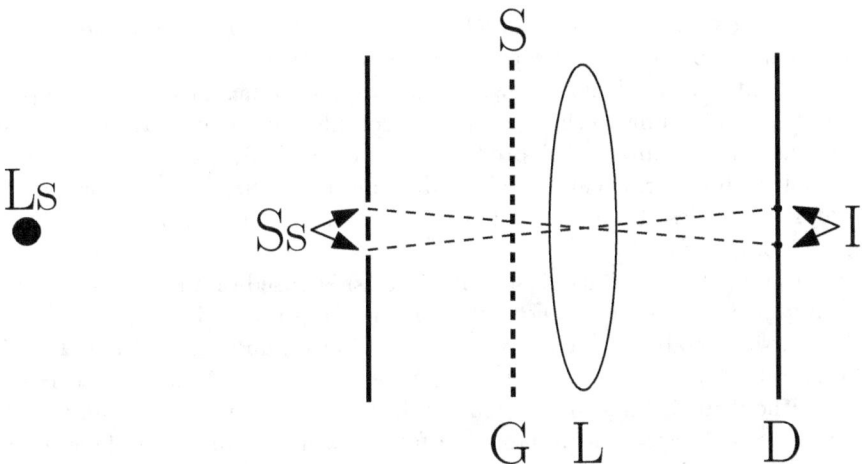

Figure 1: Schematic illustration of the basic setting of the slit-grid experiment where Ls stands for light source, Ss for slits, G for grid, S for the screen at the location of grid, L for lens, D for display screen, and I for images of slits (or pinholes). The dimensions are obviously not proportionate or to scale. Refer to § 10.1 for details.

analyze and assess further the (original) EPR argument and its consequences. However, before that we should note that the EPR argument rests on two main postulates or assumptions:
• **Postulate of completeness**: "*every element of the physical reality must have a counterpart in the physical theory*".
• **Postulate of physical reality**: "*If, without in any way disturbing a system, we can predict with certainty (i.e., with probability equal to unity) the value of a physical quantity, then there exists an element of physical reality corresponding to this physical quantity*".

10.2.1 The Essence of the Original EPR Argument

The primary purpose of the original EPR argument is to claim that quantum mechanics (represented by the wavefunction) is not complete since the wavefunction (in its description of reality) does not contain all the elements of reality. However, the argument seems to have another objective that is smearing the uncertainty principle.

The essence of the original EPR argument is summarized in the following bullet points:
• "A sufficient condition for the reality of a physical quantity is the possibility of predicting it with certainty, without disturbing the system" (see the postulate of physical reality).
• The lack of simultaneous knowledge of incompatible observable quantities (related to the uncertainty principle) is because either (1) "the description of reality given by the wave function in quantum mechanics is not complete" or (2) "these two quantities cannot have simultaneous reality".
• Considering the case of correlated (or "entangled") systems, 2 (in the previous point) is false, therefore 1 must be true (because the negation of one of two exclusive and disjoint alternatives necessitates the confirmation of the other alternative).[507]
• Hence, we conclude that the quantum mechanical description of reality is not complete.
To sum up, our inability to acquire knowledge about a pair of incompatible observables simultaneously is because the description of reality (as given by wavefunction) is not complete.

To be more clear, let have (as an example) two correlated and separated particles A and B which are supposedly not interacting. If we measure (say) the momentum of A precisely then we know the momentum of B precisely (without disturbing B) and hence the momentum of B represents an element of physical

---
[507] In fact, the presentation of the essence of this point is distorted in the EPR paper and hence for the sake of clarity we stated it in a rather different (and more direct) way.

reality (according to the postulate of physical reality). Now, if we also measure the position of A precisely then by the same token the position of B represents an element of physical reality.[508] Accordingly, we have two elements of reality for particle B. Now, if quantum mechanics provides a complete description to the physical reality, then (according to the postulate of completeness) the wavefunction should be able to provide us with (precise and simultaneous) predictions of incompatible observables (i.e. momentum and position in our example). However, because such predictions are not possible (according to the uncertainty principle of quantum mechanics) then "the quantum-mechanical description of physical reality given by wave functions is not complete".

In fact, the entire argument (as well as most of its details) is based on many arguable and challengeable claims, propositions, formulations, etc. For example, the argument is based on an entirely classical rationale (as well as extreme realism which will be discussed next) and hence it may be rejected altogether on the basis that quantum physics do not follow this classical rationale. Also, EPR seem (e.g. by noting the use of phrases like "without disturbing the system") to justify the uncertainty principle by the disturbance factor (rather than by an intrinsic and fundamental factor) which is generally disputed and rejected (see for instance point 1 of § 6.6.1). Accordingly, the argument is not worth of detailed investigation or assessment (and hence we ignore it). Our advice to those who are interested in this argument is to read the EPR paper carefully and with open and skeptical mind to find for themselves the shabby nature of this argument.

Interestingly, Bohr in his reply to the EPR paper ignored (almost totally) discussing and criticizing the EPR argument technically.[509] It seems as if Bohr felt that this argument is not "serious" and hence it is not worthwhile to discuss and criticize in a "serious way"; instead he just kept stating and citing the position of quantum mechanics (mainly from the Copenhagen viewpoint) so that EPR understand it properly.[510]

Another interesting point is that this tattered (original) EPR argument is significantly different from the modern time "EPR argument" (or arguments which look more tidy and rational) whose credit is given (unduly) to EPR. In fact, EPR (and Einstein in particular) are commonly given the credit for the entirety of this subject (and some even consider the subject of quantum entanglement as the brainchild of Einstein) despite the poor nature of this contribution and the fact that the root of this argument originates from the early days of quantum mechanics and its probabilistic interpretation.[511]

10.2.2 Realism

It is obvious that EPR adopt an extreme form of realism,[512] as can be felt, for example, from the following quotes:
- "In a complete theory there is an element corresponding to each element of reality".
- "every element of the physical reality must have a counterpart in the physical theory".

In fact, the EPR argument represents a view commonly described as "local realism", i.e. it respects the principle of locality and it sees science as an exact reflection of definite reality (noting the "element of reality").[513] In contrast, quantum mechanics may be commonly seen as non-"local realistic" theory, i.e.

---

[508] For the measurement of position, we can imagine A and B to be separating (starting from a common initial position) with equal speed in opposite directions.

[509] Refer to the Bohr reply in "N. Bohr, Phys. Rev. 48, 696, 1935".

[510] Regardless of agreeing with Bohr or not, he (considering his background) certainly understands quantum mechanics better than Einstein and his colleagues.

[511] This should reflect the unhealthy obsession with Einstein thanks to the systematic and relentless indoctrination (which is like a ritual practice in modern-day science education). In fact, the literature of quantum physics contains many nonsensical arguments and paradoxes (like this one) although most of which are not as famous and "credible" as the EPR argument. It should be surprising that an argument (or paradox) like this takes the center stage in the quantum mechanical debates and investigations since the early days of quantum theory (and it is still so). I think if the "Einstein" tag is not on this argument then very few will buy it or pay attention to it.

[512] Refer for instance to § 1.10 and § 7.14 about our view on realism and the rejection of extreme epistemological realism.

[513] In fact, the label "complete" in the EPR argument is interpreted by some to mean essentially a theory that is local and realistic, i.e. it satisfies the requirement of "local realism" in the above sense. It is worth noting that although EPR do not refer to locality explicitly, it is present implicitly in their argument.

non-local or non-realistic or both.

In our view, accepting local realism in itself is sensible and even favorable (from the perspective of rationality and considering a moderate form of realism) although it may not be experimentally supported (due, for instance, to the experiments of quantum entanglement which supposedly require the violation of locality or realism or both). Similarly, accepting non-local realism is also sensible (at least in our view); moreover it seems to be experimentally supported in its scientific aspect (e.g. by quantum entanglement experiments as well as by other scientific indications to the invalidity of the principle of locality). However, accepting local non-realism may not be sensible because if realism is lost then what is the significance and value of locality (i.e. from the perspective of rationality what is the point of keeping non-realistic locality)?

Interestingly, many physicists seem to favor locality against realism and hence if they have to choose between realism and locality they will choose locality. This may be explained in part by the fact that realism is essentially a philosophical and epistemological issue (which may not be of much interest to many scientists) while locality is essentially a scientific issue (as can be appreciated from its link to special relativity which is very close to the heart of most modern physicists and this should be another factor for the desire to keep locality since it means saving special relativity). Moreover, realism is already threatened (supposedly) by other aspects of quantum theory (especially according to some of its schools of interpretation) such as by indeterminism and the implications of measurement and wavefunction collapse (see for instance § 5.7 and § 6.11 as well as § 8.1) regardless of locality, and hence in the eye of many physicists the battle for realism is already lost (or almost). So, a common attitude among modern physicists is to try their best to keep locality with no much concern about realism. In fact, some may not even worry about locality as long as special relativity is saved (by their alleged excuses such as allowing non-locality when no information or energy can be transferred).

It is worth noting that the criterion of reality set by EPR (basically in the postulate of physical reality) can be challenged quantum mechanically (and even non-"quantum mechanically") from different aspects (especially by the interpretation of some schools). Also, the claim of EPR that it is "in agreement with classical as well as quantum-mechanical ideas of reality" is not acceptable (at least quantum mechanically especially from the widely-accepted Copenhagen school). We should also note that the association between "element of physical reality" and predictability is questionable. For example, an "element of reality" (if we have to accept this extreme form of realism) may not be accessible to any theory (and hence we have "element of physical reality" with no predictability). Also, we may have accidental correlations that can provide predictability but without any correspondence to reality in the given sense (and hence we have predictability with no "element of physical reality"). In fact, this is one of the dubious aspects of the criterion of reality set by EPR (which we indicated earlier).

We should also note (with regard to Einstein) that the (original) EPR argument is based on a form of realism that does not seem consistent with other views (about realism) expressed by Einstein in other contexts and situations. However, this does not seem strange to Einstein who has the habit of changing his position and mind according to what suits him most in the particular situation even if this leads to contradictions and inconsistencies. In fact, this sort of Einsteinian opportunism is widely tolerated in the literature where excuses are always found to justify his instability in opinion and lack of genuine belief in what he claims (or pretends) to embrace.

10.2.3 Completeness and Correctness

The EPR paper is concerned with completeness not correctness (noting that they do not declare their belief in the correctness despite the opposite claims). For example, they state: "It is the second question[514] that we wish to consider here, as applied to quantum mechanics". However, it should be noted that their notion of completeness (which is summarized in their postulate of completeness) is based on an extreme form of realism and hence it is rather different from our notion of completeness.

In the following points we discuss some issues related to completeness and correctness:

---

[514] This refers to their earlier question: "Is the description given by the theory complete?"".

1. Despite the fact that the EPR argument is essentially about completeness not correctness, the argument could have an impact (or implication) on correctness through its (implicit) questioning of the uncertainty principle which (as a fundamental principle) is part of the formalism of quantum mechanics (see for instance § 4.7.1). Anyway, the EPR argument cannot be a serious challenge to quantum mechanics and its validity (neither from the uncertainty principle aspect nor from other aspects like locality). Yes, the EPR argument could be a challenge to some schools of interpretation of quantum mechanics.
2. Although the EPR argument is essentially based on a local hidden-variable view, EPR do not mention locality or hidden variable. In fact, locality is implicit in the EPR argument (as indicated earlier), while "hidden-variable" is an idea created after the appearance of EPR argument (apparently by Bell during his discussion to the EPR argument). Yes, the idea of "hidden-variable" can be felt in the EPR argument (e.g. in their talk about "element of reality" within the context of denying completeness) and hence the EPR argument can justifiably be classified as a (local) hidden-variable theory or view.
3. According to EPR: "While we have thus shown that the wave function does not provide a complete description of the physical reality, we left open the question of whether or not such a description exists. We believe, however, that such a theory is possible".

   This may suggest that EPR believe in the non-uniqueness of science (see § 2.4.3) since they imagine the possibility of having an alternative quantum theory that is different from the existing quantum theory. However, this (in our view) is not true because the alternative theory that they anticipate and look for is no more than a completion to the existing quantum theory and hence its role is not identical to the role of the current quantum theory. In other words, the conclusion of the EPR paper about the possibility of finding a complete theory does not contradict the uniqueness of science because the new theory is supposed to be a completion of the existing theory and not as a substitute or replacement, i.e. the complete theory is or could be "unique".

   In fact, we may even claim that uniqueness of science can be felt (implicitly) during the reading of the EPR paper. For example, the entire EPR argument is based on the presumed validity of quantum mechanics (despite its claimed non-completeness) as if it is the only possible quantum formulation (i.e. for the specific quantum predictions). Also, the extreme form of realism adopted by EPR should suggest a belief in the uniqueness of science (or at least this type of realism is more consistent with such a belief). Anyway, if the EPR argument is not based (implicitly) on an assumption of uniqueness of science, it at least does not contradict such an assumption.

10.2.4 Other Issues about the Original EPR Argument

An important issue about the original EPR argument is related to the uncertainty principle. In fact, this issue caused a considerable amount of confusion (about the nature, content and objective of the argument) in the literature. In our view, although the declared objective of the EPR argument is to challenge the completeness of quantum mechanics, it seems that EPR were also interested in questioning the uncertainty principle and casting a shadow on its validity (at least in its sense as an intrinsic uncertainty which is essential to some quantum mechanical interpretations, especially Copenhagen, and possibly even formulations). In fact, from reading the EPR paper we can guess (or feel) that the idea of this argument was not sufficiently ripe or well-developed and hence they tried to gather as much as possible of points and issues against quantum mechanics (according to their view) in support of their argument, and hence the uncertainty principle emerged as one of such problematic issues of quantum mechanics (and possibly the best) that they can exploit in their argument.

Another important issue about the original EPR argument is related to the questionable aspects (in the argument and paper) which we referred to earlier. In fact, there are many aspects in the EPR argument and paper that can be questioned and challenged. For example:
• Their claim "this criterion is in agreement with classical as well as quantum-mechanical ideas of reality" is strange and untrue (or at least dubious).
• EPR seem to understand the wavefunction as a physical entity (in the sense of being a reflection of physical reality, rather than a mathematical device and tool) which is questionable.

- The "disturbance factor" which they seem to adopt as a justification for the uncertainty principle (as mentioned earlier in § 10.2.1) is questionable.
- The EPR adopt an extreme form of realism which is not supported by evidence and cannot even be defended (due to the existence of evidence or indication against it).
- There are also question marks about some technicalities in their formulations which can be criticized as a sign of looseness and lack of rigor (if not a sign of poor comprehension and lack of understanding).

10.3 Proofs and Arguments in Support of Bell's Theorem

In this appended section we present a number of proofs and arguments in support of the Bell theorem noting that there are many more (and hence this is just a sample) although many of them are not fundamentally different. Most of these proofs and arguments lead to derivation of a quantitative relationship (usually in the form of inequality) that can be used to compare the predictions of quantum mechanics and those of local hidden-variable (LHV) theories. As indicated earlier, different proofs and arguments of Bell's theorem may have different implications and consequences (and possibly different refutations) and they usually lead to different forms of inequality (or other forms of quantitative relationships) which may have different implications. They also usually differ in their vulnerability to challenges and potential refutations, as they usually differ in rigor and technicality.

As mentioned earlier (see § 9.3.2), we present in this appended section two proofs and two arguments. Our motivation for showing this diversity of proofs and arguments was outlined in § 9.3.2. For example, one motivation is because of the importance of Bell's theorem especially at the epistemological level. In fact, this theorem is at the center of many ongoing research areas and branches of quantum physics (formal as well as epistemological). Another motivation is that different proofs and arguments shed light on different aspects of the issues involved and they lead to different implications and formulations. The proofs and arguments also differ in their physical conditions and assumptions, their strength, extension, resistance to challenges and criticisms, intuitivity, rigor, and so on. Overall, a collection of proofs and arguments will be more beneficial, more assertive and more informative (even if we assume, arguably, that such a diversity is not needed).

10.3.1 A Proof of Bell's Theorem

In this appended subsection we present a technical proof of the Bell theorem and inequality (in a different form to its original form). Although this proof produces another version of the Bell inequality (which is rather less general), its advantage is that it is more illuminating and has more physical content and substance (and possibly is less abstract).[515]

The proofs of Bell's theorem and inequality (including the present proof) generally employ entangled (or correlated) quantum systems such as entangled electrons or photons (see § 5.6 and § 6.10). We use in the present discussion and proof spin-1/2 particles (e.g. electrons) as a prototype for entangled quantum systems although this will not affect the generality of the essence since all the arguments and results can be easily generalized and extended (or modified slightly) to include other types of entangled quantum systems such as polarized photons.[516] We should note that due to obvious pedagogical reasons, part of the following phrasing and reasoning is rather loose (from more strict and rigorous technical standards) to make it easier for the reader to understand the pivotal points in Bell's theorem and its proof.

Let first remind ourselves that the essence of Bell's theorem can be stated as follows: it is impossible for local hidden-variable (LHV) theories to produce all the predictions of quantum mechanics (refer to § 9.3.2 for more details about Bell's theorem). However, it is important to note that when we talk here about local hidden-variable theories we mean theories that accept the (experimentally well-established) predictions of quantum mechanics and hence these theories are "quantum in essence" although they are

---

[515] We acknowledge that in this proof we partly follow Rae and Napolitano (see the References in the back of the book).
[516] The reader, however, should note that the formulation in the two cases is not the same because electrons are spin-1/2 fermions while photons are spin-1 bosons. We should also note that as far as the violation of Bell's inequality is concerned, violation in any particular system should be enough for proving the essence of the Bell theorem.

10.3.1 A Proof of Bell's Theorem

"classically-rationalized in interpretation" by assuming the existence of local hidden variables so that locality, determinism and causality in the classical sense are preserved. So, when we compare LHV theories with quantum mechanics we actually compare LHV theories (which accept the established principles and predictions of quantum mechanics) with the dominant standard interpretation of quantum mechanics (which rejects, or at least suspects, the LHV interpretation). To put it more briefly and loosely: we actually compare LHV theories (of quantum mechanics) to non-"LHV" theories (of quantum mechanics) noting that the essence and objective of this comparison is to show that the predictions of LHV theories is not the same as the predictions of quantum mechanics (and hence LHV theories cannot provide an acceptable interpretation to quantum mechanics, while non-"LHV" could provide, though not necessarily, such an interpretation).[517]

To start with, let have a pair of entangled (or correlated) and locally-separated spin-1/2 particles (labeled A and B). If this pair is created by the decay of a spin-0 particle then the total spin of the entangled system is zero. Now, if we measure the spins of A and B simultaneously[518] along a given direction (say z direction) then it is certain that one will be $+1$ and the other will be -1 (in units of $\hbar/2$). This is an experimentally-established result and it is acceptable both classically and quantum mechanically (since the total spin is zero).[519]

Now, instead of measuring both spins along the z direction, let measure the spin of A along the z direction and the spin of B along a direction that makes an angle θ with the z direction (where $0 < \theta < \pi$ can be imagined to be the polar coordinate of spherical coordinates r, θ, ϕ and where the direction vector of the B measurement is assumed to be in the xz plane with no loss of generality). The following proof of Bell's theorem is based on this presumed experimental setting where we will show that the LHV proposal leads to predictions that contradict the predictions of quantum mechanics[520] and this should prove Bell's theorem. In fact, we will obtain an inequality that we can (and will) use to compare (theoretically) the two predictions quantitatively and demonstrate their differences in specific cases and hence prove Bell's theorem theoretically. Moreover, such an inequality can be used as a basis for experimental tests that (can in principle) determine decisively and specifically if the quantum phenomena follow the LHV predictions (and hence LHV theories are correct) or otherwise (and hence LHV theories are wrong).[521] As a result of such experimental tests, the fate of the LHV theories and interpretations can be determined (noting that many such tests have been conducted and have shown, supposedly, the falsehood of the LHV predictions and the correctness of the quantum mechanical predictions).

---

[517] We should note in this context that although local hidden-variable theories are commonly labeled as "classical" they are not really classical. In fact, it may be more appropriate to label them, for instance, as "realistic theories" or "local-realistic theories" or "deterministic theories" (depending on their substance and flavor). The "classical" label (which may be attached to them in the literature) is based on their presumable classical rationale (and that is why we described them as classically-rationalized). Anyway, this is not an important issue (as long as confusion and misunderstanding are avoided) and hence we may allow ourselves to use labels like "classical" occasionally to refer to these theories as opposite to the "strictly" quantum mechanical theories (i.e. both in content and spirit or in formalism and interpretation) which are based on the commonly-accepted and intrinsically-indeterministic rationales (as represented by characteristically-quantum features like superposition of states prior to measurement and reduction of wavefunction by measurement).

[518] For non-locality considerations, what is needed is only that the two measurements should occur within a time interval that does not allow local communication (i.e. no physical signal can be exchanged between the two particles during this time interval). In fact, this should relieve us from taking care of simultaneity and how it is defined and hence in the following discussion we can assume this is the case (noting that since we are considering an LHV scenario, instantaneous communication is not a possibility).

[519] We note that this type of experiments should (from a quantum mechanical viewpoint) be sufficient for the investigation of non-locality since the measurement on one object leads to the collapse of their wavefunction with no delay (which is a non-local influence).

[520] More precisely, we will show that if we assume that the predictions of the LHV proposal are identical to the predictions of quantum mechanics we will get a false (or contradictory) result (and hence we can conclude that this assumption is false which means that the two predictions are different and this should prove Bell's theorem).

[521] As indicated earlier, we are actually assuming (implicitly) the validity of the predictions of quantum mechanics. Otherwise, we should have three main possibilities: the predictions of LHV theories are correct (and hence quantum mechanics is wrong), the predictions of quantum mechanics are correct (and hence LHV theories are wrong), and neither is correct (and hence we need to look for another quantum theory). We should also note that if we accept the possibility that quantum mechanics may not be interpretable then the falsehood of LHV theories (and correctness of quantum mechanics) does not imply the correctness of non-"LHV" interpretation because quantum mechanics may not have any interpretation at all (i.e. neither LHV nor non-"LHV").

10.3.1 A Proof of Bell's Theorem

It should be obvious that the measurement on A (along z direction) should be either $+1$ or -1. Similarly, the measurement on B (along θ direction) should be either $+1$ or -1. This means that we can get any one of the following four cases: $++, +-, -+, --$ (where $+-$ for instance means $+1$ for A and -1 for B).[522] So, unlike the previous case where both spins are measured along the z direction (and hence we get either $+-$ or $-+$), now we can also have both spins up (i.e. $++$) and both spins down (i.e. $--$). However, we want to know how these measurements are correlated quantum mechanically. In other words, if we repeat this experiment on a very large number of (independent, identical and) identically-prepared pairs of entangled particles in this setting then how many times we will get $+-$, and similarly how many times we will get $++$, $-+$ and $--$. So, what we are actually looking for is the probabilities $P_{++}, P_{+-}, P_{-+}, P_{--}$ (where P_{-+} means the probability of getting -1 for A AND $+1$ for B, and the rest follow this pattern).

To investigate and analyze the situation formally, let start from the well-known quantum mechanical results about spin-1/2 particles. The operators $\hat{S}_x, \hat{S}_y, \hat{S}_z$ that represent the components of spin in the x, y, z directions are:[523]

$$\hat{S}_x = \frac{\hbar}{2}\begin{bmatrix} 0 & 1 \\ 1 & 0 \end{bmatrix} \qquad \hat{S}_y = \frac{\hbar}{2}\begin{bmatrix} 0 & -i \\ i & 0 \end{bmatrix} \qquad \hat{S}_z = \frac{\hbar}{2}\begin{bmatrix} 1 & 0 \\ 0 & -1 \end{bmatrix} \qquad (18)$$

Now, if we use our intuition then the operator $\hat{S}_\theta$ that represents the spin component in the θ direction is:

$$\hat{S}_\theta = \hat{S}_x \sin\theta + \hat{S}_z \cos\theta = \frac{\hbar}{2}\begin{bmatrix} 0 & 1 \\ 1 & 0 \end{bmatrix}\sin\theta + \frac{\hbar}{2}\begin{bmatrix} 1 & 0 \\ 0 & -1 \end{bmatrix}\cos\theta = \frac{\hbar}{2}\begin{bmatrix} \cos\theta & \sin\theta \\ \sin\theta & -\cos\theta \end{bmatrix} \qquad (19)$$

and hence the eigenvalues of this (matrix) operator are $\pm\hbar/2$.[524] The eigenvectors of $\hat{S}_\theta$ corresponding to the eigenvalues $+\hbar/2$ and $-\hbar/2$ are respectively:[525]

$$\begin{bmatrix} \sin\theta \\ 1-\cos\theta \end{bmatrix} \qquad \text{and} \qquad \begin{bmatrix} \cos\theta - 1 \\ \sin\theta \end{bmatrix} \qquad (20)$$

On normalizing these eigenvectors,[526] we get the following normalized eigenvectors of $\hat{S}_\theta$ (corresponding to the eigenvalues $+\hbar/2$ and $-\hbar/2$ respectively):

$$\begin{bmatrix} \cos\frac{\theta}{2} \\ \sin\frac{\theta}{2} \end{bmatrix} \qquad \text{and} \qquad \begin{bmatrix} -\sin\frac{\theta}{2} \\ \cos\frac{\theta}{2} \end{bmatrix} \qquad (21)$$

---

[522] It should be noted that if $\theta = 0$ then only $+-, -+$ are possible, and if $\theta = \pi$ then only $++, --$ are possible, and this should explain why we restricted θ to the range $0 < \theta < \pi$ despite the fact that the range of the polar coordinate of spherical system is $0 \leq \theta \leq \pi$.

[523] For the sake of accuracy and to avoid possible confusion, we kept $\hbar/2$ (remembering that ± 1 are in units of $\hbar/2$).

[524] From linear algebra, the characteristic equation is:

$$\det\left(\frac{\hbar}{2}\begin{bmatrix} \cos\theta & \sin\theta \\ \sin\theta & -\cos\theta \end{bmatrix} - \begin{bmatrix} \lambda & 0 \\ 0 & \lambda \end{bmatrix}\right) = \begin{vmatrix} \frac{\hbar}{2}\cos\theta - \lambda & \frac{\hbar}{2}\sin\theta \\ \frac{\hbar}{2}\sin\theta & -\frac{\hbar}{2}\cos\theta - \lambda \end{vmatrix} = \lambda^2 - \left(\frac{\hbar}{2}\right)^2 = 0$$

and hence $\lambda = \pm\hbar/2$.

[525] We have:

$$\frac{\hbar}{2}\begin{bmatrix} \cos\theta & \sin\theta \\ \sin\theta & -\cos\theta \end{bmatrix}\begin{bmatrix} \sin\theta \\ 1-\cos\theta \end{bmatrix} = \frac{\hbar}{2}\begin{bmatrix} \cos\theta\sin\theta + \sin\theta(1-\cos\theta) \\ \sin^2\theta - \cos\theta(1-\cos\theta) \end{bmatrix} = +\frac{\hbar}{2}\begin{bmatrix} \sin\theta \\ 1-\cos\theta \end{bmatrix}$$

and

$$\frac{\hbar}{2}\begin{bmatrix} \cos\theta & \sin\theta \\ \sin\theta & -\cos\theta \end{bmatrix}\begin{bmatrix} \cos\theta - 1 \\ \sin\theta \end{bmatrix} = \frac{\hbar}{2}\begin{bmatrix} \cos\theta(\cos\theta - 1) + \sin^2\theta \\ \sin\theta(\cos\theta - 1) - \cos\theta\sin\theta \end{bmatrix} = -\frac{\hbar}{2}\begin{bmatrix} \cos\theta - 1 \\ \sin\theta \end{bmatrix}$$

[526] We have (noting that $0 < \theta < \pi$ and hence $0 < \frac{\theta}{2} < \frac{\pi}{2}$):

$$\begin{bmatrix} \frac{\sin\theta}{\sqrt{\sin^2\theta + (1-\cos\theta)^2}} \\ \frac{1-\cos\theta}{\sqrt{\sin^2\theta + (1-\cos\theta)^2}} \end{bmatrix} = \begin{bmatrix} \frac{\sin\theta}{\sqrt{2-2\cos\theta}} \\ \frac{1-\cos\theta}{\sqrt{2-2\cos\theta}} \end{bmatrix} = \begin{bmatrix} \sqrt{\frac{1-\cos^2\theta}{2-2\cos\theta}} \\ \sqrt{\frac{(1-\cos\theta)^2}{2-2\cos\theta}} \end{bmatrix} = \begin{bmatrix} \sqrt{\frac{1+\cos\theta}{2}} \\ \sqrt{\frac{1-\cos\theta}{2}} \end{bmatrix} = \begin{bmatrix} \cos\frac{\theta}{2} \\ \sin\frac{\theta}{2} \end{bmatrix}$$

and

$$\begin{bmatrix} \frac{\cos\theta - 1}{\sqrt{(\cos\theta - 1)^2 + \sin^2\theta}} \\ \frac{\sin\theta}{\sqrt{(\cos\theta - 1)^2 + \sin^2\theta}} \end{bmatrix} = \begin{bmatrix} -\frac{1-\cos\theta}{\sqrt{2-2\cos\theta}} \\ \frac{\sin\theta}{\sqrt{2-2\cos\theta}} \end{bmatrix} = \begin{bmatrix} -\sqrt{\frac{(1-\cos\theta)^2}{2-2\cos\theta}} \\ \sqrt{\frac{1-\cos^2\theta}{2-2\cos\theta}} \end{bmatrix} = \begin{bmatrix} -\sqrt{\frac{1-\cos\theta}{2}} \\ \sqrt{\frac{1+\cos\theta}{2}} \end{bmatrix} = \begin{bmatrix} -\sin\frac{\theta}{2} \\ \cos\frac{\theta}{2} \end{bmatrix}$$

10.3.1 A Proof of Bell's Theorem

Now, it is obvious that the (normalized) eigenvectors of $\hat{S}_z$ are $\begin{bmatrix} 1 \\ 0 \end{bmatrix}$ and $\begin{bmatrix} 0 \\ 1 \end{bmatrix}$ corresponding respectively to the eigenvalues $+\hbar/2$ and $-\hbar/2$.[527] So, if we note that the θ orientation is relative to the z orientation (which is represented in its positive and negative directions by the $\hat{S}_z$ eigenvectors $\begin{bmatrix} 1 \\ 0 \end{bmatrix}$ and $\begin{bmatrix} 0 \\ 1 \end{bmatrix}$ respectively) in the xz plane then we can express the $\hat{S}_z$ eigenvectors as linear combinations of the (normalized) eigenvectors of $\hat{S}_\theta$, that is:

$$\begin{bmatrix} 1 \\ 0 \end{bmatrix} = c_{-+} \begin{bmatrix} \cos\frac{\theta}{2} \\ \sin\frac{\theta}{2} \end{bmatrix} + c_{--} \begin{bmatrix} -\sin\frac{\theta}{2} \\ \cos\frac{\theta}{2} \end{bmatrix} = \begin{bmatrix} c_{-+}\cos\frac{\theta}{2} - c_{--}\sin\frac{\theta}{2} \\ c_{-+}\sin\frac{\theta}{2} + c_{--}\cos\frac{\theta}{2} \end{bmatrix} \quad (22)$$

$$\begin{bmatrix} 0 \\ 1 \end{bmatrix} = c_{++} \begin{bmatrix} \cos\frac{\theta}{2} \\ \sin\frac{\theta}{2} \end{bmatrix} + c_{+-} \begin{bmatrix} -\sin\frac{\theta}{2} \\ \cos\frac{\theta}{2} \end{bmatrix} = \begin{bmatrix} c_{++}\cos\frac{\theta}{2} - c_{+-}\sin\frac{\theta}{2} \\ c_{++}\sin\frac{\theta}{2} + c_{+-}\cos\frac{\theta}{2} \end{bmatrix} \quad (23)$$

where the significance of the suffices (i.e. + and −) of the coefficients c's will be clarified soon. On solving these two sets of simultaneous equations (where each set is represented by the first and second rows of the above two matrix equations) we get:

$$c_{-+} = \cos\frac{\theta}{2} \qquad c_{--} = -\sin\frac{\theta}{2} \qquad c_{++} = \sin\frac{\theta}{2} \qquad c_{+-} = \cos\frac{\theta}{2} \quad (24)$$

which can be easily checked by substitution in Eqs. 22 and 23.

Now, if the spin of A is -1 (in units of $\hbar/2$) then the eigenvector of A is $\begin{bmatrix} 0 \\ 1 \end{bmatrix}$ and hence the "eigenvector of B along the z orientation" should be $\begin{bmatrix} 1 \\ 0 \end{bmatrix}$ which is given by Eq. 22. So, if the spin of B along the θ orientation is to be $+1$ (in units of $\hbar/2$) then the eigenvector of B in this case is $\begin{bmatrix} \cos\frac{\theta}{2} \\ \sin\frac{\theta}{2} \end{bmatrix}$ (which corresponds to the positive eigenvalue of $\hat{S}_\theta$) and hence the coefficient c_{-+} of $\begin{bmatrix} \cos\frac{\theta}{2} \\ \sin\frac{\theta}{2} \end{bmatrix}$ in Eq. 22 should be (the probability amplitude) representing the case "-1 spin for A and $+1$ spin for B" (i.e. along z and θ respectively) and this should explain the suffix $-+$ in c_{-+}. Similar arguments should easily explain the suffices of the other three coefficients (i.e. c_{--}, c_{++}, c_{+-}).

Accordingly, the probabilities of getting $+1$ and -1 along the z and θ orientations are:

$$P_{++} = |c_{++}|^2 = \sin^2\frac{\theta}{2} \quad (25)$$

$$P_{+-} = |c_{+-}|^2 = \cos^2\frac{\theta}{2} \quad (26)$$

$$P_{-+} = |c_{-+}|^2 = \cos^2\frac{\theta}{2} \quad (27)$$

$$P_{--} = |c_{--}|^2 = \sin^2\frac{\theta}{2} \quad (28)$$

where the first and second signs in the suffices correspond to the z (i.e. for A) and θ (i.e. for B) orientations respectively. For example, the symbol P_{+-} means the probability of having A with spin $+1$ (along the z orientation) AND B with spin -1 (along the θ orientation), and the other symbols follow this pattern.

Now, let see the consequences of these quantum mechanical results if we accept local hidden-variable theories (where we will show that if we assume the LHV predictions to be identical to these quantum mechanical results then we will get false consequences). It is obvious that according to these theories

---
[527] We have:

$$\frac{\hbar}{2}\begin{bmatrix} 1 & 0 \\ 0 & -1 \end{bmatrix}\begin{bmatrix} 1 \\ 0 \end{bmatrix} = +\frac{\hbar}{2}\begin{bmatrix} 1 \\ 0 \end{bmatrix} \qquad \text{and} \qquad \frac{\hbar}{2}\begin{bmatrix} 1 & 0 \\ 0 & -1 \end{bmatrix}\begin{bmatrix} 0 \\ 1 \end{bmatrix} = -\frac{\hbar}{2}\begin{bmatrix} 0 \\ 1 \end{bmatrix}$$

10.3.1 A Proof of Bell's Theorem

(or rather to some types of these theories) the state of a quantum object is determined in reality (by some hidden variables embedded locally in the object) and hence the role of measurement is just to reveal this reality. In other words, the outcome of measurement is determined in advance (by the local hidden variables) although our knowledge of this outcome is conditioned by conducting the measurement to unveil this outcome. Now, let accept this local hidden-variable proposal and apply it to the spin of spin-1/2 entangled particles (as in the above analysis) and get some results (according to LHV) that can be compared (possibly indirectly) to the above quantum mechanical results (which are summarized in Eqs. 25-28).

So, let have three directions (which we label as $1, 2, 3$ noting that these directions are not required to be mutually orthogonal or even linearly independent) and let consider the spin components of a spin-1/2 particle in these directions. According to the local hidden-variable proposal these components are determined in advance and hence the particle in reality (or in itself) has a given state (i.e. $+1$ or -1) for each one of these components before conducting any measurement although (according to the rules of quantum mechanics) only one component can be determined at a time since simultaneous measurement of more than one component is banned. For example, a particle can be in a state $+-+$ when its spin component in the 1st direction is $+1$, its spin component in the 2nd direction is -1, and its spin component in the 3rd direction is $+1$.

In fact, any particle can be in one of 8 distinct states representing all the possible combinations of the three signs, i.e.

$$+++ \qquad ++- \qquad +-+ \qquad +-- \qquad -++ \qquad -+- \qquad --+ \qquad ---$$

Accordingly, any set of such particles can be divided into 8 disjoint (or mutually exclusive) subsets (some of which could be empty) where each one of these subsets contains a number n of these particles (where n is a non-negative integer function of the three signs). For example, if the subset of particles with state $-+-$ contains 9 particles then we can write $n(-+-) = 9$. We may also consider other types of subsets for which less than 3 components are considered. For example, if the subset of particles whose 1st spin component is -1 and their 2nd spin component is $+1$ (while their 3rd spin component is irrelevant to us) contains 25 particles then we can write $n(-+\pm) = 25$. Accordingly, we can write:

$$n(-+\pm) = n(-++) + n(-+-) \tag{29}$$

and hence we can conclude that $n(-++) = 25 - 9 = 16$. To avoid any misunderstanding or confusion we repeat that all this is based on the proposal of local hidden-variable where the states have a reality independent of measurement.

Now, let have a large number (say N) of entangled (or rather correlated) pairs of such particles (remembering that the total spin of each pair is zero). It should be *obvious* that (according to the local hidden-variable proposal) we can identify two spin components of any entangled particle in any pair simultaneously by conducting a measurement of one of these components on the particle itself and conducting a measurement of the other component on its twin (i.e. the other particle in the entangled pair).[528] For example, if A and B are such entangled particles and we measure (simultaneously) the 1st spin component of A and the 2nd spin component of B and we got $+1$ for the former and -1 for the latter then we know that A is in state $++\pm$ (regardless of the third component which could be $+$ or $-$ as indicated) where the knowledge of $+1$ for the 1st spin component comes from the measurement on A while the knowledge of $+1$ for the 2nd spin component comes from the measurement on B (in conjunction with the "entanglement") by getting -1 for the 2nd spin component of B. We should similarly know that B is in state $--\mp$ where the knowledge of -1 for the 1st spin component comes from the measurement on A (by getting $+1$ for the 1st spin component of A) while the knowledge of -1 for the 2nd spin component comes from the measurement on B. It should be *obvious* (see footnote [528]) that such simultaneous measurements (and knowledge) do not violate the quantum mechanical rules if we accept the local hidden-variable proposal because we are conducting a single measurement on each of the "entangled" particles at a time (and getting knowledge of

---

[528] This *obviousness* may be disputed. Anyway, the simultaneous identification of two components is reasonable since the two particles (according to LHV) are actually independent of each other and they are not interacting. Also see point 2.

10.3.1 A Proof of Bell's Theorem

only one component at a time), and the extra knowledge (i.e. knowing both components simultaneously for each particle) that we get is the benefit of "entanglement" or correlation (as well as the presumption of LHV and independence of reality) and not because we conducted two measurements on the particle (and hence disturbed it twice).

Now, let test these results carefully using basic logic and arithmetic as well as the previously-obtained quantum mechanical results (represented mainly by Eqs. 25-28). So, let assume that we prepared a set (say S1) of very large number (say N) of entangled pairs of such particles[529] and we conducted two simultaneous measurements (say on the 1st and 2nd spin components) on each pair in this set according to the above procedure (i.e. by measuring one component for each particle in the pair). As a consequence, we obtain $n_1(++\pm)$, $n_1(+-\pm)$, $n_1(-+\pm)$ and $n_1(--\pm)$ where the subscript 1 (i.e. in n_1) refers to S1 and where

$$N = n_1(++\pm) + n_1(+-\pm) + n_1(-+\pm) + n_1(--\pm) \tag{30}$$

Accordingly, we can write for instance:

$$n_1(++\pm) = n_1(+++) + n_1(++-) \tag{31}$$

Now, if we prepare another set (say S2) that is identical to S1 (including the number of entangled pairs N) and we conducted two simultaneous measurements (say on the 1st and 3rd spin components this time) on each pair in this set according to the above procedure, then we can similarly write for instance:

$$n_2(+\pm+) = n_2(+++) + n_2(+-+) \tag{32}$$

where the subscript 2 (i.e. in n_2) refers to S2. If we finally prepare a third set (say S3) that is identical to S1 and we conducted two simultaneous measurements (say on the 2nd and 3rd spin components this time) on each pair in this set according to the above procedure, then we can similarly write for instance:

$$n_3(\pm-+) = n_3(+-+) + n_3(--+) \tag{33}$$

where the subscript 3 (i.e. in n_3) refers to S3.

Now, since the three sets (i.e. S1, S2 and S3) are identical (noting that according to the LHV proposal they should have identical reality) then they should be statistically identical and hence we can drop the subscripts of n in Eqs. 31-33. Accordingly, we can write (using Eqs 31-33):

$$\begin{aligned}n(++\pm) - n(+\pm+) + n(\pm-+) &= \cancel{n(+++)} + n(++-) - \cancel{n(+++)} \\ &\quad -\cancel{n(+-+)} + \cancel{n(+-+)} + n(--+) \\ &= n(++-) + n(--+)\end{aligned}$$

It is obvious that $n(++-) + n(--+) \geq 0$ and hence we can write:

$$n(++\pm) - n(+\pm+) + n(\pm-+) \geq 0 \tag{34}$$

which can be seen as a form of Bell's inequality.

Now, let assume that the LHV predictions are identical to the quantum mechanical predictions (as given by Eqs. 25-28). Accordingly, we can substitute form Eqs. 25-28 into the last inequality. However, to be able to do this we need to express the last inequality in terms of the probabilities of Eqs. 25-28.[530] To do so, we **first** divide the inequality by N and hence we get:

$$P(++\pm) - P(+\pm+) + P(\pm-+) \geq 0 \tag{35}$$

---

[529] N in the following should be understood as the number of particles (rather than pairs) in the ensemble (despite the opposite suggestion of our expression which is used for simplicity). In fact, N can be understood as the number of pairs but in this case the sum of n's (in Eq. 30 for instance) should be understood as representing half of the entangled particles (with some required restrictions and clarifications which we ignore to avoid unnecessary complications). Anyway, this is a trivial issue (noting that N can be discarded in this argument although we used it for more clarity).

[530] As indicated, the purpose of expressing the last inequality (which is based on LHV) in terms of the probabilities of Eqs. 25-28 (which are quantum mechanical) is to see if the two (i.e. LHV and quantum mechanics) are compatible or not. In other words, if the inequality obtained from this expression (which mixes LHV with quantum mechanics) leads to any false result then we can conclude that LHV and quantum mechanics are not compatible, i.e. the predictions of LHV cannot be identical to the predictions of quantum mechanics (and this is the essence of Bell's theorem).

10.3.1 A Proof of Bell's Theorem

Second, we note that the term $P(++\pm)$ represents a θ between the first and second directions (which we label as θ_{12}) and hence we can write this term as $P^{\theta_{12}}(++)$ where the signs correspond (respectively) to the indices of θ_{12}. Similar considerations apply to the second and third terms and hence the inequality of Eq. 35 becomes:

$$P^{\theta_{12}}(++) - P^{\theta_{13}}(++) + P^{\theta_{23}}(-+) \geq 0 \tag{36}$$

Third, the pair of signs in each term in Eq. 36 are for the same particle while each sign in Eqs. 25-28 is for one particle in the pair, and hence if we should put the pairs of signs in a form consistent with the form of Eqs. 25-28 then we should reverse the second sign of each pair in Eq. 36. Accordingly, Eq. 36 becomes (using the subscripted notation of Eqs. 25-28):

$$P^{\theta_{12}}_{+-} - P^{\theta_{13}}_{+-} + P^{\theta_{23}}_{--} \geq 0 \tag{37}$$

On substituting from Eqs. 25-28 into Eq. 37 we get:

$$\cos^2\frac{\theta_{12}}{2} - \cos^2\frac{\theta_{13}}{2} + \sin^2\frac{\theta_{23}}{2} \geq 0 \tag{38}$$

The last inequality (which can also be seen as a form of Bell's inequality) is based on the LHV proposal combined with the predictions of quantum mechanics (as if the two predictions are identical), and hence any violation of this inequality should prove Bell's theorem. This is because this inequality is derived by the LHV assumption in conjunction with the predictions of quantum mechanics (as represented by Eqs. 25-28) and hence if this inequality is violated then these predictions must be incompatible with the predictions of LHV and hence the predictions of quantum mechanics and those of LHV must be different (as claimed by Bell's theorem; see footnote [520] on page 230).

So, all we need now to prove Bell's theorem is to find an instance in which this inequality is invalid. In fact, there are many such instances. For example, if all the directions are in the xz plane and the 1st direction is along the positive z axis (i.e. due north) and the 2nd direction is along the positive x axis (i.e. due east) while the 3rd direction is along the rising $z = x$ line (i.e. north-east) then $\theta_{12} = \pi/2$ and $\theta_{13} = \theta_{23} = \pi/4$.[531] Hence from Eq. 38 we get:

$$\cos^2\frac{\pi}{4} - \cos^2\frac{\pi}{8} + \sin^2\frac{\pi}{8} \geq 0 \tag{39}$$

$$\frac{1}{2} - 0.85355 + 0.14645 \geq 0 \tag{40}$$

$$-0.20711 \geq 0 \tag{41}$$

which is obviously false. As we see, we get a false result (based on the assumption of identicality of the two predictions) and hence the assumption of the identicality of LHV predictions and quantum mechanical predictions should be false, and this should prove the Bell theorem. In fact, this should even invalidate the local hidden-variable proposal since the predictions of quantum mechanics (as represented by Eqs. 25-28) are already verified and hence the predictions of LHV should be false.

To sum up, the inequality of Eq. 37 is derived on the basis of the LHV scenario. So, if we assume that the LHV predictions are identical to the quantum mechanical predictions then we can substitute from Eqs. 25-28 (which represent the predictions of quantum mechanics) into this inequality. On doing so we got an absurdity (of Eq. 41). Therefore, we conclude that our assumption of identicality is false, and this proves the Bell theorem.

In the following points we discuss some issues related to this proof:
1. There are many proofs for the Bell theorem. Hence, if some proofs are challenged (as we will do next by proposing some possible challenges to the current proof) and possibly refuted, it should not affect the validity of the theorem in general although it may affect the specific version of the inequality derived in that proof. Yes, if all the existing proofs are challenged and (supposedly) refuted then this will be a threat to the theorem, but this seems unlikely (or difficult to do) and hence the theorem is well established in general (although it could still be subject to questioning and skepticism).

---

[531] We are considering only the magnitude of these angles since their sense does not matter in the calculations of Eq. 38.

10.3.1 A Proof of Bell's Theorem

2. As indicated in footnote [528], the possibility of identifying two spin components simultaneously may be disputed and hence this proof may be challenged by claiming that entangled objects should be considered as a single quantum system. However, we think this possibility is reasonable because according to LHV the two particles are independent of each other and they do not interact in any way. So, they are not really entangled in the quantum mechanical sense to consider them as a single system (as we claimed earlier with regard to quantum-mechanically entangled systems). Yes, they are correlated (classically) through their history.

3. Because the predictions of quantum mechanics are verified in countless experiments and observations on countless types of quantum systems, it may be claimed (as we did earlier) that the Bell theorem itself (with its obvious theoretical consequences obtained from the derived inequality) should be sufficient to refute the LHV theories with no need for any further experimental verification or test. However, this may be challenged by the fact that the particular correlation (as demonstrated and quantified loosely by the inequality) should still require verification and test since this particular correlation has not been tested (at least directly) before the emergence and employment of Bell's theorem. However, this may be the case with some proofs and arguments of Bell's theorem and some formulations of Bell's inequality in which no specific (or previously-established) predictions of quantum mechanics enter in the proof/argument and formulation.[532] But there are some proofs (like the present proof) in which specific predictions of quantum mechanics (i.e. those represented by Eqs. 25-28) enter in the proof and in the formulations and since these predictions are (supposedly) verified previously (otherwise they will not be accepted and adopted) then the mere theoretical violation of the inequality (as demonstrated for instance by the result of Eq. 41) should be sufficient for the refutation of LHV. Anyway, this is not a big issue and we accept that specific experimental tests for the derived Bell inequalities are (rather) necessary to remove any doubt or suspicion about the invalidity of LHV. Also see point 12 of § 9.3.2 as well as the last paragraph of appended § 10.3.4.

So to sum up (with regard to the experimental tests required for/by the Bell theorem) we can say: the violation of the Bell inequality (through theoretical calculations) should prove Bell's theorem (i.e. LHV and quantum mechanics differ in prediction), but the determination of which is right (i.e. LHV or quantum mechanics or neither) may (where this "may" depends on the particular inequality and how it is obtained) require experimental tests to determine the specific pattern of the outcome and with which predicted pattern it agrees (if any).

4. Some types of the Bell theorem proofs (like the above proof) should depend (in their validity when used as basis for experimental tests) on assuming full (or very high) detection rate (or level) of the particles involved.[533] Hence, the validity of the results of experimental tests based on these types of proofs should generally (or even critically) depend on the detection efficiency (i.e. for the results to be conclusive the detection efficiency should be full or very high). In fact, the early Bell experiments have generally poor detection efficiency, but this was improved substantially in subsequent experiments where some of the recent experiments have (supposedly) reached 100% efficiency (or almost).

Anyway, we should note that not all types of proofs require full (or very high) efficiency and hence some types tolerate partial efficiency. In most cases, high (not necessarily full or very high) detection efficiency should be enough for the experimental validity of the proofs (i.e. for the validity of experimental results based on these proofs). So, in general each proof should be analyzed (with respect to the particular experiment conducted in conjunction with that proof) to assess its efficiency demand (for the experimental results to be valid or conclusive), and this means that the issue of detection efficiency is case dependent.

5. There are some local hidden-variable theories that are not completely deterministic and hence the above proof (and its alike) may not apply to them because in this proof total determinism is assumed.[534]

---

[532] The proof of appended § 10.3.2 may be seen as an instance of this type. However, we should note that it can also be compared to particular predictions of quantum mechanics (which are supposedly verified already) when it is applied in particular cases of theoretical calculations (as demonstrated for instance by our calculations in Eq. 60). Also, the general argument of appended § 10.3.3 may be an instance of this type.

[533] For example, the above proof requires the test of all the N pairs in the S1, S2 and S3 sets; otherwise the required probabilities of Eq. 35 cannot be obtained (at least with sufficient accuracy).

[534] In principle, the rationale of the above proof does not apply to these theories because the proof assumes that these LHV

Accordingly, other types of proof are required for proving Bell's theorem (with regard to this type of LHV theories) and refuting them.
6. The assumption that we made in the above proof that we can prepare identical sets (i.e. S1, S2 and S3) my be disputed. However, it seems reasonable on a statistical basis.

10.3.2 Another Proof of Bell's Theorem

In this appended subsection we give another technical proof of the Bell theorem and inequality. In fact, this proof produces the Bell inequality in its original form as found in Bell's paper (see "J.S. Bell, Physics 1(3), 195, 1964") although we use different symbols. This proof also follows (in general) the line of the proof of Bell himself.

Let $\mathbf{a}, \mathbf{b}, \mathbf{c}$ be three arbitrary unit vectors in the 3D (Euclidean) space, and let have an entangled (or correlated) system made (for instance) of two spin-1/2 particles π_1 and π_2 where their spin is measured in units of $\hbar/2$ (and hence the spin of each particle along any direction is given as $+1$ or -1). If this system is created by the decay of a spin-0 particle then the spin of the system is zero. So, if the spin of π_1 along $\mathbf{a}$ is ± 1 then the spin of π_2 along $\mathbf{a}$ is ∓ 1 and hence the expectation value (or average) μ for the product of their spins (as an outcome of measurement on a number of identical entangled systems of this type) should be $\mu(\mathbf{a}, \mathbf{a}) = -1$ where the argument $(\mathbf{a}, \mathbf{a})$ determines the directions of the spin measurement of π_1 and π_2 respectively. Similarly, if the spin of π_1 along $\mathbf{a}$ is ± 1 then the spin of π_2 along $-\mathbf{a}$ is also ± 1 and hence the expectation value for the product in this case should be $\mu(\mathbf{a}, -\mathbf{a}) = +1$. More generally, the expectation value *according to quantum mechanics* for the product of the spin of π_1 along a given vector (say $\mathbf{a}$) and the spin of π_2 along a given vector (say $\mathbf{b}$) is given by:[535]

$$\mu_{\text{qm}}(\mathbf{a}, \mathbf{b}) = -\mathbf{a} \cdot \mathbf{b} = -\cos\theta \tag{42}$$

where μ_{qm} is the expectation value according to quantum mechanics, θ is the angle between $\mathbf{a}$ and $\mathbf{b}$ (noting that it is not necessarily the polar angle as in the previous proof) and the dot represents the dot product.

Now, let assume that the spin state of π_1 and π_2 is determined by local hidden variable(s) symbolized by λ (and hence the measurements of the spin of π_1 and π_2 are independent of each other). Also, let A be a function that determines the result of spin measurement on π_1 (and hence $A = \pm 1$) and B be a function that determines the result of spin measurement on π_2 (and hence $B = \pm 1$). It should be obvious that A and B depend on the orientation of measurement (as determined by the unit vectors) and on λ and hence we can write $A(\mathbf{a}, \lambda)$ and $B(\mathbf{b}, \lambda)$.[536] So, according to the above results we can write:

$$B(\mathbf{a}, \lambda) = -A(\mathbf{a}, \lambda) \tag{43}$$

Now, Eq. 42 determines the expectation value according to quantum mechanics and hence we need to determine the expectation value μ_{LHV} (i.e. for the product of the spin of π_1 along a given vector, say $\mathbf{a}$, and the spin of π_2 along a given vector, say $\mathbf{b}$) according to the local hidden-variable (LHV) assumption and compare the two to see if they are identical or not.[537] So in brief, all we need to do now is to determine μ_{LHV} and do the comparison.[538]

---

theories are totally deterministic since the reality of the states of particles is (assumed) totally determinate in advance.

[535] This result can be derived formally (see for instance the solution of Problem 4.50 of Griffiths book "Introduction to Quantum Mechanics" second edition). The solutions manual of Griffiths book is available on the internet in pdf format. It is also available in paper form from some publishers.

[536] The use of $\mathbf{a}$ (with A) and $\mathbf{b}$ (with B) is just an instance.

[537] We note that if we have parallel or anti-parallel situation then quantum mechanics and LHV agree and hence we did not distinguish earlier between them in symbolism (i.e. we used μ for the expectation value without subscript since in these cases $\mu_{\text{qm}} = \mu_{\text{LHV}}$).

[538] We note that the determination of μ_{LHV} will be done indirectly through an inequality which restricts μ_{LHV}. We also note that the comparison between μ_{qm} and μ_{LHV} will be done indirectly by substituting μ_{qm} (which represents the predictions of quantum mechanics) in the LHV inequality and hence if the inequality that we obtain from this substitution or merge is violated then we can conclude that the two predictions cannot be identical, i.e. $\mu_{\text{LHV}} \neq \mu_{\text{qm}}$ in general (also see footnote [520] on page 230). It should be obvious that μ_{LHV} (like μ_{qm}; see Eq. 42) is a function of the vectors of A and B and hence we can write (for instance) $\mu_{\text{LHV}}(\mathbf{a}, \mathbf{b})$ and $\mu_{\text{LHV}}(\mathbf{a}, \mathbf{c})$.

10.3.2 Another Proof of Bell's Theorem

As a general background knowledge, the expectation value of a function[539] (or "random variable") is the integral of "the function times its probability density" over the (independent) variable of that function. For example, if $f(x)$ is a function of x and its probability density is $\rho(x)$ then its expectation value is:

$$\langle f \rangle = \int_{\text{all } x} f \, \rho \, dx \tag{44}$$

Accordingly, the expectation value μ_{LHV} of the product $A(\mathbf{a}, \lambda) B(\mathbf{b}, \lambda)$ [which is a function of the variable λ since $\mathbf{a}$ and $\mathbf{b}$ are fixed constants] is obtained by integrating "$A(\mathbf{a}, \lambda) B(\mathbf{b}, \lambda)$ times its probability density $\rho(\lambda)$" over the variable λ, that is:

$$\mu_{\text{LHV}}(\mathbf{a}, \mathbf{b}) = \int_{\text{all } \lambda} A(\mathbf{a}, \lambda) B(\mathbf{b}, \lambda) \, \rho(\lambda) \, d\lambda = -\int_{\text{all } \lambda} A(\mathbf{a}, \lambda) A(\mathbf{b}, \lambda) \, \rho(\lambda) \, d\lambda \tag{45}$$

where we used Eq. 43 in the last step. Accordingly:

$$\mu_{\text{LHV}}(\mathbf{a}, \mathbf{b}) - \mu_{\text{LHV}}(\mathbf{a}, \mathbf{c}) = \left[-\int_{\text{all } \lambda} A(\mathbf{a}, \lambda) A(\mathbf{b}, \lambda) \, \rho(\lambda) \, d\lambda \right] - \left[-\int_{\text{all } \lambda} A(\mathbf{a}, \lambda) A(\mathbf{c}, \lambda) \, \rho(\lambda) \, d\lambda \right] \tag{46}$$

$$= -\int_{\text{all } \lambda} \left[A(\mathbf{a}, \lambda) A(\mathbf{b}, \lambda) - A(\mathbf{a}, \lambda) A(\mathbf{c}, \lambda) \right] \rho(\lambda) \, d\lambda \tag{47}$$

$$= -\int_{\text{all } \lambda} \left[A(\mathbf{a}, \lambda) A(\mathbf{b}, \lambda) - A(\mathbf{a}, \lambda) A^2(\mathbf{b}, \lambda) A(\mathbf{c}, \lambda) \right] \rho(\lambda) \, d\lambda \tag{48}$$

where line 2 is justified by the linearity of integration while line 3 is justified by the fact that $A^2 = 1$ (since $A = \pm 1$).

Now, if we take the modulus of both sides of the last equation we get:

$$\left| \mu_{\text{LHV}}(\mathbf{a}, \mathbf{b}) - \mu_{\text{LHV}}(\mathbf{a}, \mathbf{c}) \right| = \left| \int_{\text{all } \lambda} \left[1 - A(\mathbf{b}, \lambda) A(\mathbf{c}, \lambda) \right] A(\mathbf{a}, \lambda) A(\mathbf{b}, \lambda) \, \rho(\lambda) \, d\lambda \right| \tag{49}$$

$$\leq \int_{\text{all } \lambda} \left| \left[1 - A(\mathbf{b}, \lambda) A(\mathbf{c}, \lambda) \right] A(\mathbf{a}, \lambda) A(\mathbf{b}, \lambda) \, \rho(\lambda) \right| d\lambda \tag{50}$$

$$= \int_{\text{all } \lambda} \left| \left[1 - A(\mathbf{b}, \lambda) A(\mathbf{c}, \lambda) \right] A(\mathbf{a}, \lambda) A(\mathbf{b}, \lambda) \right| \rho(\lambda) \, d\lambda \tag{51}$$

$$= \int_{\text{all } \lambda} \left| \left[1 - A(\mathbf{b}, \lambda) A(\mathbf{c}, \lambda) \right] \right| \rho(\lambda) \, d\lambda \tag{52}$$

$$= \int_{\text{all } \lambda} \left[1 - A(\mathbf{b}, \lambda) A(\mathbf{c}, \lambda) \right] \rho(\lambda) \, d\lambda \tag{53}$$

$$= \int_{\text{all } \lambda} \rho(\lambda) \, d\lambda - \int_{\text{all } \lambda} A(\mathbf{b}, \lambda) A(\mathbf{c}, \lambda) \, \rho(\lambda) \, d\lambda \tag{54}$$

$$= 1 - \int_{\text{all } \lambda} A(\mathbf{b}, \lambda) A(\mathbf{c}, \lambda) \, \rho(\lambda) \, d\lambda \tag{55}$$

$$= 1 + \mu_{\text{LHV}}(\mathbf{b}, \mathbf{c}) \tag{56}$$

where line 2 is justified by the triangle inequality for integrals, line 3 is justified by the fact that $\rho \geq 0$ because it is a probability density function, line 4 is justified by the fact that $|A(\mathbf{a}, \lambda) A(\mathbf{b}, \lambda)| = |\pm 1| = 1$, line 5 is justified by the fact that $A(\mathbf{b}, \lambda) A(\mathbf{c}, \lambda) = \pm 1$ and hence $\left[1 - A(\mathbf{b}, \lambda) A(\mathbf{c}, \lambda) \right] \geq 0$ (because it is either 0 or 2), line 6 is justified by the linearity of integration, line 7 is because ρ is (assumed) normalized, and line 8 is from Eq. 45. So in brief, we get:

$$\left| \mu_{\text{LHV}}(\mathbf{a}, \mathbf{b}) - \mu_{\text{LHV}}(\mathbf{a}, \mathbf{c}) \right| \leq 1 + \mu_{\text{LHV}}(\mathbf{b}, \mathbf{c}) \tag{57}$$

---

[539] We mean continuous function (as opposite to discrete).

which is the Bell inequality in its original form (with some differences in symbolism).

Comparing the predictions of Eq. 42 (which represents quantum mechanics) with the predictions of this inequality (which represents LHV) it can be easily shown that the two predictions are different which establishes Bell's theorem. Such comparison can be made (as we did in the proof of § 10.3.1) by showing (through substituting from Eq. 42 into the inequality of Eq. 57) that μ_{qm} does not satisfy this inequality (in general) and hence μ_{qm} cannot be identical to μ_{LHV}. For instance, if $\mathbf{a}, \mathbf{b}, \mathbf{c}$ are coplanar and $\mathbf{a}, \mathbf{b}$ are orthogonal while $\mathbf{c}$ divides their angle in half then $\theta(\mathbf{a}, \mathbf{b}) = \pi/2$ and $\theta(\mathbf{a}, \mathbf{c}) = \theta(\mathbf{b}, \mathbf{c}) = \pi/4$ and hence according to Eq. 42 we have:

$$\mu_{\text{qm}}(\mathbf{a}, \mathbf{b}) = -\cos\frac{\pi}{2} = 0 \qquad \mu_{\text{qm}}(\mathbf{a}, \mathbf{c}) = \mu_{\text{qm}}(\mathbf{b}, \mathbf{c}) = -\cos\frac{\pi}{4} = -\frac{1}{\sqrt{2}} \tag{58}$$

These values obviously do not satisfy Bell's inequality of Eq. 57 because:

$$\left|0 - \left(-\frac{1}{\sqrt{2}}\right)\right| \not\leq 1 + \left(-\frac{1}{\sqrt{2}}\right) \tag{59}$$

$$\left(\frac{1}{\sqrt{2}} \simeq 0.707\right) \not\leq \left(1 - \frac{1}{\sqrt{2}} \simeq 0.293\right) \tag{60}$$

This should conclude the proof of Bell's theorem.

10.3.3 General Argument in Support of Bell's Theorem

In the following we present a general argument in support of Bell's theorem. However, it should be noted that this argument is neither rigorous nor technical (or mathematical). Nevertheless, it is useful in endorsing Bell's theorem and providing a logical support for its qualitative content in a qualitative way since it rationalizes the Bell theorem in a comprehensible way (rather than an abstract mathematical way).

Let have two objects, A and B, which are entangled (or rather correlated) at time t_0 (e.g. the time of creating a pair of spin-1/2 particles by the decay of a spin-0 particle). Moreover, let the (spin) states of these objects evolve in time (whether according to the quantum mechanical conceptualization or according to the local hidden-variable conceptualization) and we are free to test the (spin) state of any one of these objects (say A) at any future time (say t_2).

Now, according to the local hidden-variable scenario when we test A at t_2 the state of B at t_2 (assuming a global time)[540] is already determined and hence the test of A at t_2 is irrelevant as far as the state of B at t_2 is concerned. In fact, in principle we can know the state of B at t_2 even before t_2 and regardless of any test on A or on any other object. This is because the state of B at any time is determined by the hidden variable which is local to B and independent of any non-local factor such as A and the test on it. This is completely unlike the situation according to the quantum mechanical scenario (in its commonly-accepted interpretation and predictions) where the state of B at any time (before test) is not independent of the state of A because A and B are entangled (in a quantum mechanical sense which means that they have a single wavefunction as an entangled system). This means that the state of B at t_2 is determined *by* the test and not independent of it.

To make the difference more clear, let the object A be destroyed at $t_1 < t_2$. According to the local hidden-variable scenario the state of B at t_2 is unaffected (i.e. it is the same as the state of B at t_2 if A is not destroyed at t_1), while according to the quantum mechanical scenario the situation is very different (since the state of B at t_2 will be determined by the event of destruction, whether it is regarded as collapse or not, and hence the state of B at t_2 is determined by the state of B, as well as the state of A,[541] at t_1 when the destruction occurs).

---

[540] The implication (or rather the negative impact) of "global time" on special relativity should be obvious. The issue of global time as a quantum mechanical implication has been investigated earlier (see for instance § 5.6 and § 9.2.1).
  Anyway, global time may not be needed here since all we need is a specific time (which could be frame-dependent).
[541] Or rather by the state of the entangled system which is made of A and B.

In fact, the essence (and culprit) of all this is that the unitary evolution of quantum mechanics (which is determined and governed by the Schrodinger equation or any equivalent equation) is not the same as the classical evolution (which is essentially determined and governed by classical mechanics such as Newton's laws or at least by the rationale of classical mechanics as discussed earlier in § 10.3.1), and hence the state at a given time (say t_2) of a physical system that is subject to the former type of evolution is not necessarily the same as the state at t_2 of that system if it is subject to the latter type of evolution.

We should finally note that the rationale of this argument may be restricted to the totally deterministic type of local hidden-variable theories (as discussed in point 5 of § 10.3.1). However, the essence of the argument may avoid this restriction if we base this rationale on the difference of the type of evolution (as explained in the last paragraph).

10.3.4 Geometric Argument in Support of Bell's Theorem

In this appended subsection we present a simple geometric argument in support of Bell's theorem. Although this argument may not be sufficiently rigorous (and may also be restricted to totally deterministic local hidden-variable theories), it is useful in itself for giving a rather intuitive geometric insight (as well as being useful for providing extra "partial" support to Bell's theorem like some other arguments and proofs which also may not be sufficiently rigorous). In the following, we assume the 3D space to be coordinated by standard spherical coordinates but for simplicity we only consider the xz plane of this system. So, although we may use the terminology of plane geometry (like semi-circle) it should be understood that the actual geometry is solid in 3D (and hence the semi-circle[542] is actually a hemisphere).

Let have an entangled (or correlated) double-particle quantum system of zero total spin like the one described and used in the proofs of § 10.3.1 and § 10.3.2. According to the local hidden-variable theories (or at least the totally deterministic ones) the spins of the two particles are determined in reality and hence we can represent them by two vectors equal in magnitude and opposite in direction. This is depicted graphically in the left frame of Figure 2 where the solid part represents the spin of one particle (say A) and the dashed part represents the spin of the other particle (say B). Now, if we measured the spin component of A along the z orientation and found this component to be up then the spin vector of A should be in the upper z hemisphere (represented by the upper semi-circle abc in the right frame of Figure 2); otherwise it will not have a component in the positive z direction. If we then measured the spin component of B along the θ orientation[543] and found this component to be also up then the spin vector of B should be in the upper θ hemisphere (represented by the north-east semi-circle def in Figure 2). Now, this can be the case only if the spin vector of A is in the sector aOd (and hence the spin vector of B is in the sector cOf).

Accordingly, the probability of P_{++} (i.e. having spin-up for A along z and spin-up for B along θ) should be proportional to the size of the cOf sector (i.e. relative to the size of semi-circle). Now, if we note that $0 \leq \theta \leq \pi$ and the size of the semi-circle is proportional to π while the size of the sector cOf is proportional to θ then we should have $P_{++} = \theta/\pi$. This argument applies exactly to the P_{--} by just relabeling "up" as "down" (or by reversing the orientations of axes of measurement) and hence $P_{--} = \theta/\pi$. Now, if we note that the event ++ (i.e. A up along z and B up along θ) and the event +− (i.e. A up along z and B down along θ) are complementary then we should have $P_{+-} = 1 - P_{++} = 1 - \theta/\pi$. Similarly, −− (i.e. A down along z and B down along θ) and −+ (i.e. A down along z and B up along θ) are complementary events and hence we should have $P_{-+} = 1 - P_{--} = 1 - \theta/\pi$. So in brief, the probabilistic predictions of the LHV theories are:

$$P_{++} = \frac{\theta}{\pi} \tag{61}$$

$$P_{+-} = 1 - \frac{\theta}{\pi} \tag{62}$$

$$P_{-+} = 1 - \frac{\theta}{\pi} \tag{63}$$

---

[542] Or rather: semi-disk.
[543] The θ orientation is represented by the vector Oe in the right frame of Figure 2.

10.4 Aspect's Experiment

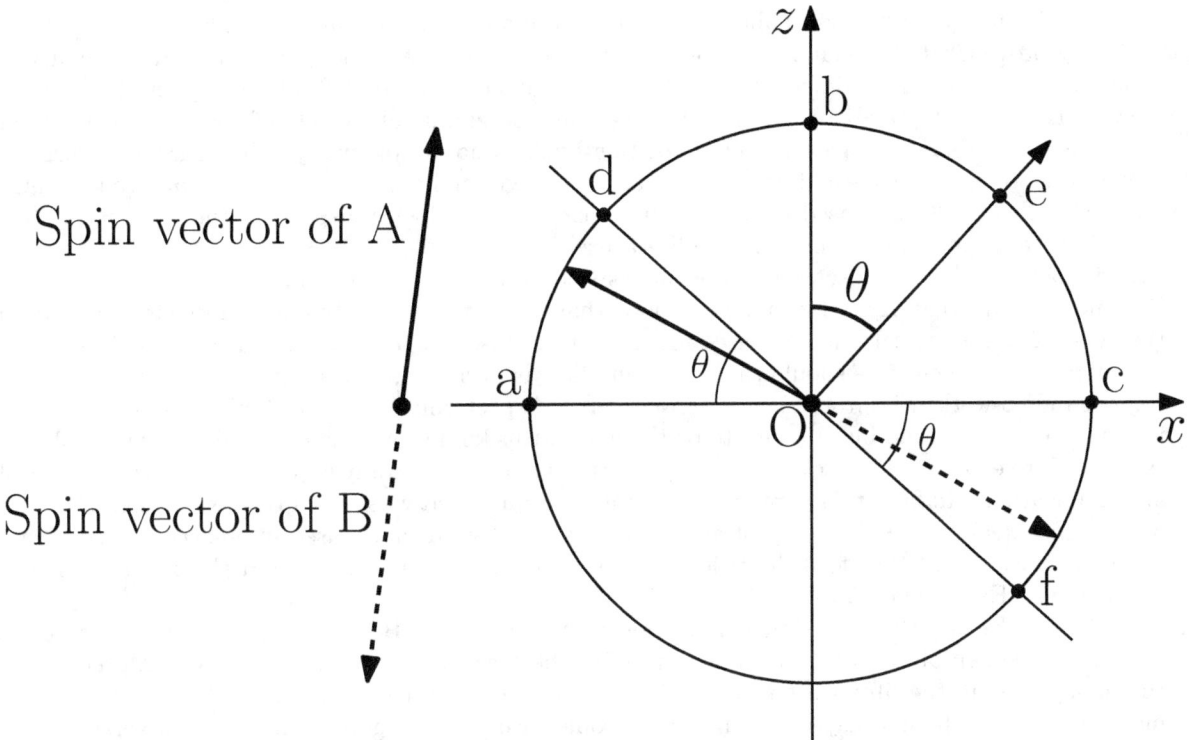

Figure 2: Schematic illustration of the basic setting of the geometric argument in support of Bell's theorem. Refer to § 10.3.4 for details.

$$P_{--} = \frac{\theta}{\pi} \qquad (64)$$

As we see, these predictions are different from the predictions of quantum mechanics which we derived earlier in § 10.3.1 (see Eqs. 25-28). Therefore, the LHV theories cannot reproduce all the predictions of quantum mechanics and this should establish the Bell theorem. As discussed earlier (see for instance point 3 of § 10.3.1), this should also refute the LHV with no need for any further experimental test (assuming that the predictions of Eqs. 25-28 are already verified).

10.4 Aspect's Experiment

This experiment is supposedly the first of its kind (or at least one of the first) that demonstrates the violation of the Bell inequality and provides evidence in support of the predictions of quantum mechanics and against the predictions of local hidden-variable theories.[544] The technical details of this experiment and how it works are lengthy and messy and out of our scope (as well as being irrelevant to our objectives). Therefore, the interested readers should refer to the literature for such details.[545] In fact, there are many similar experiments like this one especially in the recent times where different types of entangled quantum particles (as well as different experimental settings and techniques) are used.[546]

---

[544] We note that Aspect's experiment is not the first of its kind although it is the one that is widely celebrated and recognized as the first "conclusive" experiment in the LHV theories debate. Also see point 1.

[545] See for instance "A. Aspect, P. Grangier, and G. Roger, Phys. Rev. Lett. 49(2), 91, 1982" and "A. Aspect, J. Dalibard,, and G. Roger, Phys. Rev. Lett. 49(25), 1804, 1982". Also see Rae and Napolitano in the References for a basic description of this experiment.

[546] As indicated earlier (see § 9.3.2), this kind of experiments became a standard educational practice in undergraduate labs.

10.4 Aspect's Experiment

There are also many controversies about the significance and value of this experiment and its alike and if they really prove what they claim to prove. For example, experimental loopholes related to locality and efficiency of detection were claimed against these experiments (see point 3; also see point 1 of § 9.3.3). Anyway, it is almost impossible to verify and assess all the empirical aspects of these experiments and their criticisms by someone (like the author of this book) who is not involved in their procedures and technicalities. So, our analysis and assessment (as presented partly and briefly in the following points as well as earlier in § 9.3.3) are based mainly on the theoretical and logical aspects of these experiments (as well as what was reported or claimed in the literature).

In the following points we discuss briefly some aspects of the Aspect experiment:

1. It is noteworthy that Aspect conducted more than one experiment but for simplicity (considering the type of these experiments) we use singular form. Also, Aspect conducted his experiments with collaborators and hence we should say something like "experiments of Aspect *et al*" or "experiments of Aspect and co-workers" but for simplicity we refer to Aspect only. Also, we think Aspect's experiment is not the first of its kind but it seems to be the first of this kind that uses temporally-varying polarizers (whose relative orientation varies "randomly" in time) and is commonly seen as conclusive in its results and implications. In fact, there are many variants and improved versions of this experiment and hence we can consider this as a category of experiments started with (and based in essence on) the Aspect pioneering work, and therefore the title of this section should more appropriately be something like "Aspect-type Experiments".

2. As indicted already, this experiment is obviously a test for the Bell inequality (or rather a test for a modified version of it) and hence it is a test for the local hidden-variable theories. Moreover, it is commonly seen in the literature as a test for the EPR argument (see § 9.3.1 and § 10.2). This view may be justified (though may not be totally in some cases) with regard to the modern versions of the EPR argument but not with regard to the original EPR argument. Noting the difference in the setting and some other details between the two experiments (i.e. the original EPR thought experiment and the Aspect experiment) this could be true with regard to the spirit (not the letter) of the original EPR argument and with regard to some of its aspects. In fact, even some modern versions of the EPR argument may not fit exactly within the procedure and rationale of some Aspect-type experiments (or Bell's experiments to be more general).

3. There are many criticisms and challenges directed to this experiment and its results and conclusions from technical and procedural perspectives (as well as from other more fundamental perspectives). Although most (or many) of these concerns were addressed in later improved versions[547] of this experiment or in other experiments of this type (i.e. experiments whose purpose is to investigate the Bell-type inequalities and the validity of local hidden-variable theories), not all criticisms and concerns are addressed and resolved or settled. So, our view (or feeling) is that these experiments generally provide high confidence in the reported results and conclusions but they are not utterly conclusive (and hence we should remain open-minded about other possibilities).

Also refer to § 9.3.3.

---

[547] We note that these improvements are in formulation or/and in procedure.

Epilogue

From our investigation in this book we can draw some general conclusions and judgments outlined in the following points (noting that some of these can be a subject of debate and deliberation):

1. Quantum theory is one of the greatest and most successful scientific theories but it is also one of the strangest. It is phenomenal both in its practical success (i.e. empirically) and in its theoretical and conceptual failure (i.e. epistemologically). The empirical success (associated with the conceptual failure) of quantum theory may make it ideal for technology but not for science in its epistemological dimension (i.e. as an attempt to understand the world especially at the microscopic level). In fact, most of the successes of quantum theory so far are applicational and technological in nature rather than scientific and theoretical (in the sense of understanding and having deep insight). We may even claim that quantum theory contributed in some cases to confusion and misunderstanding, and this is particularly and vividly reflected in the literature of the interpretation of quantum mechanics and the oddity of schools, opinions, hallucinations, .. etc. in this regard.

2. The literature of quantum theory (particularly on the interpretation side) is full of contradictions, errors, naive and nonsensical arguments, inconsistencies, illusions, hallucinations, ... etc. In fact, no branch of physics (if we exclude the relativity theories) can match the mess of quantum theory. This is due partly to the absence of a solid logical and epistemological framework that is required to analyze and interpret scientific theories rationally and consistently.

3. From what we saw in this book we can conclude that modern physics is far from being perfect or complete, and this does not seem to represent the general attitude and feeling among modern physicists. In fact, the position of the majority of physicists today is similar to the position of some physicists in the end of the 19th century where they believed that physics is about to be perfected and completed very soon and hence it almost reached its final destination. In our view, science in its current state is an inhomogeneous mixture of paradigms and theories and hence it may not be wise to expect science to provide a completely consistent picture (neither as formalism nor as interpretation). In fact, the failure to have a satisfactory interpretation to quantum mechanics is just an instance and symptom of this more fundamental problem. The exceptional prominence of this instance is because of the distinct nature and features of the quantum world and quantum phenomena.[548]

4. We should insist on the necessity and importance of epistemology and philosophy of science for scientists and hence these subjects should be seen as a necessity for developing healthy and productive science. In fact, this requirement should be considered in the curricula of science education by including these subjects in the training programs of scientists. The classical term "natural philosophy" should be seen as a paradigm and model not only for old physics but even for modern physics.

5. Despite the fact that almost each one of the proposed interpretations of quantum mechanics has its own merit (since it addresses certain problems and usually propose good solutions to these problems) no one of the existing interpretations is entirely satisfactory and can address and solve all the existing problems (let alone potential problems that are likely to emerge in the future). So, should this lead us to the conclusion that no rational and satisfactory interpretation can be found (or even "exist") for quantum mechanics? Or should we keep trying and trying until we (hopefully) find this fugitive interpretation?

   Regarding the issue of interpretability, our view is that quantum mechanics (despite its empirical brilliance) is non-interpretable. Everyone has the right to keep arguing endlessly with or against this

---

[548] In our view, we should inspect and analyze the epistemological crisis of quantum theory within a broader context, i.e. the context of science in general. In other words, the crisis of quantum mechanics is a miniature version or instance of the crisis in modern physics. So far, there is no universal physical theory; what we actually have is a collection of inhomogeneous (and sometimes contradicting) partial theories dealing with different subjects and starting from different axiomatic and theoretical frameworks. So, we have one theory for electrodynamics, another one for quantum mechanics, and so on. In fact, even the so-called unification theories are not actually universal (or universalization) theories but they are theories of stitching-together and homogenizing of this inhomogeneous collection. We can safely claim that the justification of this process of stitching-together and homogenizing (as the only option for the development and evolution of science) is implicitly based on the assumption of uniqueness of science which we strongly reject (see § 2.4.3).

or that epistemology and interpretation of quantum theory, and we can continue this seemingly-futile investigation and debate for as long as we wish. However, our advice to everyone is to save time and energy and stop trying to find a consistent and acceptable interpretation to quantum mechanics. Based on our belief in the non-uniqueness of science (see § 2.4.3), it may be more productive and fruitful to try to create a new quantum theory (that is interpretable as well as empirically successful and possibly superior) instead of spending a huge amount of resources on the subject of interpreting the current quantum mechanics by designing arguments, conducting experiments, going through endless quarrels and disputes, and so on. However, it seems that there is a general erroneous belief among modern physicists that science is unique and hence we have no choice but to embrace quantum mechanics and accept it as it is as if there is no prospect for alternative.

So to conclude, those who are interested in practical applications should continue using this theory (at least for the time being and as long as it is working) without worrying about its epistemology and interpretation, while those who are interested in the epistemology and philosophy of the quantum world should stop looking for an interpretation to this theory (to avoid more waste of time and resources). Meanwhile, those scientists who believe in the non-uniqueness of science should look for another quantum theory that is both empirically brilliant (and possibly even more brilliant than quantum mechanics) and epistemologically useful (or interpretable) hoping that this is achievable within the existing theoretical frameworks and conceptual infrastructures.

6. It is very difficult (if not impossible) to "understand" and appreciate quantum theory (and its formalism in particular) without working on the technicalities of quantum mechanics within proper theoretical or practical contexts and applications.[549] Unfortunately, some (if not many) of those who put themselves in the position of assessing and interpreting quantum physics lack such experiences and hence their assessments and interpretations (which are usually based on general readings and superficial inspection) reflect their stereotypes, prejudices and misunderstandings. I think an ideal individual for assessing and interpreting quantum physics is a technically-knowledgeable quantum physicist with strong philosophical sense and background knowledge. Unfortunately, this category of individuals is a minority among physicists and philosophers of science (who are interested in these issues).

7. Regarding realism (which can be seen as the central issue in the epistemology of quantum physics), we summarize our views and conclusions in the following bullet points:

• Realism is a philosophical (or ontological) choice but it is an epistemological necessity because with no realism (as embedded essentially in the principles of reality and truth and should be extended to other rationality aspects like causality) there is no knowledge (and no epistemology) because with no realism there is no rationality.[550] In our view, epistemological realism (for quantum theory) is at least a possibility (and hence it should be adopted or at least kept alive) even if we accept that the quantum theory may have implications that seem inconsistent with realism. Anyway, if quantum mechanics is a threat to total realism it cannot (or should not) be a threat to realism in its entirety and hence we should maintain realism classically and maintain it, when possible, even quantum mechanically, i.e. we discard realism minimally and as a last resort and keep it when and where it is possible (noting that the insistence on realism is not out of dogmatic fanaticism or philosophical puritanism but it is out of sheer pragmatism).

• In our view, if (hypothetically) we reached a definite conclusion that realism is not compatible with quantum theory (i.e. it is impossible to find a realistic interpretation to this theory) then we could or should blame the theory rather than realism (i.e. we accept quantum mechanics as a calculus or recipe for dealing with the quantum world and abandon the attempts to make sense of it by resorting to non-realistic interpretations). In fact, keeping realism is more fundamental and important to our scientific and epistemological progress and for healthy evolution of science (and knowledge in general) than

---

[549] In fact, this demand applies to science (and even knowledge) in general but it is much more intense in quantum theory due to its exceptional nature.

[550] As indicated earlier, science is not about reality as it is (which is a philosophical and ontological issue) but about reality as we see it (which is an epistemological issue). We should also remember that realism is an epistemological necessity in its moderate form not in its extreme form (which we reject). Also see (for instance) § 1.10 and § 2.5.2 as well as point 4 of § 8.6.

having a satisfactory interpretation to quantum mechanics. This could even be exploited beneficially as a motive for developing a new quantum theory that is compatible with realism (or at least improving the current quantum theory to become so).

• Even if we assume that we are forced (for whatever reason) to adopt a non-realistic interpretation to quantum physics and hence abandon realism in the microscopic world, we can still maintain realism in the macroscopic world (and the theories that represent this world like classical mechanics). This seemingly-inconsistent position may be justified by claiming that due to the scale factor (see § 1.3.2) the epistemology that we developed during our evolution can be influenced by this factor, and hence as the physics of the classical and quantum worlds are different (where in one world classical physics applies while in the other world quantum physics applies) the epistemological rules of the two worlds (which are based in part on our scientific knowledge) could also be different.[551] So in brief, as a last resort for maintaining realism (classically) we may even accept different epistemological theories and principles towards classical and quantum phenomena as we accepted different scientific theories towards these phenomena.[552]

• Finally, science in our view should be based on experimental realism rather than theoretical idealism (noting that we embrace only moderate form of realism). Quantum mechanics (as a scientific theory) is certainly experimental and its interpretation (if it has any acceptable interpretation) should be realistic. Hence, if realism is not achievable (i.e. in a supposedly-acceptable interpretation) then it may be better to stay with no interpretation (rather than adopting a non-realistic "acceptable" interpretation). In simple words, realism with no interpretation is better than interpretation with no realism.

---

[551] In fact, the scale factor can justify the claim that the current human epistemology is suitable to the classical world but not to the quantum world since all our inherited intellectual structures evolved within the classical world (noting that all our quantum and microscopic experiences are obtained in the last "moment" of our evolution and hence they did not have time to permeate into our intellectual structure). Hence, a new epistemology that can cope with the challenges of the quantum world may be required and should be looked for (and this may require a lengthy evolutionary process). Alternatively, a new "quantum theory" that is consistent with the current "classical epistemology" should be sought and developed (if we assume or accept that it is possible).

[552] It is worth remembering that "classical" in this context is opposite to "quantum" (in accord with the default). So, realism in its classical and strict sense should be kept in some "non-classical" (but non-quantum) theories like Lorentz mechanics (and hence we should keep challenging special relativity, for instance, by the principles of reality and truth). The reason is that these theories are actually classical from this perspective (i.e. the quantum perspective) and hence the classical realism should apply to them. In other words, any necessary changes and amendments to realism due to quantum non-realism should be restricted to the quantum theory and quantum world.

References

J.S. Bell. *Speakable and unspeakable in quantum mechanics*. Cambridge University Press, first edition, 1987.

J. Binney; D. Skinner. *The Physics of Quantum Mechanics*. Oxford University Press, first edition, 2013.

B.H. Bransden; C.J. Joachain. *Quantum Mechanics*. Pearson Education, second edition, 2000.

B.H. Bransden; C.J. Joachain. *Physics of Atoms and Molecules*. Pearson Education, second edition, 2003.

W. Demtröder. *Atoms, Molecules and Photons An Introduction to Atomic-, Molecular- and Quantum Physics*. Springer, second edition, 2010.

A.P. French; E.F. Taylor. *An Introduction to Quantum Physics*. Chapman & Hall, first edition, 1979.

D.J. Griffiths. *Introduction to Quantum Mechanics*. Prentice Hall, Inc., first edition, 1995.

M. Jammer. *The Philosophy of Quantum Mechanics: The Interpretations of Quantum Mechanics in Historical Perspective*. John Wiley & Sons, Inc., first edition, 1974.

D.A.B. Miller. *Quantum Mechanics for Scientists and Engineers*. Cambridge University Press, first edition, 2008.

Open University Team. *Quantum Mechanics (for SM355 course)*. Open University, first edition, 1998.

A.C. Phillips. *Introduction to Quantum Mechanics*. John Wiley & Sons Ltd, first edition, 2003.

A.I.M. Rae; J. Napolitano. *Quantum Mechanics*. CRC Press, sixth edition, 2016.

T. Sochi. *The Mechanics of Lorentz Transformations*. CreateSpace, first edition, 2018.

T. Sochi. *General Relativity Simplified & Assessed*. Amazon Kindle Direct Publishing, first edition, 2020.

Note: as well as the above references, we also consulted during our work on the preparation of this book many other books, research and review papers and general articles about this subject (some of which are cited in the text).

Index

3D
    momentum vector, 6
    space, 6, 66, 237, 240
    vector, 77

Absolute frame, 55, 146, 150
Accuracy, 18, 50, 70, 92, 93, 96, 104, 135, 231, 236
Action at a distance, 9, 22, 25, 26, 28, 30, 31, 119, 147–149
Afshar experiment, 143
After measurement (stage), 87, 106–108, 122, 169, 198
Angular momentum, 88, 89, 93, 104, 117, 118, 120, 153, 155, 161, 176, 178
    quantum number, 89, 104
Annihilation, 23, 45, 86, 137, 147
Arbitrarily-precise measurement, 136
Arbitrary
    accuracy, 93
    precision, 95
Art, 9, 31, 38, 39
Artificial satellite, 61
Aspect (Alain), 29, 221, 241, 242
Aspect experiment, 55, 127, 149, 168, 181, 191, 192, 214, 221, 222, 241, 242
Aspect-type experiments, 190, 191, 242
Asterisk (or star symbol), 6, 11
Astronomical, 24, 164
Astronomy, 13, 15, 44, 74, 164
Astrophysics, 13, 74, 164
At measurement (stage), 102, 125, 194
Atom, 9, 13, 16, 63, 69, 72, 74, 75, 83, 87, 89, 90, 95, 97, 103, 104, 120, 129, 131, 140, 152, 207
Atomic
    ground state, 95, 104
    model, 62, 104, 139, 161, 171, 177, 178
    orbit, 65, 76, 171
    particle, 104
    physics, 83, 131, 163
    radius, 72
    scale, 8, 10
    size, 10
    system, 10, 95
    transition, 97, 123
Atomism, 10
Average, 6, 75, 82, 88, 95, 237
Axiomatic framework (of quantum mechanics), 52, 56, 70, 80–82, 85, 90, 103, 120, 123, 130, 131, 133, 151, 170, 173, 176, 188, 194, 201, 209

Ballistic
    missile, 61
    trajectory, 76
Barrier (of potential), 115–117, 144–146
Basic
    elements (of quantum mechanics), 103
    quantum theory (mechanics), 78, 126, 132
Beable, 45, 189, 190
Before measurement (stage), 81, 87, 102, 106, 107, 118, 122, 125, 169, 194, 198
Bell
    argument, 120, 133, 215, 217, 218, 220
    experiments, 149, 168, 214, 215, 217–222, 236, 242
    inequality, 29, 55, 190, 191, 215–220, 222, 229, 234–237, 239, 241, 242
    paper, 237
    theorem, 29, 55, 59, 127, 147, 149, 175, 181, 190–192, 214–222, 229, 230, 234–237, 239–241
Bell (John), 45, 59, 142, 190, 191, 216, 217, 220, 228, 237
Bell-CHSH inequality, 218
Bell-type
    arguments, 218
    inequalities, 218, 242
Between measurements (stage), 82, 84, 102, 125, 130–133, 194
Big Bang, 74
Biological, 34, 35, 40, 62, 63, 199, 202
    sciences, 63, 133
Biology, 32, 62, 63
Bohm (David), 191
Bohm theory, 59, 175, 183, 189–191, 203, 218
Bohr (Niels), 99, 141, 142, 186, 188, 206, 215, 226
Bohr atomic model, 104, 139, 161, 171, 177, 178
Born (Max), 215
Born interpretation, 56, 77, 128–131, 170, 176, 181, 200
Boson, 65, 89, 172, 229
Boundary, 117, 199
Branching (of worlds), 57, 193–199
Brownian motion, 71
Bubble
    chamber, 163
    paradox, 207

Calculus, 8, 11
Canonically conjugates, 11
Cartesian, 6, 90
Cathode ray, 163
Celebrity, 207
    culture, 198
Certainty, 26, 50, 80, 94, 105, 136, 154, 166, 191, 214, 218, 221, 225
Chaotic system, 106
Charge, 13, 16, 68, 89, 93, 132, 155–157, 172
    density, 132
Chargeability, 16
Chargeable, 16
CHSH inequality, 218
Classic-quantum duality, 109, 113, 139, 161
Classical, 8, 65, 66, 68–72, 75, 104, 164
    electrodynamics, 125
    entanglement, 120
    evolution, 240
    indeterminism, 106, 133, 189, 210
    interference, 114
    measurement, 69, 70, 93, 107, 121, 151, 185
    mechanics, 8, 37, 38, 41, 46, 47, 70, 72
    particle, 68, 69, 74, 112, 115, 116, 143
    physics, 8, 11, 13, 16–19, 32, 72
    superposition, 107
    theory, 8, 65, 69, 70, 72
    wave, 17, 66–68, 73, 78, 83, 109, 112, 116, 117, 125, 134, 145, 156, 157, 161, 171, 172

Clauser (John), 218, 221
Collapse (of wavefunction), 11, 12, 15, 19, 28–31, 33, 81, 120–123, 130, 150–154, 185
Collimators, 163
Common sense, 16, 17, 40, 48, 70, 111, 136, 147, 160, 165, 180, 183, 194, 210
Communist, 53, 54
Commutation (of operators), 138
Commutator (of operators), 6, 88
Complementarity, 14, 99–102, 109, 120, 125, 141–143, 172, 199
Complementary, 11, 92, 94, 96, 100, 101, 134, 136, 142, 149, 152, 214, 240
Completeness (of quantum mechanics), 214, 215, 225–228
Complex
    conjugate, 6, 11
    function, 11, 66, 77, 79, 128
    number, 11, 66, 77–79, 81, 82, 114, 128, 165, 177
    phase factor, 78
    variable, 11, 66
Compliance with
    logic, 34, 40, 41
    realism, 43, 127, 175
    the established theories, 47
    the principle of causality, 44, 53
    the principles of reality and truth, 42, 43, 47, 53
Comprehensibility, 56
Compton
    effect, 104, 109
    scattering, 57
Comte (Auguste), 29
Conductivity (in metals), 95, 96
Configuration space, 66, 129
Conjugate (complex), 6, 11, 199
Conjugate observables, 11
Consciousness, 20, 32, 33, 43, 69, 70, 114, 122, 123, 126, 151–153, 158, 179, 186, 196, 199, 209, 211, 212
Conservation principles, 26, 30, 72, 92, 126, 137, 138, 154, 155, 161, 169, 171, 178
Consistent histories interpretation, 200, 201
Conspiracy theories, 181
Continuity, 65, 79
Continuous, 65, 79, 81, 87, 94, 104, 105, 129, 177, 238
Continuum mechanics, 73, 177
Copenhagen interpretation, 55, 57, 77, 82, 96, 101, 121, 127, 141, 143, 170, 173, 176, 181, 182, 184–189, 191, 194–203, 205–208, 211, 226–228
Coplanar, 239
Correctness, 40, 41, 45–47, 52, 61, 62, 98, 152, 166, 192, 205, 209, 220, 227, 228, 230
Cosmological, 74, 164, 179
Cosmology, 13, 15, 44, 74, 164
Countable, 10, 65, 79
Counter-
    intuitive, 20, 22, 38, 148, 152, 165, 172, 187, 191, 194, 200, 221, 222
    intuitivity, 148, 165, 172, 183, 190, 194, 197
Creation, 14, 23, 38, 44, 45, 54, 55, 86, 137, 145, 147, 172
Cultural, 12, 16, 60, 199, 202, 210
Culture, 14, 40, 198, 210

Dark
    energy, 14, 44, 45, 54, 55, 164
    matter, 14, 44, 45, 54, 164
Davisson-Germer experiment, 109, 168

de Broglie
    equation, 67, 94, 109, 156, 171
    hypothesis, 67, 109, 131, 156, 171
de Broglie (Louis), 45, 191
Decoherence, 198–200, 209
Decoherent histories interpretation, 200
Deduction, 23, 24
Definiteness, 26, 68, 79, 132, 166, 215
Delayed
    cause, 20, 22, 23, 26, 27, 124, 158, 179, 200, 209
    choice, 15, 20, 158, 159
Delocalization, 16
delta function, 80, 115
Detectable, 9, 97, 110, 112, 147, 148
Detection efficiency, 18, 222, 236, 242
Determinant, 6
Determinism, 20, 26, 70, 71, 105–107, 133, 134, 166, 188, 189, 192, 195, 198, 202, 208, 215, 230, 236
DeWitt (Bryce), 198
Dialectical materialism, 54
Differential
    calculus, 113
    equation, 83, 84
    operator, 6, 81, 82, 86
Diffraction, 16, 96, 109, 114, 224
    pattern, 114, 224
Dimensionality, 11, 79
Dirac equation, 84, 90, 132
Discontinuity, 79
Discontinuous, 122, 151
Discrete, 10, 65, 72, 79, 81, 87, 94, 97, 104, 129, 177, 238
    energy levels, 97
Discreteness, 10, 65, 94
Dispersion, 67
Distinguishability, 65, 68
Divine
    entity, 38
    intervention, 181
Divinity, 39, 53
Divisibility, 68
Dot product, 6, 237
Double-slit experiment, 8, 17, 18, 20–22, 55, 99, 112, 114, 124, 134, 141, 143, 153, 155–159, 168, 171, 224
Doubly-ionized lithium, 83
Dualism, 69
Duality, 14, 19, 22, 94, 96, 99–102, 109–113, 115, 120, 137, 138, 141, 145, 146, 156, 157, 161, 171, 176
Dynamics (of quantum mechanics), 102, 194, 197–199

Effect of observation and knowledge, 18–20, 114, 124, 126, 156–158, 179
Ehrenfest theorem, 72, 85, 97, 98, 141, 162, 163
Eigenequation, 10
Eigenfunction, 10, 11, 78, 81, 82, 85–87, 93, 169, 185
Eigenstate, 79, 81, 82, 86–88, 90, 91, 108, 121, 125, 128, 129, 152, 153, 169, 222
Eigenvalue, 10, 81–83, 85–88, 93, 108, 121, 128, 129, 169, 177, 231, 232
    equation, 10, 83, 85, 86
Eigenvector, 86, 177, 231, 232
Einstein (Albert), 135, 142, 191, 206, 214, 215, 224, 226, 227
Einsteinian opportunism, 227
Electrodynamics, 8, 125, 243
Electromagnetic
    field, 44, 137

signal, 25, 28, 147
wave, 17, 44, 45, 49, 66, 68, 79, 111, 134, 157, 171
Electromagnetism, 8, 11, 45, 49, 85
Electron, 13–15, 17–20, 65, 66, 68, 69, 72, 76, 83, 89, 90, 93–95, 109–111, 114, 115, 118–120, 123, 129, 136, 145, 152, 156, 157, 161, 163, 164, 171, 172, 177, 208, 218, 222, 229
Element of physical reality, 225, 227
Elementary
linear algebra, 11, 113
objects, 65
particles, 75, 111, 161
quantum mechanics (theory), 8, 85
Eligibility criteria, 25, 34, 38, 39, 41–43, 45, 47, 50, 53, 55, 60, 181, 182, 197
Empirical
failure, 205
incompleteness, 213
success, 14, 46, 52, 60, 61, 94, 132, 137, 160, 166, 167, 178, 183, 193, 205, 209, 212, 243
Empiricism, 23, 48, 165
Energy, 6, 16, 17, 20, 28–30, 66, 67, 72, 81–87, 89, 92, 93, 95, 97, 104, 109, 115–117, 126, 129, 132, 134, 135, 137–139, 144–146, 154, 155, 157, 164, 165, 171, 172, 182, 199, 210, 227, 244
-frequency relation, 67, 109
level, 87, 97, 104
operator, 83, 85–87
spectrum, 87
Entangled
electrons, 118, 119, 222, 229
objects, 15, 28, 117–120, 136, 149, 214, 222, 223, 236
particles, 153, 231, 233, 234, 241
photons, 118, 119
systems, 117–119, 136, 153, 222, 229, 230, 236, 237, 239
Epistemological
choice, 16, 219
failure, 14, 52, 160, 176, 182
idealism, 32
incompleteness, 213
necessity, 16, 26, 27, 30–32, 36, 40, 43, 44, 173, 174, 244
principles of science, 34, 35, 37, 38, 42
realism, 16, 32, 36, 43, 226, 244
rules, 203, 245
success, 166
Epistemology, 1, 8, 14, 31, 32, 34, 35, 42, 59, 60, 64, 70, 75, 77, 84, 91, 92, 95, 132, 139, 157, 160, 161, 165, 167, 173, 174, 182, 192, 193, 201, 203, 206, 208, 210–212, 243–245
EPR
argument, 136, 149, 186, 214–218, 220, 224–228, 242
paper, 214, 224–228
paradox, 207
Euclidean
geometry, 41
space, 11, 237
Eulerian mechanics, 37
Everett (Hugh), 196, 198
Everett interpretation, 193
Evolution, 11, 12, 24, 40, 59–62, 78, 80, 82, 84, 85, 102, 105, 125, 130, 131, 153, 160, 170, 178, 179, 194, 195, 198, 199, 202, 240, 243–245
Evolutionary, 13, 40, 148, 198, 202, 245
Exchange
energy, 65, 126

force, 65, 126
symmetry, 65, 68, 126, 192
Excited
energy levels, 104
state, 95, 105, 152
Existence of
interpretation, 51, 59, 60
reality, 35, 36, 42, 132, 158
Expectation value, 6, 7, 11, 72, 75, 82, 86, 88, 108, 181, 237, 238
Extrinsic consistency, 206
Fermion, 65, 89, 90, 92, 229
Feynman (Richard), 155, 179
Fine structure, 83
Finite
dimensionality, 11, 79
height (of potential), 116, 145
integral, 79
lifetime, 105
mass, 68
potential, 115
size, 68
speed, 9, 25, 29, 30, 147, 148, 153, 154
square well potential, 87
time, 27, 45, 147
wavelength, 109
width (of potential), 116
Finity (of speed), 147, 148
First order
derivative, 79, 84
differential equation, 83, 84
Formalism, 1, 8, 9, 51, 77, 103, 113, 170
duality, 109, 113
Formalism-interpretation confusion, 109, 170
Fourier analysis, 17, 94, 96, 134–136
Frame of reference, 52, 55, 67, 109, 110, 170, 213
Franck-Hertz experiment, 104, 168
Free particle, 80, 87, 154, 171
Freewill, 26, 62–64, 134
Frequency, 7, 16, 17, 67, 105, 109, 134, 172
Fundamental principle, 39, 82, 85, 91, 96, 98, 101, 132, 171, 228

Galileo, 147
Gell-Mann (Murray), 185
General
argument (for Bell theorem), 236, 239
argument (for uncertainty principle), 136
physics, 8
relativity, 14, 28, 49, 150, 213
Geometric
argument (for Bell theorem), 240, 241
insight, 240
point, 68
shape, 16
Global, 30, 118, 119, 122, 146, 151, 174, 187, 197, 213, 239
collapse, 119, 146
frame, 30, 213
time, 30, 119, 239
Gravity, 49, 61, 164, 165, 190, 213
Ground state, 95, 104
Group velocity, 22, 67, 172

Half-integer

angular momentum, 89
   quantum number, 72, 89
   spin, 89, 90
Hamiltonian, 6, 82, 84, 155
   mechanics, 37
   operator, 6, 82–86
Harmonic oscillator, 95, 97, 126
Harmony, 49, 183
hat (symbol), 6, 82, 86, 87
Hawking (Stephen), 198
Hawking radiation, 137
Heisenberg (Werner), 45, 93, 113, 129, 188, 215
Helium, 63, 83
   atom, 63
Hemisphere, 240
Hermitian operator, 81, 86
Hermiticity, 86
Hidden-variable (interpretation, theory, formulation), 6, 7, 29, 37, 55, 59, 77, 127, 128, 132, 133, 147, 174, 175, 181, 185, 187–195, 202, 215–222, 228–230, 232, 233, 235–237, 239–242
Hilbert space, 11, 66, 79, 81
History, 200
   of evolution, 59
   of quantum (theory, physics, mechanics), 29, 61, 73, 167, 173, 177, 217
   of science, 29, 37, 46, 60, 61, 178, 214
Holt (Richard), 218
Homogeneous differential equation, 83, 84
Horne (Michael), 218
Humanism, 12, 14, 72, 76, 206
Humanist, 12, 13, 71
Humanities, 32
Hydrodynamics, 177
Hydrogen, 83, 131
   atom, 83, 131
Hyper-fine structure, 83

Idealism, 20, 31, 32, 36, 123, 245
Identicality, 65
Ideological, 40, 53, 54
Imaginary
   part, 66
   unit, 6, 82, 83
Immeasurability, 30
Immeasurable, 28, 67
Incompatibility, 101, 102, 141, 165, 183, 205, 212, 216
Incompatible, 11, 79, 81, 82, 88, 89, 92, 93, 99–101, 109, 111–113, 125, 134–136, 142, 153, 172, 175, 184, 191, 225, 226, 235
Incompleteness (of quantum mechanics), 205, 213–215, 228
Independence of
   observer, 15, 16, 35
   reality, 33, 158, 234
Indeterminacy, 26, 32, 63, 64, 84, 87, 100, 105, 106, 169, 185–187, 189, 203, 204, 210
Indeterminism, 14, 26, 27, 36, 63, 64, 68, 70, 94, 96, 105–108, 121, 132–134, 140, 150, 155, 166, 167, 184–190, 195, 202, 213–215, 227
Indistinguishability, 65, 105, 111, 126, 133, 172, 192
Induction, 23, 24
Inert, 16, 100
Inertness, 15, 16, 49, 52, 69, 123, 137, 212
Infinite
   accuracy, 93

dimensional, 79
dimensionality, 11, 79
discontinuity, 79
function value, 79
integral, 79
lifetime, 105
number of realities, 152
number of worlds, 195
speed, 9, 21, 25, 28–30, 67, 119, 146–148, 172, 207
square well potential, 87, 95, 126
time interval, 134
wavelength, 109
width (of potential), 116
Infinitely
   -many branches, 198
   -precise measurement, 94, 136
   high (potential barrier), 115
Infinitesimal
   time interval, 84
   volume element, 6, 80, 81
Infra-sound, 44
Inner product, 11
Instantaneous, 25, 28, 118, 119, 122, 129, 146, 150, 151, 153, 187, 197, 207, 230
Integer
   angular momentum, 89
   function, 233
   quantum number, 72, 89
   spin, 89, 90
Integral, 79, 81, 82, 86, 88, 108, 238
   calculus, 113
Intensity, 11, 13, 17, 105, 107, 111, 114, 224
Interference, 16, 17, 19–21, 66, 109, 113, 114, 134, 171, 200
   pattern, 18–22, 67, 99, 114, 143, 144, 158, 224
Interpretation, 1, 8, 9, 34, 51, 53, 57–59, 170, 176, 180, 184, 188, 193, 199–201
Interpretativity, 50, 58, 167, 192
Intrinsic
   consistency, 202, 206
   indistinguishability, 65
   uncertainty, 11, 64, 70, 82, 106, 108, 111, 124, 133, 135, 136, 150, 186, 187, 228
Intuition, 16, 23, 48, 135, 136, 147, 157, 162, 165, 170, 179, 180, 194, 231
Intuitivity, 38, 71, 113, 123, 191, 198, 229
Invariance, 155
Ionized, 83, 105

jj coupling, 89

Kinetic energy, 6, 82–84, 87, 115, 144, 145, 157
   operator, 6, 87

Lagrangian mechanics, 37
Lamb shift, 83
Large
   box (normalization), 80
   quantum number, 11, 48, 72, 96–98, 203
Law, 32
   of parsimony, 37
Laws of motion, 61, 72, 105, 163
Legal, 210
Lifetime, 95, 102, 105
Light, 6, 9, 18, 19, 21, 27, 28, 44, 67, 96, 109, 112, 117, 133, 147, 149, 207, 224, 225

Linear
- algebra, 10, 11, 113, 124, 231
- combination, 81, 84, 121, 124, 232
- differential equation, 83, 84, 125
- operator, 86, 125
- theory, 124, 125

Linearity of
- integration, 238
- quantum operators, 124
- quantum theory (mechanics), 124, 125, 131
- Schrodinger equation, 80, 124, 131

Linearly independent, 233
Literature, 9, 39
Lithium, 83
Local
- hidden-variable (interpretation, theory, formulation), 6, 7, 55, 133, 175, 181, 189, 191, 192, 215–222, 228–230, 232, 233, 235–237, 239–242
- realism, 226, 227

Localization, 16, 21, 68, 110, 116, 145, 172
Logic, 34, 39–43, 53, 165, 182, 183, 202, 216, 234
Logicality, 35, 39, 40, 42, 53, 147, 165
Lorentz
- factor, 17, 115
- mechanics, 8–10, 13, 28, 30, 31, 38, 41, 46, 51, 52, 55, 72, 73, 78, 98, 131, 141, 147–149, 162, 163, 170, 180, 213, 245
- transformations, 10, 147, 170

Lorentzian, 9, 13, 17, 38, 46, 84, 85, 90, 115, 132
LS coupling, 89

Macroscopic, 8, 11, 13, 14, 164, 165
Magnetic dipole moment, 89, 90, 172
Many-
- consciousness interpretation (hypothesis), 196
- histories interpretation, 198
- minds interpretation (hypothesis), 196, 198, 199
- worlds interpretation, 56, 57, 125, 127, 143, 152, 183, 185, 187, 193–200, 202

Marginal principle, 56, 91, 96, 98, 99, 101, 139, 141, 143
Mass, 6, 16, 17, 68, 84, 109, 115, 116, 156, 157
- -energy relation, 16, 115

Massive, 17, 70, 111, 112, 115
Massless, 17, 111
Mathematics, 10, 38, 59, 85, 113, 165, 177, 202
Matrix
- mechanics, 11, 37, 66, 77, 78, 113, 129
- operator, 86, 231

Matter wave, 17, 21, 28, 66, 67, 72, 97, 109–111, 128, 131, 146, 156, 157, 171, 172
Maxwell equations, 43
Mean value, 6, 82, 95
Measurability, 50, 136
Measurable, 9, 17, 66, 81, 82, 86, 87, 93, 114
Measurement, 9, 11, 12, 14, 15, 69, 120, 150
- operator, 121, 125
- postulate, 12, 31, 52, 81, 84, 122, 130, 133, 151, 152, 194, 195, 209

Megascopic, 164
Memory, 22
Metaphysical, 9, 25, 44, 45, 53–55, 62, 109, 137, 148, 164, 172, 173, 183, 190, 191, 193, 194, 199
Metaphysics, 9, 44, 54, 173
Microscopic, 8, 13, 14, 164, 165
Modern
- physics, 14, 25, 33, 52, 61, 119, 137, 142, 157, 170, 180, 193, 199, 205, 217, 222, 243
- science, 48, 49, 61, 147, 173, 180, 192

Modulus, 11, 79, 80, 105, 108, 114, 121, 134, 169, 238
- squared, 79, 105, 108, 114, 121, 169

Molecular
- physics, 83
- scale, 10

Molecule, 10, 74, 75, 89, 95, 97, 140
Momentum, 6, 9, 15, 16, 27, 66–68, 81, 82, 84, 86–88, 92–97, 100, 102, 105, 109, 116, 122, 129, 134, 136, 137, 139, 154, 155, 169, 172, 198, 214, 225, 226
- operator, 6, 86, 87
- space, 129

Moral, 210
Multi universe interpretation, 193
Multiple worlds interpretation, 193

nabla operator, 6, 82
Natural
- evolution, 24
- history, 44
- philosophy, 243

Nature, 12, 38, 40, 56, 189
Negative
- energy, 115, 116, 144, 145
- kinetic energy, 144, 145

Neutron, 65, 66, 90, 95
- star, 66, 95

Newton
- laws, 11, 61, 62, 105, 163, 240
- second law, 85, 97, 169
- third law, 56, 190

Newtonian mechanics, 37
Non-
- Euclidean geometry, 41
- Hermitian operator, 86
- interpretable, 50, 51, 162, 167, 171, 179, 201, 243
- linear operator, 125
- linear system, 106
- linear term, 125, 131, 209
- linear theory, 124, 125
- linearity, 124, 125, 131, 153
- local, 29, 30, 122, 129, 146, 148–151, 174, 175, 183, 187, 190–192, 197, 199–202, 215, 216, 218, 219, 222, 227, 230, 239
- locality, 25, 28, 30, 31, 129, 146–148, 150, 174, 183, 190–192, 197, 202, 213, 217, 222, 227, 230
- physical wave, 17, 66, 111
- uniqueness of interpretation, 60
- uniqueness of knowledge, 37
- uniqueness of science, 32, 36, 37, 60, 108, 113, 133, 140, 166, 173, 175, 202, 204, 220, 221, 228, 244

Normalizable, 79–81
Normalization, 79, 80, 121, 124, 171
Normalized, 11, 80, 81, 88, 231, 232, 238
Nucleon, 14
Nucleus, 14, 69, 89, 95, 171

Objectivity, 15, 32, 33, 49, 50, 52, 70, 169, 186
Observable, 6, 9, 11, 85–88
Observation, 9, 11, 15, 16, 19, 20, 157, 158, 169, 170, 185, 186, 190, 192, 193, 196, 197, 199, 203, 208–213, 224, 236
Observer, 9, 13–16, 18, 20, 32, 169, 180, 185–188, 190, 193, 195, 196, 198–200, 203, 210–212

Occam's razor, 37
Oddity, 22, 110, 183, 184, 193, 194, 197–199, 243
Old quantum theory, 74, 139, 161, 178
Ontological
    choice, 31, 36, 42, 174
    idealism, 32, 36
    necessity, 27, 36, 173
    oddity, 194
    realism, 16, 32
Ontology, 42, 198
Operator, 6, 10, 72, 73, 79, 81–88, 94, 108, 121, 122, 124, 125, 138, 161, 165, 169, 177, 231
Optical
    fibers, 163
    parametric down conversion, 118, 119
    waves, 117
Optimality criteria, 48–50, 58, 162, 167
Original EPR argument, 214, 215, 224–228, 242
Orthogonal, 233, 239
Orthogonality, 86
Overlap integral, 82

Paradoxes, 206–212
Parallel worlds interpretation, 193
Parametric down conversion, 118, 119
Parapsychology, 33, 114, 158
Partial
    causality, 26, 27
    correctness, 47
    differential equation, 83
    incorrectness, 47
    interpretation, 56
    realism, 32
Particle, 8–10, 14, 16, 17, 68, 69, 109–113
Particle-wave duality, 17–19, 57, 74, 93, 96, 97, 99, 100, 107, 109–113, 137, 138, 144, 145, 155–157, 159–162, 171, 177
Passive, 9, 15, 33, 46, 47, 100, 114, 123, 151, 158, 185
Passivity, 15, 49, 151
Pauli equation, 132
Phase
    factor, 78
    relationships, 200
    velocity, 22, 67, 172
Philosophy, 31, 32, 38, 39, 58, 71, 142, 174, 180, 192, 203, 210, 243, 244
    of science, 1, 71, 243
Phlogiston theory, 178
Photoelectric effect, 57, 104, 109
Photon, 15–18, 20, 90, 94, 104, 109, 111, 112, 118, 119, 123, 136, 156–158, 163, 171, 172, 218, 222, 224, 229
Physical
    reality, 14, 33, 43–45, 68, 105, 114, 128, 134, 135, 140, 151, 158, 165, 166, 169, 176, 177, 202, 208, 214, 225–228
    space, 66
    wave, 17, 28, 66, 67, 111, 134, 171
Physicality, 25, 28, 44, 45, 53, 55, 62, 66, 67, 128, 147, 169, 172, 176, 181, 190, 197
Pilot wave, 19, 21, 67, 68, 110–112, 128, 146, 156, 157, 171, 172
Planck
    constant, 6, 92, 93, 109, 110
    quantization hypothesis, 104
Plum pudding model, 62

Podolsky (Boris), 214, 224
Poincare (Henri), 16
Polar
    angle, 7, 237
    coordinate, 230, 231
Polarity, 172
Polarization, 44, 118, 119, 223
Polarized photon, 218, 222, 229
Political, 61
Position, 6, 16, 21, 22, 27, 60, 65, 66, 68, 74, 76, 77, 79, 81, 82, 85–88, 92–97, 100, 101, 104, 105, 129, 136, 153, 181, 207, 214, 224, 226
    operator, 6, 85–87
    vector, 6, 82, 86, 87
Position-momentum form (of uncertainty principle), 92–94, 97, 116, 134
Postulate of
    completeness (of EPR argument), 225–227
    measurement, 12, 31, 52, 81, 84, 122, 130, 133, 151, 152, 194, 195, 209
    physical reality (of EPR argument), 225–227
    projection, 81, 151
    reduction, 81, 153
    special relativity, 28, 29
Postulates of
    quantum mechanics, 85, 130
    von Neumann, 80
Potential
    barrier, 115–117, 145
    energy, 6, 82–85, 87, 89, 115
    energy operator, 6, 87
    well, 97, 115, 116
Precision, 21, 50, 74, 92, 94, 95, 99, 134–136
Predictability, 26, 27, 102, 173, 189, 192, 193, 227
Predictivity, 50, 167, 192
Principle of
    causality, 22–27, 36, 44, 53, 71, 132, 146–148, 150, 151, 158, 173, 174, 201, 209
    complementarity (complementarity principle), 56, 96, 99–102, 110, 112, 138, 141–144, 146, 156, 158, 185–187, 224
    conservation of angular momentum, 118, 120, 155
    conservation of charge, 155
    conservation of energy, 84, 92, 115, 116, 137, 139, 145, 154, 155, 165, 171
    conservation of momentum, 92, 137, 139, 155
    conservation of parity, 126, 155, 169
    conservation of probability, 126, 154, 169
    conservation of spin, 153, 154
    consistency, 39
    correspondence (correspondence principle), 10, 11, 48, 72, 73, 76, 96–99, 101, 102, 110, 139, 140, 161–163, 187, 203
    economy, 37, 38, 48, 49, 190, 195
    exclusion (exclusion principle), 89, 90, 92, 103, 126
    existence of reality, 35, 36, 42, 132, 158
    intuitivity, 38, 48–50
    locality, 25, 27–30, 118, 119, 122, 133, 148, 151, 174, 175, 187, 191, 197, 200, 201, 207, 214, 215, 218, 226, 227
    logic, 39
    logicality, 35, 42
    non-uniqueness of science, 32, 36, 37, 60, 108, 113, 133, 140, 166, 173, 175, 202, 204, 220, 221, 228, 244
    rationality, 36

uncertainty (uncertainty principle), 11, 15, 27, 64, 66,
  70, 74, 86, 87, 92–96, 99–102, 106, 108, 116, 124,
  125, 127, 128, 134–138, 140, 142, 145, 149, 154,
  155, 161, 162, 165, 171, 178, 181, 185–187, 192,
  214, 215, 225, 226, 228, 229
uniqueness of reality, 35, 132, 139, 193, 195
uniqueness of truth, 35, 36, 47, 98, 139, 140

Principles of
  quantum theory, 61, 91, 101, 132
  reality and truth, 12, 31, 35–37, 42–44, 47, 53, 60, 98,
    127, 185–187, 193, 195, 201, 203, 244, 245

Privileged frame, 30, 150
Probabilisticity, 68, 70, 107, 108, 127, 128, 132, 166, 185–187,
  189, 213, 214
Probability, 6, 11, 70, 80–82, 97, 102, 103, 105, 108, 113–116,
  121, 122, 129, 154, 155, 163, 169, 177, 197, 198,
  225, 231, 232, 240
  amplitude, 11, 66, 67, 77, 79, 80, 108, 128, 129, 131,
    134, 146, 170, 172, 181, 184, 185, 232
  density, 7, 11, 78–81, 87, 114, 134, 238
  distribution function, 79
  wave, 17, 66, 97, 111, 112, 128, 134, 146, 153, 157
Projectile, 139
Projection (of wavefunction), 81, 121, 151
Projection postulate, 81, 151
Proton, 89, 93
Psychology, 32
Ptolemaic model, 37, 178

Quantitativity, 50
Quantization, 68, 94, 95, 97, 104, 105, 171
Quantized, 65, 89, 90, 95, 104, 105
Quantum, 8, 10, 65, 66, 68–72, 75, 104, 164
  communication, 29, 61, 62, 119, 147
  computing, 58, 59, 61, 62, 119, 214
  cryptography, 58, 59, 147
  electrodynamics, 8
  encryption, 29, 119
  entanglement, 28–30, 117–120, 125, 136, 146–150, 158,
    172, 191, 212, 213, 226, 227
  field theory, 8, 84, 132, 171
  gravity, 165, 213
  indeterminism, 105, 106, 108, 132–134, 188, 189, 210
  information, 59, 214, 217
  interference, 113, 114, 134
  logic, 41, 42
  measurement, 15, 33, 52, 69–71, 87, 88, 93, 94, 105, 106,
    120–125, 133, 135, 150–153, 169, 185, 186
  mechanics, 1, 7, 8, 80, 102, 130, 154
  number, 11, 48, 72, 89, 96–98, 104, 203
  operator, 6, 72, 81, 85–87, 94, 124, 161
  particle, 8, 18, 65, 68, 69, 81, 89, 111, 115, 116, 133,
    145, 163, 241
  penetration (of barriers), 115, 116, 145, 146
  physics, 1, 8–11, 72, 73, 75, 167
  scattering, 115, 116, 144–146
  soup, 137
  superposition, 106, 107, 132, 140, 195, 200
  system, 8–11
  theory, 1, 8, 9, 77, 91, 101, 104, 127, 160, 205
  tunneling, 115–117, 144–146, 212
  wave, 66–68, 78, 114, 134, 157, 172

Radioactive, 115, 207

Rationalization, 22, 31, 40, 57, 58, 94, 96, 99, 112, 116, 121,
  134, 135, 137, 141, 144, 146, 151, 154, 155, 158,
  184
Real
  number, 66, 79, 81, 86, 114, 128, 165
  part, 66
  particle, 14, 137
  variable, 66
  wave, 17, 67, 68, 157
Realism, 14, 16, 31–33, 36, 174, 175, 201–204, 226–229, 244,
  245
Realisticity, 49
Rectangular Cartesian, 6, 90
Reduced Planck constant, 6, 82, 83, 92
Reduction (of wavefunction), 11, 81, 82, 114, 121, 123, 130,
  151, 153, 174, 185, 230
Reduction postulate, 81, 153
Reflection, 115, 117
Refractive index, 117
Relative state interpretation, 193
Relativistic, 9, 17, 28–30, 84, 147, 148, 150, 170, 172, 174,
  201
Relativity theories, 119, 149, 165, 174, 182, 183, 187, 190,
  191, 197, 200, 210, 213, 217, 243
Religion, 32, 38, 54
Renaissance, 147, 192
Resolution, 16, 18, 93, 186, 192
Rosen (Nathan), 214, 224
Rules of
  classical physics, 206
  common sense, 40, 210
  intrinsic consistency, 202
  logic, 12, 34, 35, 37, 39–43, 60
  ordinary logic, 42
  physics, 108
  probability, 70, 102, 108
  quantum mechanics, 168, 233
  quantum physics, 9, 161
  reality and truth, 35
  sanity, 183
  syllogism, 39
  thinking, 39

Sandwich integral, 86, 88, 108
Scale factor, 13–15, 19, 22, 65, 71, 72, 76, 137, 161, 203, 206,
  208, 245
Scattering, 65, 115–117, 123, 144–146
Schrodinger
  cat paradox, 106, 183, 186, 207–212
  equation, 9, 11, 12, 21, 73, 74, 77–80, 82–85, 89, 90,
    102, 103, 105, 113, 122, 124, 125, 130–132, 151,
    152, 162, 163, 177, 178, 183, 194, 195, 198, 240
Schrodinger (Erwin), 191
Scientific
  interpretation, 53, 55–60, 181, 182, 197
  theory, 1, 8, 12, 13, 15, 25, 34, 35, 37–39, 45, 46, 48, 49
Second
  order differential equation, 83
  postulate of special relativity, 28, 31
Self-interference, 18, 20–22, 114, 126, 141, 155, 157, 159, 179,
  212
Semi-
  circle, 240
  classical, 14, 61, 73, 83, 85, 113, 139, 161, 162, 176
  deterministic, 21, 22

disk, 240
random, 19, 21, 22
randomness, 21, 22
SG type experiments, 6, 90
Shimony (Abner), 218
Simplicity, 37, 48, 71, 162, 170, 187
Simultaneity, 25, 30, 146, 150, 213, 230
Simultaneous, 11, 24, 94, 99, 112, 134, 135, 156, 176, 214, 224–226, 232–234
Simultaneously-measurable, 82, 86, 93
Single-
particle interference, 20
valued function, 79
Singly-ionized helium, 83
Slit-grid experiment, 55, 56, 100, 127, 141, 143, 144, 181, 186, 224, 225
Social, 12, 16, 40, 60, 61
sciences, 63, 133
Sociological, 63
Sound (wave), 44, 66, 112
Soviet Union, 54
Space, 6, 13–15, 25, 30, 74, 77, 78, 89, 104, 119
-reflection symmetry, 155
-rotation symmetry, 155
-time, 25, 30, 60, 78, 89, 119, 137, 163
-translation symmetry, 155
Spatial, 9, 11, 17, 21, 25, 28, 30, 31, 60, 65, 66, 68, 78–80, 82, 84, 85, 89, 90, 95, 102, 110, 118, 130, 137, 147–149, 153, 207
action at a distance, 9, 25, 28, 30, 31, 147, 148
dependency, 11, 78, 79, 84, 85, 89
intensity, 17
non-locality, 25, 30
Special
relativistic, 28, 29, 147, 148, 150, 170
relativity, 9, 10, 27–31, 41, 51, 52, 119, 146–150, 174, 180, 183, 187, 191, 206, 207, 213, 227, 239, 245
Spectral lines, 104
Speed, 6, 9, 13, 17, 21, 25–30, 38, 67, 74, 84, 111, 119, 146–150, 153, 154, 157, 172, 207, 226
limit, 25–27, 30, 149
of light, 6, 9, 21, 27, 28, 147, 149
restriction, 28, 30
Spherical coordinates, 6, 230, 240
Spin, 6, 7, 14, 68, 76–79, 85, 86, 88–91, 103, 104, 118–120, 122, 129–132, 152–154, 161, 171, 172, 176, 178, 197, 198, 216, 218, 222, 223, 229–234, 236, 237, 239, 240
-0, 230, 237, 239
-1, 90, 229
-1/2, 89, 90, 118, 119, 122, 153, 216, 218, 222, 229–231, 233, 237, 239
-down, 6, 90, 118, 119, 153, 154, 197, 198, 218
-orbit coupling, 171
-up, 6, 90, 118, 119, 153, 154, 197, 198, 218, 240
operators, 6, 86, 231
quantum number, 89, 104
Splitting (of worlds), 194, 196, 197, 199
Square well potential, 83, 87, 95, 126
Star, 29, 54, 66, 95, 115, 186
Stationary (state), 78, 83, 134, 152
Step (of potential), 117
Stern-Gerlach
apparatus, 6, 90
experiment, 6, 90, 104, 153, 168

Sub-atomic
particle, 104
scale, 8, 10
system, 10, 95
Sub-microscopic, 8
Subjective, 14, 32, 108, 114, 122, 123, 128, 151, 177, 182, 188, 206, 218
Subjectivity, 20, 49
Substitution rules, 86, 161
Suicide paradox, 211
Superluminal speed, 9, 25–30, 67, 119, 146–149, 154, 207, 214
Superluminality, 29–31, 147, 148, 213
Superposition, 11, 16, 17, 78, 79, 81, 82, 86–88, 90, 91, 102, 105–109, 117, 121, 124, 125, 131–133, 140, 150, 151, 162, 166, 172, 184–186, 193, 197, 200, 201, 207, 209, 211, 230
Superscopic, 164
Symmetry, 65, 68, 126, 155, 192

Technology, 61, 243
Tempo-spatial, 11, 60, 79, 82, 85, 89, 102, 130, 131
Temporal, 9, 11, 25, 26, 28, 30, 31, 78, 80, 84, 137, 147, 175
action at a distance, 9, 25, 26, 28, 31, 147
dependency, 11, 78, 80, 85, 89
non-locality, 25, 30
Testability, 45, 46, 53–56, 181, 197
Testable, 45, 55, 197, 200
Theological, 45, 53–55
Theology, 44
Thermodynamics, 97, 98
Thermonuclear fusion, 115
Thompson atomic model, 62
Thoroughness, 56
Thought experiment, 106, 135, 141, 168, 177, 186, 200, 206, 208, 209, 211, 218, 242
Time, 6, 10, 11, 13–15, 25, 30, 74, 77, 78, 83, 89, 104, 119
-dependent Schrodinger equation, 9, 78, 82–85, 102, 130
-dependent wavefunction, 9, 11, 78, 80, 83
-energy form (of uncertainty principle), 92, 93, 134, 135, 138, 154
-independent Schrodinger equation, 9, 83–85
-independent wavefunction, 9, 11, 77, 78, 80
-of-flight, 158
-translation symmetry, 155
reversal, 78, 155
Tone, 169
Total energy, 82–84, 87, 115, 116
Trajectory, 16, 19, 20, 72, 74, 76, 78, 99, 110, 139, 143, 163, 171, 186
Transactional interpretation, 55, 127, 143, 181, 199, 200
Triangle inequality for integrals, 238
Tunnel effect, 115
Turbulent system, 106

Ultra-sound, 44
Uncertainty, 6, 11, 21, 26, 27, 32, 36, 64, 87, 92, 108, 134, 136–138, 204
Unification, 37, 73, 197, 199, 212, 243
Uniqueness of
cause, 24
interpretation, 59, 60
reality, 35, 44, 132, 139, 193, 195
science, 128, 139, 140, 220, 228, 243
truth, 35, 36, 47, 98, 139, 140
Unitary evolution, 82, 102, 133, 194, 240

Unity, 11, 78, 81, 154, 225
Universe, 14, 38, 63, 66, 74, 164, 179, 180, 199, 203

Validity
    criteria, 25, 34, 45, 47, 53, 57, 182
    domain, 48, 75, 76, 139, 162, 163
Vector space, 11, 79
Virtual particles, 137, 172
Visible light, 18
Volume, 6, 11, 16, 79–81, 108
von Neumann (John), 80, 191, 220

Wave, 8, 13, 14, 16, 17, 66–68, 109, 112, 113
    equation, 67, 83, 109, 161
    mechanics, 11, 37, 66, 73, 77, 78, 83, 85, 113, 129, 177
Wavefunction, 7, 9, 11, 12, 15, 19, 28–31, 33, 77–80, 120–123, 128–130, 150–154
Wavelength, 7, 16–18, 67, 68, 94, 105, 109, 111, 156, 157, 172
Wavepacket, 9, 17, 19, 21, 22, 57, 67, 68, 95, 96, 110–112, 134, 146, 156, 157, 171, 172, 177
Weinberg (Steven), 198
Well-defined, 79
Wheeler (John), 198
White dwarf, 66
Wigner (Eugene), 70, 211
Wigner friend paradox, 211, 212
World line, 78, 163

X-ray, 157

Zeeman effect, 83
Zero-point energy, 95, 126

Author Notes

- All copyrights of this book are held by the author.
- This book, like any other academic document, is protected by the terms and conditions of the universally recognized intellectual property rights. Hence, any quotation or use of any part of the book should be acknowledged and cited according to the scholarly approved traditions.
- This book is totally made and prepared by the author including all the graphic illustrations, indexing, typesetting, book cover, and overall design.